Dr. Wen-Fong Ke
James R. Clay

Algebraic StructuresandCombinatorial Designs

Dr. Wen-Fong Ke
James R. Clay

Algebraic StructuresandCombinatorial Designs

A Comprehensive Guide to Block Designs, Nearrings,and Geometric Configurations – 2nd Edition

LAP LAMBERT Academic Publishing

Imprint

Cover image: www.ingimage.com

Publisher:
LAP LAMBERT Academic Publishing
is a trademark of
Dodo Books Indian Ocean Ltd. and OmniScriptum S.R.L publishing group

120 High Road, East Finchley, London, N2 9ED, United Kingdom
Str. Armeneasca 28/1, office 1, Chisinau MD-2012, Republic of Moldova, Europe
Managing Directors: Ieva Konstantinova, Victoria Ursu
info@omniscriptum.com

Printed at: see last page
ISBN: 978-620-8-46016-7

Algebraic Structures and Combinatorial Designs

A Comprehensive Guide to Block Designs, Nearrings, and Geometric Configurations – 2nd Edition

Wen-Fong Ke
National Cheng Kung University, Tainan
James R. Clay

To

Carol

Preface

After the publication of "Nearrings: Geneses and Applications," Jim began to write this book. His sudden death in 1996 left this work unfinished. Carol suggested that I should complete the missing sections and find a publisher for it. Twenty years have past and it is finally done.

The central theme of the book is the equation $ax = bx + c$. One learns this equation in school very early. Simple as it is, different areas of beautiful mathematics appear here. Jim applied it in 1968 to a class of algebraic objects, namely the planar nearrings. It was a fruitful endeavor in that planar nearrings have many applications, in other areas of mathematics and beyond.

Part of this book overlaps with "Nearrings: Geneses and Applications," in particular, the long chapter on planar nearrings. This present book, however, focuses on the various issues surrounding the equation $ax = bx + c$, emphasizing its combinatorial, geometric and algebraic aspects.

In the first chapter, nine problems are offered, each represents an idea for developing a topic. It then introduces the counting techniques, affine planes, project planes, Möbius planes, and mutually orthogonal Latin squares.

Chapter II introduces the equation $ax = bx + c$, for fields, division rings, nearfields, and finally planar nearrings. In a field or a division ring, the equation $ax = bx + c$ can be solved if a is not equal to b. This is generally not true in nearfields and nearrings. Planarity is a property that sieves out planar nearfields and planar nearrings. The chapter then discusses the details of nearfields and planar nearrings.

The balanced incomplete block designs (BIBDs) is an active field of research in combinatorics. Statisticians use BIBDs for designing large and elaborate agricultural experiments. Finite planar nearrings are a good source for producing BIBDs. Chapter III is a detailed discussion of applications of nearrings in BIBDs.

The last chapter contains the geometric aspects of planar nearrings. Circular BIBDs mimic the circles in the complex plane, and are proven to be rich in applications. We present a characterisation of field generated circular planar nearrings. Double planar nearrings may be used to construct triangles and with suitable conditions. The three angles of such a triangle can be measured and the three angles will sum up to 180°! The last example is to construct an affine configuration with one pencil on $\mathbb{Z}_{p^2}$, p an odd prime.

The appendix contains some basic algebraic objects (groups, rings, fields) for the convenience of the readers.

Through the years, there have been voices from the nearring community, enquiring the fate of this book. I would like to thank the LAP LAMBERT Academic Publishing, especially Ian Fowler, for making it possible to publish the book. I especially feel grateful for being a part of the great Clay family: Jim and Carol; Tricia, Tina, and Terri; Benjamin, Annalisa, Nickolas, Chelsea, Thea, Brian, Michael, Joshua, Adam, and Marc. It was a great time there in Tucson for my wife, Lichen, our daughter, Ming, and myself from 1988 to 1992.

Wen-Fong Ke
Tainan, 2016

Note to the Second Edition (2025):

This edition retains the original content published in 2017. The title and cover have been updated for presentation purposes; no changes have been made to the text or research findings.

Contents

List of Symbols **xi**

I Introduction **1**

1 Introductory ideas 1
2 Counting 101 14
3 Basic field theory applications 31

II The equation $ax = bx + c$ **75**

4 Division rings, or skew fields 76
5 Nearfields 84
6 Planar nearrings 107
7 How to make an integral planar nearring 114

III Planar nearrings and combinatorics **123**

8 Balanced incomplete block designs, part 1 123
9 Nets 129
10 Mutually orthogonal Latin squares 134
11 Duality 137
12 Transversal designs 148
13 Tactical configurations and association schemes 151
14 Partially balanced incomplete block designs 158
15 Balanced incomplete block designs, part 2 162

16 Elements of graph theory 164
17 Applications to coding theory 169
18 Finite affine planes and finite projective planes are BIBDs . . . 192

IV Planar nearrings and geometry **199**
19 Circular planar nearrings 199
20 Double planar nearrings 215
21 Rays, angles, segments, triangles, spheres, and circles 219
22 Group of units of $(\mathbb{Z}_{p^2}, +, \cdot)$ 231
23 Affine configurations 233

Appendix **241**
A1 Elements of set theory 241
A2 Elements of field and ring theory 253
A3 Elements of semigroup and monoid theory 297
A4 Elements of group theory 309

Bibliography **337**

Index **341**

List of Symbols

After each entry will appear a reference to a numbered item in the text. The symbol is discussed just before or in this numbered item.

Set Notation

$\cup\mathscr{A}$	union of the family of sets $\mathscr{A}$	A1.1
$\cap\mathscr{A}$	intersection of the family of sets $\mathscr{A}$	A1.1
$A \setminus B$	elements of the set A which are not in the set B	A1.1
$a \in A$	element a is a member of the set A	A1.1
$a \notin A$	it is false that $a \in A$	A1.1
$\mathbb{N}$	set of natural numbers	A1.1
$\mathbb{Q}$	set of rational numbers	A1.1
$\mathbb{R}$	real numbers	A1.1
$\mathbb{C}$	complex numbers	A1.1
$\{x \mid p(x)\}$	all x such that $p(x)$ is true	A1.1
$\{a_1, \ldots, a_n\}$	set with members $a_1, \ldots, a_n$	A1.1
$(a_1, \ldots, a_n)$	an element of $A_1 \times \cdots \times A_n$	A1.1
(a, b)	an element of $A \times B$	A1.1
$A \times B$	cartesian product of set A with set B	A1.1
$A \cup B$	union of sets A and B	A1.1
$A \cap B$	intersection of sets A and B	A1.1
$A \subseteq B$	set A is a subset of the set B	A1.1

$A \subset B$	set A is a proper subset of the set B	A1.1
$\varnothing$	empty set, null set	A1.1
A^n	set of all n-tuples $(a_1, \ldots, a_n)$, $a_i \in A$	A1.1
A^ω, $A^\mathbb{N}$	set of all infinite sequences $(a_1, \ldots, a_n, \ldots)$, $a_i \in A$	A1.1
$\mathscr{P}(A)$	power set of A; family of all subsets of A	A1.1
2^A	another notation for $\mathscr{P}(A)$	A1.1
$\lvert A \rvert$	cardinality of a set A	A1.4
$\mathscr{C}(\mathbb{R})$	all continuous $f : \mathbb{R} \to \mathbb{R}$	A2.5
$M(X)$	all mappings $f : X \to X$	A3.2
$E(S)$	all endomorphisms of S	A3.12
$A(S)$	all automorphisms of S	A3.19
$\mathrm{Inn}M$	all inner automorphisms of M	A3.28
$\mathrm{Aff}F$	all $[b,m](x) = b + mx$	A4.2
$\mathbb{R}_+$	all positive real numbers	A4.2
$\mathscr{P}_2(N)$	the family of two element subsets of N	13.2

Relation Notation

A/R	equivalence classes of R on A	A1.1
$[a]_R$	equivalence class of $a \in A$ for R	A1.1
aRb	$(a,b) \in R$ where R is a relation	A1.1

Function Notation

B^A	set of all functions $f : A \to B$	A1.3
$f \vert X$	restriction of f to X	A1.3
$f(A)$	image of A under f	A1.3
$f^{-1}(A)$	inverse image of A under f	A1.3
$\mathrm{Im}f$	the range of f	A1.3

$f \circ g$	function f composed with function g; also, polynomial f composed with polynomial g	A1.3
f^{-1}	the inverse of a bijective $f: A \to B$	A1.3
$f: A \to B$	f is a function with domain A and codomain B	A1.3
1_A	identity function on the set A	A1.3
$\mathscr{C}(\mathbb{R})$	all continuous $f: \mathbb{R} \to \mathbb{R}$	A2.5
$M(X)$	all mappings $f: X \to X$	A3.2
$E(S)$	all endomorphisms of S	A3.12
$A(S)$	all automorphism of S	A3.19
Inn$M$	all inner automorphisms of M	A3.28
Aff$F$	all $[b,m](x) = b + mx$	A4.2
$a^\bullet$	$a^\bullet: R \to R$ and $a^\bullet(x) = ax$, $a, x \in R$	A4.30
$\Phi^\bullet$	$\{a^\bullet \mid a \in \Phi\}$	A4.31
$[a,b,c,d]$	linear fractional transformation	3.14
$(*, z_2, z_3, z_4)$	cross ratio	3.19
$[b,m]$	$[b,m](x) = b + mx$	5.16

Ring and Field Theory Notation

$\mathbb{Z}$	ring of integers	A1.1
$\mathbb{Q}$	field of rational numbers	A1.1
$\mathbb{R}$	field of real numbers	A1.1
$\mathbb{C}$	field of complex numbers	A1.1
$\mathscr{C}(\mathbb{R})$	all continuous $f: \mathbb{R} \to \mathbb{R}$	A2.5
$\mathbb{Z}_n$	integers modulo n	A2.2
$I + r$, $r + I$	coset of an ideal I	A2.8
R/I	quotient ring of ring R and ideal I	A2.10

(a)	principal ideal generated by a	A2.14
$\ker\phi$	kernel of ϕ	A2.22
$(a_0, a_1, \ldots)$	a polynomial	A2.24
$\mathbf{0}$	zero polynomial	A2.26
$\deg a$	degree of the polynomial a	A2.26
F^*	$F \setminus \{0\}$	A3.2
$\mathscr{U}(R)$	group of units of R	A3.14
T	all $e^{i\theta}$	A4.2
$a+ib$	a complex number	A4.7
$\lvert a+ib\rvert$	absolute value of $a+ib$	A4.7
$e^{i\theta}$	$\cos\theta + i\sin\theta$	A4.7
$a^\bullet$	$a^\bullet : R \to R$ and $a^\bullet(x) = ax$, $a, x \in R$	A4.30
$\Phi^\bullet$	$\{a^\bullet \mid a \in \Phi\}$	A4.31
$\overline{z}$	conjugate of a complex number z	3.20
$\lvert z\rvert$	absolute value or length of a complex number z	3.20
$\mathscr{I}(z)$	imaginary part of a complex number z	3.20
$\mathscr{R}(z)$	real part of a complex number z	3.20
$\overline{q}$	conjugate of a quaternion q	4.7
$\Delta(A)$	determinant of A	4.8
$F(x)$	field of rational functions over F	5.16

Integral Planar Nearring Theory Notation

$a =_m b$	$ax = bx$ for all x	6.2
1_a	a's right identity	6.5
B_a	1_a's elements	6.5
Φ	the group of fixed point free automorphisms of a Ferrero pair	7.1

(N, Φ)	Ferrero pair	7.2
$\Phi(a)$	the orbit of Φ acting on a	7.2
N^*	$N \setminus \{0\}$, that is, the nonzero elements of N	8.1
Ng, Ng_i	a basic block that is a subgroup of $(N, +)$	13.5
Ne, Ne_j	a basic block that is not a subgroup of $(N, +)$	13.5

Semigroup and Monoid Theory Notation

1_A	identity function on the set A	A1.3
$a * b$, ab	a times b	A2.1
$\mathscr{C}(\mathbb{R})$	all continuous $f : \mathbb{R} \to \mathbb{R}$	A2.5
$M(X)$	all mappings $f : X \to X$	A3.2
$E(S)$	all endomorphisms of S	A3.12
$\mathscr{U}(M)$	group of units of M	A3.14
$A(S)$	all automorphism of S	A3.19
AB	$\{ab \mid a \in A,\ b \in B\}$	A3.25
Inn$M$	all inner automorphisms of M	A3.28

Group Theory Notation

$a - b$	$a + (-b)$	A2.9
$\mathscr{U}(R)$	group of units of R	A3.14
AB	$\{ab \mid a \in A,\ b \in B\}$	A3.25
$-b$	the additive inverse of b	A4.2
$GL(2, K)$	group of 2 by 2 nonsingular matrices over K	A4.2
$Z(G)$	center of G	A4.4
C_n	$(\mathbb{Z}_n, +)$ written multiplicatively	A4.7
$\|c\|$	order of a group element c	A4.10
gH	left coset of H	A4.11
$[G : H]$	index of H in G	A4.16

$\ker \phi$	the kernel of ϕ	A4.26
$a^\bullet$	$a^\bullet : R \to R$ and $a^\bullet(x) = ax$, $a, x \in R$	A4.30
$\Phi^\bullet$	$\{a^\bullet \mid a \in \Phi\}$	A4.31
$LF(K)$	group of linear fractional transformations over K	3.15
$\Phi(a)$	orbit of Φ at a	5.13

Geometric Notation

$C(a;r)$	circle with center a and radius r	3.1
$L(m,b)$	line with slope m and y-intercept b	3.4
$L(\infty,a)$	the line $x = a$	3.4
$\mathscr{L}(m)$	all lines with slope m	3.4
$\mathscr{L}_F$	all lines in $F \times F$	3.4
$\mathscr{L}_q$	all lines in $\mathbb{Z}_q \times \mathbb{Z}_q$	3.4
$\mathscr{L}'(\alpha)$	all $L \cup \{\alpha\}$ with $L \in \mathscr{L}(\alpha)$	3.10
$\mathscr{L}'$	all projective lines derived from the lines $\mathscr{L}$ of an affine plane	3.10
$\mathscr{L}$	the family of lines of a geometric structure	9.1
$(N \times N, \mathscr{L}_N)$	a geometric or combinatorial structure made from N	9.2
$(P, \mathscr{B}, I)$	a general incidence structure	11.1
xIB	$(x,B) \in I$, that is, x is related to B by I	11.1
$(x_1;x_2;x_3)$	a point of a special projective plane	11.6
$[x_1,x_2,x_3]$	a line of a special projective plane	11.6
$p(a,b,c)$	a formula for making some $(x_1;x_2;x_3)$	11.6
$L(a,b,c)$	a formula for making some $[x_1,x_2,x_3]$	11.6
$\mathbf{a} \bullet L$, $L \bullet \mathbf{a}$	a formula for testing $\mathbf{a}IL$	11.6
$\mathbf{L}(\mathbf{a},\mathbf{b})$	the line determined by **a** and **b**	11.6

$\mathbf{p}(A,B)$	the point common to lines A and B	11.6

Combinatorics or Counting Notation

$\binom{n}{k}$	binomial coefficient	2.10
$\binom{n}{m_1,m_2,\ldots,m_k}$	multinomial coefficient	2.12
$P(n,k)$	number of permutations of n things taken k at a time	2.7
$C(n,k)$	number of combinations of n things taken k at a time	2.10
(v,b,r,k,λ)	parameters of a BIBD	8.1
$\mathscr{B}^*$	$\{N^*a+b \mid a,b \in N,\ a \neq 0\}$	8.1
F'	$F' = F \cup \{\infty\}$	9.1
$\delta(x,y)$	a function for a transversal design	12.1
$(X,\mathscr{B},I,\mathscr{G})$	a transversal design	12.1
$\mathscr{A}$	part of an association scheme	13.2
$(N,\mathscr{T},\in,\mathscr{A})$	an association scheme	13.2
p^h_{ij}	parameters of an association scheme	13.2
PBIBD	partially balanced incomplete block design	14.1
$(V,\mathscr{E},I)$	a graph	16.1
$e\{x,y\}$	the edge of a graph between vertices x and y	16.1
$e\{x,y\} \in \mathscr{E}$	there is an edge between x and y	16.1
$e\{x,y\} \notin \mathscr{E}$	there is not an edge between x and y	16.1
$V_\Gamma(x)$	the set of vertices sharing an edge with x	16.1
$v_\Gamma(x)$	the number of vertices sharing an edge with x	16.1
v_Γ	the number of vertices sharing an edge with any vertex of a graph	16.1
$\Gamma_1 \times \Gamma_2$	the cartesian product of graph Γ_1 with Γ_2	16.4

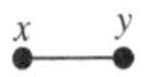	an edge between nodes x and y	16.5
	also an edge between nodes x and y	16.5
$\mathbf{a}$, $\mathbf{b}$	elements of $\mathbb{Z}_2^n$, that is, code words	17.1
$\mathbf{a} = a_1a_2\ldots a_n$	a sequence of 0s and 1s; a codeword	17.1
$w(\mathbf{a})$	the weight of $\mathbf{a}$; the number of 1s in $\mathbf{a}$	17.1
$d(\mathbf{a},\mathbf{b})$	the distance between $\mathbf{a}$ and $\mathbf{b}$	17.1
D	the minimal distance between codewords of a code	17.1
$\mathscr{C}(n)$	a code; a subset of $\mathbb{Z}_2^n$	17.1
$\mathscr{C}^A(n)$	the row code of an incidence matrix A	17.1
$\mathscr{C}_B(n)$	the column code of an incidence matrix A	17.1
W	the weight of a constant weight code	17.1
$B(\mathbf{a},r)$	the sphere or circle with center $\mathbf{a}$ and radius r	17.12
$A(n,D,W)$	the cardinality of the biggest code in Z_2^n with constant weight W and minimal distance D	17.12
δ	$\delta = D/2$	17.12

Miscellaneous Notation

$\gcd\{m,n\}$	greatest common divisor of m and n	A4.30
$\operatorname{lcm}\{m,n\}$	least common multiple of m and n	A4.31
$[x]$	the greatest integer less than or equal to x	17.3

Chapter I

Introduction

This chapter first introduces the reader to some basic combinatorial structures. It is a goal of this work to present some algebraic and geometric methods for constructing these combinatorial structures. Some of these will have abundant geometric structures, also, and another goal will be to study these. We will need some elementary but powerful combinatorial tools, so these will be developed next. Algebraic structures are useful (a) to construct combinatorial structures, and (b) to construct geometric structures, which in turn are used to construct additional combinatorial structures. Some of these will be introduced in the next chapter. The ones we have chosen are all related by a "planar condition," which is to say that each reasonable equation of the form $ax = bx + c$ has a unique solution.

1 Introductory ideas

In this section we introduce some key ideas. This will be done with a series o problems. Now these problems were chosen to illustrate in a rather dramat way the power of the methods we intend to develop. It would be a ve rare and courageous student who could solve these problems when they stated. So, it is not intended that the reader should solve them until further tools are presented. However, should you feel the desire to try, please do so. Now, let us begin!

Problem 1.1. Consider the set $S = \{0, 1, 2, \ldots, 11, 12\}$. This set has 1287 distinct subsets each containing exactly 5 elements. Select from these 1287 five-element subsets exactly 39 having the following properties: (a) each point $x \in S$ belongs to exactly 15 of these 39 subsets, and (b) each pair of distinct points $x, y \in S$, $x \neq y$, belongs to exactly 5 of these 39 subsets.

To solve Problem 1.1 would give you an example of a *balanced incomplete block design*, or BIBD, for short. And just what is a BIBD? First, we will explain exactly what one is, then we will talk a little about them, and then give an example.

Definition 1.2. Let V be a finite nonempty set, and suppose $\mathscr{B}$ is a nonempty family of nonempty subsets of V. Then $(V, \mathscr{B})$ is a *balanced incomplete block design* if there are parameters (constants) k, r, and λ with the following properties.

(a) If $B \in \mathscr{B}$, then B has exactly k elements.

(b) If $u \in V$, then there are exactly r distinct sets $B_1, B_2, \ldots, B_r \in \mathscr{B}$ each containing u.

(c) If $x, y \in V$, $x \neq y$, then there are exactly λ distinct sets $D_1, D_2, \ldots, D_\lambda \in \mathscr{B}$ each containing both x and y.

Perhaps it will be easier to think of the elements of V as *points* and each subset $B \in \mathscr{B}$ of V as a *line*. Then property (a) says that all the lines have the same number of points, property (b) says that each point is on the same number of lines, and property (c) says that each pair of distinct points are on the same number of lines. From time to time, we shall refer to the elements of V as points and the members of $\mathscr{B}$ as lines, and use the associated language of geometry.

But there is another language in use regarding BIBDs. In the 1930s, statisticians, designing large and elaborate agricultural experiments, needed more economical ways to gather data, and the idea of a BIBD grew out of this need. The use of BIBDs in large experiments of all kinds has continued to this day. In addition, BIBDs have found use in communication theory in detecting and correcting errors made in transmitting data. To the statistician, the elements of V are *varieties* and the subsets $B \in \mathscr{B}$ are *blocks*. Now you know why the symbols V, $\mathscr{B}$, and B are used. Actually, the mathematician has found that BIBDs are of interest because of their many interesting properties, and a mixture of the geometric and the statistical language is frequently used. For example, we will not hesitate to use such expressions as "each point $x \in V$ belongs to exactly r distinct blocks."

"Enough of this small talk," you say. "The time has come to see an honest BIBD." Well, here you are.

Example 1.3. Let $V = \{0,1,2,3,4,5,6\}$. We will now list the 14 blocks in $\mathscr{B}$.

$$\begin{array}{llll}
B_1 = \{0,2,4,1\}, & B_5 = \{4,6,1,5\}, & B_8 = \{0,6,5,3\}, & B_{12} = \{4,3,2,0\}, \\
B_2 = \{1,3,5,2\}, & B_6 = \{5,0,2,6\}, & B_9 = \{1,0,6,4\}, & B_{13} = \{5,4,3,1\}, \\
B_3 = \{2,4,6,3\}, & B_7 = \{6,1,3,0\}, & B_{10} = \{2,1,0,5\}, & B_{14} = \{6,5,4,2\}. \\
B_4 = \{3,5,0,4\}, & & B_{11} = \{3,2,1,6\}, &
\end{array}$$

Notice:

0 belongs to $B_1, B_4, B_6, B_7, B_8, B_9, B_{10}, B_{12}$;

1 belongs to $B_1, B_2, B_5, B_7, B_9, B_{10}, B_{11}, B_{13}$;

etc.

Also,

0, 1 belong to B_1, B_7, B_9, B_{10}; 0, 2 belong to B_1, B_6, B_{10}, B_{12};

etc.

So, $\mathscr{B}$ has the parameters $k = 4$, $r = 8$, and $\lambda = 4$. It will be rather easy for the reader to check that the remaining 5 points of V each belong to exactly 8 blocks, but it will be somewhat tedious to check that the remaining 19 pairs $x, y \in V$, $x \neq y$, each belong to exactly 4 blocks. We would encourage the reader to do so, however. In doing so, one begins to develop skills and intuition for future developments.

Exercise 1.4. Verify that each $u \in V$ of Example 1.3 actually belongs to 8 blocks of $\mathscr{B}$.

Exercise 1.5. Verify that each pair $x, y \in V$, $x \neq y$, of Example 1.3 actually belongs to 4 blocks of $\mathscr{B}$.

Exercise 1.6. If $v = |V|$, the number of points in V, and $b = |\mathscr{B}|$, the number of blocks in $\mathscr{B}$, find a relationship among the parameters v, b, r, k, and λ of Example 1.3.

Now we are ready for our second idea.

Problem 1.7. Consider the set $S = \{0, 1, 2, \ldots, 11, 12\}$. This set has exactly 715 distinct subsets each containing exactly 4 elements. Select from these 715 four-element subsets exactly 39 having the following properties: (1) each point $x \in S$ belongs to exactly 12 of these 39 subsets, (2) each pair of points $x, y \in S$, $x \neq y$, belong to exactly 3 of these 39 subsets, and (3) each triple $x, y, z \in S$ of three distinct points belong to at most one of these 39 subsets.

Notice that a solution to Problem 1.7 would also give a BIBD, but a rather special one. This leads to a modification of Definition 1.2.

Definition 1.8. A BIBD $(V, \mathscr{B})$ is *circular* if $\lambda > 1$ and every three distinct points $x, y, z \in V$ belong to at most one block $B \in \mathscr{B}$. In this case, the blocks $B \in \mathscr{B}$ are called *circles*.

Of course, the idea of a circular BIBD comes from the fact that every three distinct points on the classical euclidean plane belong to at most one circle.

We now provide an easy example of a circular BIBD. Since each block will contain exactly three points, it is not a particularly good example of a circular BIBD. (Better ones will come later.) But it is another nice example of a BIBD. Notice the similarities and differences from Example 1.3.

Example 1.9. Let $V = \{0,1,2,3,4,5,6\}$. We will now list the 14 blocks in $\mathscr{B}$:

$$\begin{array}{llll} B_1 = \{2,4,1\}, & B_5 = \{6,1,5\}, & B_8 = \{6,5,3\}, & B_{12} = \{3,2,0\}, \\ B_2 = \{3,5,2\}, & B_6 = \{0,2,6\}, & B_9 = \{0,6,4\}, & B_{13} = \{4,3,1\}, \\ B_3 = \{4,6,3\}, & B_7 = \{1,3,0\}, & B_{10} = \{1,0,5\}, & B_{14} = \{5,4,2\}. \\ B_4 = \{5,0,4\}, & & B_{11} = \{2,1,6\}, & \end{array}$$

Notice:

0 belongs to $B_4, B_6, B_7, B_9, B_{10}, B_{12}$;
1 belongs to $B_1, B_5, B_7, B_{10}, B_{11}, B_{13}$;
etc.

Also,

0, 1 belong to B_7 and B_{10}; 0, 2 belong to B_6 and B_{12};
etc.

So, $\mathscr{B}$ has the parameters $k = 3$, $r = 6$, and $\lambda = 2$.

Exercise 1.10. Verify that Example 1.9 really does form a BIBD.

Exercise 1.11. If $v = |V|$ and $b = |\mathscr{B}|$, then find a relationship among the parameters v, b, r, k, and λ of Example 1.9.

Problem 1.12. Consider the set $\mathbb{Z}_{25} = \{0,1,2,\ldots,24\}$. Select from the 53,130 distinct 5-element subsets exactly 30 having the following properties: (a) each point $x \in \mathbb{Z}_{25}$ belongs to exactly 6 of these 30 sets; (b) each pair of distinct points $x, y \in \mathbb{Z}_{25}$ belong to exactly one of these 30 sets; and (c) for any one of these 30 subsets, call it L, and any point $x \in \mathbb{Z}_{25}$ with $x \notin L$, there is exactly one subset of these 30 subsets, call it M, such that $x \in M$ and $L \cap M = \varnothing$.

A solution to Problem 1.12 would give you a BIBD with very special properties. This motivates the following concept.

Definition 1.13. A finite *affine plane* is a BIBD $(P, \mathscr{L})$ with the following properties.

(a) $\lambda = 1$.

(b) If $L \in \mathscr{L}, x \in P, x \notin L$, then there is a unique $M \in \mathscr{L}$ such that $x \in M$ and $L \cap M = \varnothing$.

(c) If $L \in \mathscr{L}$, then $|L| \geq 2$.

(d) If $L \in \mathscr{L}$, then there is a point $x \in P$ such that $x \notin L$.

Example 1.14. Let $P = \{0, 1, 2, \ldots, 8\}$. We will now list the 12 "lines" in $\mathscr{L}$.

$$\begin{aligned} K_1 &= \{0,3,6\}, & K_2 &= \{1,4,7\}, & K_3 &= \{2,5,8\}, \\ L_1 &= \{0,1,8\}, & L_2 &= \{3,4,2\}, & L_3 &= \{6,7,5\}, \\ M_1 &= \{0,2,7\}, & M_2 &= \{3,5,1\}, & M_3 &= \{6,8,4\}, \\ N_1 &= \{0,4,5\}, & N_2 &= \{3,7,8\}, & N_3 &= \{6,1,2\}. \end{aligned}$$

Exercise 1.15. Verify that Example 1.14 really is an affine plane.

Exercise 1.16. If $v = |P|$ and $b = |\mathscr{L}|$, then find a relationship among the parameters v, b, r, k, and λ of Definition 1.13.

Problem 1.17. Consider the set $\mathbb{Z}_{31} = \{0, 1, 2, \ldots, 30\}$. Select from the 736,281 distinct 6-element subsets exactly 31 having the following properties: (a) each point $x \in \mathbb{Z}_{31}$ belongs to exactly 6 of these 31 sets; (b) each pair of distinct points $x, y \in \mathbb{Z}_{31}$ belongs to exactly one of these 31 sets, and (c) for any two of these 31 subsets, say A and B, with $A \neq B$, we have exactly one point in $A \cap B$, that is, $|A \cap B| = 1$.

Again, a solution to Problem 1.17 would give a BIBD $(P, \mathscr{L})$ with significant properties which are included in the following definition.

Definition 1.18. A BIBD $(P,\mathscr{L})$ with the following properties is a finite *projective plane.*

(a) $\lambda = 1$.

(b) If $L,M \in \mathscr{L}$ and $L \neq M$, then $|L \cap M| = 1$.

(c) If $L \in \mathscr{L}$, then $|L| \geq 3$.

(d) If $L \in \mathscr{L}$, then there is a point $x \in P$ such that $x \notin L$.

Example 1.19. Let $P = \{0,1,2,\ldots,8,a,b,c,d\}$. We will now list the 13 "lines" in $\mathscr{L}$.

$$\begin{aligned}
K_1 &= \{0,3,6,a\}, & K_2 &= \{1,4,7,a\}, & K_3 &= \{2,5,8,a\},\\
L_1 &= \{0,1,8,b\}, & L_2 &= \{3,4,2,b\}, & L_3 &= \{6,7,5,b\},\\
M_1 &= \{0,2,7,c\}, & M_2 &= \{3,5,1,c\}, & M_3 &= \{6,8,4,c\},\\
N_1 &= \{0,4,5,d\}, & N_2 &= \{3,7,8,d\}, & N_3 &= \{6,1,2,d\},
\end{aligned}$$
$$I = \{a,b,c,d\}.$$

Exercise 1.20. Verify that Example 1.19 really is a projective plane.

Exercise 1.21. If $v = |P|$ and $b = |\mathscr{L}|$, then find a relationship among v, b, r, k, and λ of Example 1.19.

Notice how Example 1.19 and Example 1.14 are related. In Example 1.19, the line I is not so special.

Exercise 1.22. Choose any line X in Example 1.19 other than I. Eliminate the influence of X in P and in each of the lines of Example 1.19. That is, take the points of X away from P and each line Y of Example 1.19. One then has a set $P' = P \setminus X$, containing exactly nine points, and a family of 12 lines each of which has exactly three points. Show that what you have left is an affine plane.

There is one more type of BIBD which we shall introduce here.

Problem 1.23. Consider the set $P = \{0, 1, 2, \ldots, 24, \infty\}$ where ∞ is any symbol not in $\{0, 1, 2, \ldots, 24\}$. Select exactly 130 six-element subsets having the following properties: (a) each $x \in P$ belongs to exactly 30 of these 130 subsets; (b) each pair of distinct points $x, y \in P$ belongs to exactly 6 of these 130 subsets; and (c) if $x, y, z \in P$ are three distinct points of P, then there is exactly one of these 130 subsets containing x, y, and z.

A solution to Problem 1.23 would give you an example of a BIBD which is also a *Möbius plane*. Such a BIBD is also circular.

Definitions 1.24. A BIBD $(P, \mathscr{C})$ is a finite *Möbius plane*, or *inversive plane* if it has parameters $(v, b, r, k, \lambda) = (n^2+1,\ n(n^2+1),\ n(n+1),\ n+1,\ n+1)$ for some integer $n \geq 2$, and with the additional property that for every three distinct points $x, y, z \in P$, there is a unique $C \in \mathscr{C}$ such that $x, y, z \in C$. The blocks $C \in \mathscr{C}$ are called *circles*.

Example 1.25. Let $P = \{0, 1, 2, x, 1+x, 2+x, 2x, 1+2x, 2+2x, \infty\}$ We now list the 30 circles in $\mathscr{C}$.

$\{0,1,2,\infty\}$,	$\{0,x,2x,\infty\}$,	$\{1+2x,x,1,2+x\}$,
$\{x,1+x,2+x,\infty\}$,	$\{1,1+x,1+2x,\infty\}$,	$\{2x,0,2,2+2x\}$,
$\{2x,1+2x,2+2x,\infty\}$,	$\{2,2+x,2+2x,\infty\}$,	$\{2+2x,1+x,2,x\}$,
$\{0,1+x,2+2x,\infty\}$,	$\{0,1+2x,2+x,\infty\}$,	$\{1+2x,1,0,2x\}$,
$\{x,1+2x,2,\infty\}$,	$\{x,1,2+2x,\infty\}$,	$\{0,2+2x,x,1+2x\}$,
$\{2x,1,2+x,\infty\}$,	$\{2x,1+x,2,\infty\}$,	$\{2,2+x,1+x,1\}$,
$\{x,2,2x,1\}$,	$\{2+x,2+2x,1+2x,1+x\}$,	$\{1,2x,1+x,2+2x\}$,
$\{1+x,0,1+2x,2\}$,	$\{x,2x,2+2x,2+x\}$,	$\{0,x,2+x,2\}$,
$\{2+x,1,2+2x,0\}$,	$\{1+x,1+2x,2x,x\}$,	$\{2,1+2x,2+x,2x\}$,
$\{2x,2+x,0,1+x\}$,	$\{2+2x,2,1,1+2x\}$,	$\{1,1+x,x,0\}$.

Exercise 1.26. Verify that Example 1.25 really is a Möbius plane, if you have the courage. After we develop some theory, this will be relatively easy.

Exercise 1.27. Suppose $v = |P|$ and $b = |\mathscr{C}|$. Find a relationship among v, b, r, k, and λ of Example 1.25.

Exercise 1.28. For Example 1.25, remove ∞ from P and from each line containing it. Then remove a suitable set of circles so that the remaining $P' = P \setminus \{\infty\}$ and the remaining blocks form an affine plane.

Exercise 1.29. For Example 1.25, remove ∞ and appropriate circles to obtain a circular BIBD. What are the parameters of your BIBD?

We next introduce an idea which has a very rich history, which has numerous applications, and which is a topic of active research today.

Problem 1.30. Let $P = \{0,1,2,3,4,5,6\}$. Construct a 7 by 7 matrix

$$A = \begin{pmatrix} a_{11} & a_{12} & \dots & a_{17} \\ a_{21} & a_{22} & \dots & a_{27} \\ \vdots & \vdots & \ddots & \vdots \\ a_{71} & a_{72} & \dots & a_{77} \end{pmatrix}$$

where each $a_{ij} \in P$. This is not very difficult, and there are 7^{49} such matrices. Of course, we are looking for one with special properties. Our matrix must have the following properties:

(a) For each i, $1 \le i \le 7$, $\{a_{i1}, a_{i2}, \dots, a_{i7}\} = P$.
(b) For each j, $1 \le j \le 7$, $\{a_{1j}, a_{2j}, \dots, a_{7j}\} = P$.

So, each $x \in P$ occurs exactly once in each row and each column of A.

You may find it relatively easy to construct such a matrix A. But after you construct one, you must then construct another such matrix

$$B = \begin{pmatrix} b_{11} & b_{12} & \dots & b_{17} \\ b_{21} & b_{22} & \dots & b_{27} \\ \vdots & \vdots & \ddots & \vdots \\ b_{71} & b_{72} & \dots & b_{77} \end{pmatrix}.$$

Now that you have two such matrices, you must test your second one to see if it "goes with" the first. Look at the matrix

$$A*B=\begin{pmatrix}(a_{11},b_{11}) & (a_{12},b_{12}) & \dots & (a_{17},b_{17})\\ (a_{21},b_{21}) & (a_{22},b_{22}) & \dots & (a_{27},b_{27})\\ \vdots & \vdots & \ddots & \vdots\\ (a_{71},b_{71}) & (a_{72},b_{72}) & \dots & (a_{77},b_{77})\end{pmatrix}.$$

Now check to see if each $(x,y)\in P\times P$ is an entry in the matrix $A*B$. That is, can you find each and every $(x,y)\in P\times P$ in $A*B$ where $x,y\in P$? If not, then try another such B.

Perhaps you do find such an A and B. But we still want more. We want six such matrices $A_1,A_2,\dots,A_6$ so that if $A,B\in\{A_1,A_2,\dots,A_6\}$, and $A\neq B$, then $A*B$ has each $(x,y)\in P\times P$ as an entry. If you can find such matrices without the knowledge developed in later chapters, then you are to be congratulated.

Definitions 1.31. Let $P=\{0,1,2,\dots,n-1\}$. An n by n matrix

$$A=\begin{pmatrix}a_{11} & a_{12} & \dots & a_{1n}\\ a_{21} & a_{22} & \dots & a_{2n}\\ \vdots & \vdots & \ddots & \vdots\\ a_{n1} & a_{n2} & \dots & a_{nn}\end{pmatrix}$$

is a *Latin square of order n* if each $x\in P$ occurs exactly once in each row and each column of A. If

$$B=\begin{pmatrix}b_{11} & b_{12} & \dots & b_{1n}\\ b_{21} & b_{22} & \dots & b_{2n}\\ \vdots & \vdots & \ddots & \vdots\\ b_{n1} & b_{n2} & \dots & b_{nn}\end{pmatrix}$$

is also a Latin square of order n, then A and B are *orthogonal* if

$$A * B = \begin{pmatrix} (a_{11}, b_{11}) & (a_{12}, b_{12}) & \dots & (a_{1n}, b_{1n}) \\ (a_{21}, b_{21}) & (a_{22}, b_{22}) & \dots & (a_{2n}, b_{2n}) \\ \vdots & \vdots & \ddots & \vdots \\ (a_{n1}, b_{n1}) & (a_{n2}, b_{n2}) & \dots & (a_{nn}, b_{nn}) \end{pmatrix}$$

has exactly one entry for any $(x,y) \in P \times P$. A family $\mathscr{A} = \{A_1, A_2, \dots, A_k\}$ of Latin squares of order n are *mutually orthogonal* if $A_i, A_j \in \mathscr{A}$, $A_i \neq A_j$, imply that A_i and A_j are orthogonal.

Example 1.32. For $P = \{0,1,2,3,4\}$, we have four mutually orthogonal Latin squares of order 5 given by the following four matrices.

$$\begin{pmatrix} 0 & 1 & 2 & 3 & 4 \\ 4 & 0 & 1 & 2 & 3 \\ 3 & 4 & 0 & 1 & 2 \\ 2 & 3 & 4 & 0 & 1 \\ 1 & 2 & 3 & 4 & 0 \end{pmatrix}, \begin{pmatrix} 0 & 1 & 2 & 3 & 4 \\ 3 & 4 & 0 & 1 & 2 \\ 1 & 2 & 3 & 4 & 0 \\ 4 & 0 & 1 & 2 & 3 \\ 2 & 3 & 4 & 0 & 1 \end{pmatrix}, \begin{pmatrix} 0 & 1 & 2 & 3 & 4 \\ 2 & 3 & 4 & 0 & 1 \\ 4 & 0 & 1 & 2 & 3 \\ 1 & 2 & 3 & 4 & 0 \\ 3 & 4 & 0 & 1 & 2 \end{pmatrix}, \begin{pmatrix} 0 & 1 & 2 & 3 & 4 \\ 1 & 2 & 3 & 4 & 0 \\ 2 & 3 & 4 & 0 & 1 \\ 3 & 4 & 0 & 1 & 2 \\ 4 & 0 & 1 & 2 & 3 \end{pmatrix}.$$

Exercise 1.33. Construct an affine plane from Example 1.32. (This may prove to be very difficult at this time. You may want to return to this exercise when you learn how to construct mutually orthogonal Latin squares.)

Exercise 1.34. Verify that the four matrices given in Example 1.32 are mutually orthogonal Latin squares.

Exercise 1.35. Construct a family of nine mutually orthogonal Latin squares of order 10. (This is a joke. We only mention it to dramatize that such a simply presented problem is impossible!) So, the exercise really is to construct a family of two orthogonal Latin squares of order 3.

Exercise 1.36. Construct two Latin squares of order 6. Are your examples orthogonal?

Exercise 1.37 (Euler). Given 36 officers from 6 regiments and 6 ranks, can they be arranged into a square so that each row and each column contains exactly one officer of each rank and exactly one officer from each regiment? Hint: If your solution to Exercise 1.36 gave two orthogonal Latin squares, then you have essentially solved this exercise. Another hint: If your solution to Exercise 1.36 gave two orthogonal Latin squares, then you should check your work very carefully.

Problem 1.38. Suppose you have 20 golf balls and you are to distribute these 20 golf balls among six baskets, labelled 1, 2, 3, 4, 5, and 6. How many ways can this be done? How many ways can this be done if each basket is to have at least one golf ball? In each case, exactly what are the ways to distribute the golf balls?

Let us look at Problem 1.38 more carefully. Consider first a simpler case. Suppose we are to distribute 3 golf balls between two baskets. If each basket is to contain at least one golf ball, then there are only two ways. The first way is for basket one to have 1 and and basket two to have 2. The second way is for basket one to have 2 and basket two to have 1. If there is no restriction on the number of balls in a basket, then there are two more possibilities. One can put all three balls in basket one, or one can put all three balls in basket two.

Returning now to Problem 1.38, first consider the case where there is no restriction to the number of balls in a basket. If we put n_1 balls in basket 1, n_2 balls in basket 2, etc., then

$$n_1 + n_2 + n_3 + n_4 + n_5 + n_6 = 20.$$

So, how many ways can 20 be expressed as a sum of 6 not necessarily distinct nonnegative integers? If there is the restriction that each basket must contain at least one ball, then we would ask, "how many ways can 20 be expressed as a sum of 6 not necessarily distinct positive integers?"

For example, $(0,4,4,0,10,2)$ and $(10,0,0,4,4,2)$ would be two distinct solutions if there is no restriction on the number of balls in a basket. Similarly, $(1,3,3,1,10,2)$ and $(10,1,1,3,3,2)$ would be two distinct ways if each basket is to have at least one ball.

In general, suppose one must distribute n golf balls among k baskets labelled $1, 2, \ldots, k$, where $k \leq n-1$. How many ways can this be done? If each basket is to contain at least one golf ball, then how many ways can this be done? In each case, exactly what are the ways to distribute the golf balls? In either case, basket 1 gets n_1 golf balls, basket 2 gets n_2 balls, ..., and basket k gets n_k balls, where

$$n_1 + n_2 + \cdots + n_k = n.$$

So $(n_1, n_2, \ldots, n_k)$ is a solution to the equation

$$x_1 + x_2 + \cdots + x_k = n. \tag{1:1}$$

The first question can then be answered by solving (1:1) among the nonnegative integers, and the second question can also be answered by solving (1:1) among the positive integers.

We give now partial answers to Problem 1.38 and replace Problem 1.38 with a more general problem. There are 11,628 solutions $(n_1, \ldots, n_6)$ to the equation $x_1 + \cdots + x_6 = 20$ among the positive integers, and there are 53,130 solutions $(n_1, \ldots, n_6)$ to $x_1 + \cdots + x_6 = 20$ among the nonnegative integers.

Problem 1.39. For positive integers n and k, where $k \leq n-1$, how many solutions $(n_1, \ldots, n_k)$ are there to (1:1) among the positive integers? How many solutions $(n_1, \ldots, n_k)$ are there to (1:1) among the nonnegative integers? In each case, what are the solutions?

Problem 1.40. From the various exercises given earlier concerning the parameters v, b, r, k, and λ of a BIBD, one could naturally conjecture that there are relationships among these parameters. That is certainly the case. Formulate conjectures about such relationships. Then try to prove that your conjectures are correct.

2 Counting 101

How do you count? Most educated people have only three tools. The first is like counting on your fingers. You simply enumerate the items you wish to count. You say, "1, 2, 3, 4, ..." The second tool adds a little sophistication. Suppose you want to count the number of objects in a certain collection C. Perhaps this collection C can be partitioned nicely into pairwise disjoint sets $C_1, C_2, \ldots, C_k$. Now you count the number in each C_i, $i = 1, 2, \ldots, k$, and get numbers $n_1, n_2, \ldots, n_k$. Then $n = n_1 + n_2 + \cdots + n_k$ is the number of items in C. As an example, you want to count the money in the cash register. In this collection, there are checks, currency, and coins. First you count the amount in checks to get n_1. Then you count the amount in currency to get n_2, and finally, you count the coins to get n_3. You are assured that $n = n_1 + n_2 + n_3$ is the amount of money in this collection. The third tool is very important and very powerful. We will use this third tool to develop several elementary but powerful techniques of counting. So, what is this third tool?

Suppose you walk into a theatre, with the lights on, and notice that it has 20 rows of seats and that there are 15 seats in each row. You know immediately that there are $20 \cdot 15 = 300$ seats in the theatre. In general, if there are m rows of seats and each row has n seats in it, then there are exactly mn seats to choose from. You could choose your seat by first selecting a row, and then selecting a seat in that row. This leads to

2.1. The First Principle of Advanced Counting. If you have m ways to make a first choice, and following your first choice, no matter what it was, you have n ways to make a second choice, then you have mn ways of making a sequence of a first choice followed by a second choice.

This leads to

2.2. The General Principle of Advanced Counting. You are to make a sequence of k choices, one after the other. You have n_1 ways to make your first choice. No matter what your first choice was, you have n_2 ways of making a second choice. Having made your first choice and second choice in one of the $n_1 n_2$ possible ways, you now have n_3 ways of making a third choice, no matter how you made your first and second choices. You continue in this way. Having made a sequence of first choice, second choice, ..., $(k-1)$th choice, of which there were n_1, n_2, ..., n_{k-1} possible ways of doing so, you have n_k ways to make your kth choice. Then there are

$$n_1 n_2 n_3 \dots n_k$$

distinct ways of making this sequence of first choice, second choice, ..., and finally the kth choice.

We have found it convenient to record a sequence of k choices as described in **2.2** with a k-tuple $(c_1, c_2, \dots, c_k)$. With n_1 possible choices for c_1, n_2 possible choices for c_2, ..., and n_k possible choices for c_k, one gets $n_1 n_2 \cdots n_k$ possible k-tuples $(c_1, c_2, \dots, c_k)$ representing the history of the sequence of k choices.

Our first application of **2.2** is to count selections with replacement.

Theorem 2.3 (Selection with Replacement). *Suppose one makes a sequence of k selections from a set of n objects, and after each selection is made, that object is returned to be available for the next selection. Then there are n^k such possible sequences of k selections.*

Proof. Upon each choice for selection, there are n choices. Hence, we have $n_1 = n_2 = \cdots = n_k = n$, and so by **2.2**, $n_1 n_2 \cdots n_k = n^k$ is the number of possible sequences of k choices. □

In Problem 1.30, there are $7^2 = 49$ choices to be made, one for each (i, j)-position of the matrix $A = (a_{ij})$. Each choice must come from the seven element set $P = \{0, 1, 2, 3, 4, 5, 6\}$, so there are 7 possibilities for each of the 49 positions. Hence, there are 7^{49} such matrices A.

Theorem 2.4. *Let R be a finite commutative ring with identity with $|R| = q$ elements. For each of the $q - 1$ nonzero elements $a_n \in R$, there are q^n distinct polynomials*

$$a_0 + a_1 x + a_2 x^2 + \cdots + a_n x^n$$

of degree n in $R[x]$.

Proof. Each coefficient a_i may be chosen from among the q distinct elements of the ring R, and there are $n + 1$ such coefficients $a_0, a_1, \ldots, a_n \in R$ to be chosen. Hence, there are q^{n+1} such polynomials. □

Corollary 2.5. *If F is a finite field and if $|F| = q$, then for a polynomial $a \in F[x]$ of degree n, the quotient ring $F[x]/(a)$ has exactly q^n elements.*

Proof. The quotient ring $F[x]/(a)$ consists of the cosets

$$r_0 + r_1 x + \cdots + r_{n-1} x^{n-1} + (a),$$

where each $r_i \in F$. (See Theorem A2.32.) By Theorem 2.4, there are q^n such polynomials $r_0 + r_1 x + \cdots + r_{n-1} x^{n-1}$. □

Corollary 2.6. *For a finite field F with $|F| = q$, and for an irreducible $a \in F[x]$ of degree n, the field $F[x]/(a)$ has exactly q^n elements.*

So, for a prime p, the task of finding a field with exactly p^n elements can be reduced to finding an irreducible polynomial $a \in \mathbb{Z}_p[x]$ with $\deg a = n$.

Our second application of **2.2** is to count selections without replacement. Compare the next theorem with Theorem 2.3.

Theorem 2.7 (Selection without replacement). *Suppose one makes a sequence of k selections from a set of n objects, where $k \leq n$. After each selection, suppose the object is not returned to the set. Then there are exactly*

$$P(n,k) = n(n-1)\cdots(n-(k-1)) = \frac{n!}{(n-k)!} \tag{2:2}$$

possible sequences $(c_1, c_2, \ldots, c_k)$.

Proof. Our first choice c_1 has $n_1 = n$ possibilities. So our second choice c_2 has $n_2 = n-1$ possibilities. If we have $n_i = n-(i-1)$ choices for the ith choice c_i, then we will have $n_{i+1} = n-i$ choices for the $(i+1)$th choice c_{i+1}. Hence, we have $n_1 n_2 \cdots n_k$, i.e., $n(n-1)\cdots(n-(k-1))$ possible sequences $(c_1, c_2, \ldots, c_k)$. □

Corollary 2.8. *There are $n!$ possible ways to choose a sequence*

$$(c_1, c_2, \ldots, c_n)$$

from a set of n objects if we do not replace the object after each choice.

Proof. In Theorem 2.7, $k = n$, so $n-(k-1) = 1$. □

Definitions 2.9. A sequence of k choices from a set of n objects without replacement, as in Theorem 2.7, is a *permutation of n objects taken k at a time*. It is also called an *arrangement of k objects from a set of n objects*. If $n = k$, then such a sequence is a *permutation of n objects*.

For a permutation $(c_1, c_2, \ldots, c_n)$ of a set S of n objects, there is the original set $\{c_1, c_2, \ldots, c_n\}$. That is, $\{c_1, c_2, \ldots, c_n\} = S$. For a permutation of n objects taken k at a time $(c_1, c_2, \ldots, c_k)$, there is the subset $\{c_1, c_2, \ldots, c_k\}$ of k objects of S. So $\{c_1, c_2, \ldots, c_k\}$ is one of possibly many subsets of S having exactly k of the n objects of S as members. Of course, it may be that $(d_1, d_2, \ldots, d_k) \neq (c_1, c_2, \ldots, c_k)$ but yet $\{d_1, d_2, \ldots, d_k\} = \{c_1, c_2, \ldots, c_k\}$. For example, $(6,5,2) \neq (5,6,2)$ but yet

$\{6,5,2\} = \{5,6,2\}$. Since a set $\{c_1, c_2, \ldots c_k\}$ has $k!$ permutations, there are $k!$ permutations of n objects taken k at a time for each subset $\{c_1, c_2, \ldots, c_k\}$ of S with cardinality k. If A and B are two different subsets of S with cardinality k, then none of the $k!$ permutations of A is equal to any of the $k!$ permutations of B. For if $(a_1, a_2, \ldots, a_k)$ is a permutation of the k elements of A and if $(b_1, b_2, \ldots, b_k)$ is a permutation of the k elements of B, then $A = \{a_1, a_2, \ldots, a_k\}$ and $B = \{b_1, b_2, \ldots, b_k\}$. But $A \neq B$, so some $a_i \in A$ is not in B. That is, there is an $a_i \in A$ such that $a_i \notin B$. (Likewise, there is a $b_j \in B$ such that $b_j \notin A$.) Hence, $(a_1, a_2, \ldots a_k) \neq (b_1, b_2, \ldots, b_k)$ since they differ in at least the ith position.

But just how many subsets of cardinality k are there in a set of n objects? (Of course, we are assuming that $k \leq n$ or there would be 0.) Let $\binom{n}{k}$ denote the answer to this question. So, $\binom{n}{k}$ denotes the number of subsets of S which have cardinality k, where $|S| = n$. We will derive a formula for $\binom{n}{k}$ by counting the number of permutations of the n objects of S taken k at a time in two ways.

Our first way of counting the number of permutations of the n objects of S taken k at a time is as we have done in Theorem 2.7. So there are $n!/(n-k)!$ such ways.

For our second way, we apply the First Principle, namely **2.1**. There are $\binom{n}{k}$ ways of choosing k-element subsets of S. For each of these $\binom{n}{k}$ choices, there are $k!$ permutations of that subset. This follows from Corollary 2.8. So, by **2.1**, there are $\binom{n}{k} \cdot k!$ ways of making permutations of the n objects of S taken k at a time. Our first choice was to select one of the $\binom{n}{k}$ subsets of S of cardinality k. Our second choice was to choose the $k!$ permutations of this subset of cardinality k. Each such permutation of this subset is a permutation of the n elements of S taken k at a time.

On the one hand, we get $n!/(n-k)!$, and on the other hand, we get $\binom{n}{k} \cdot k!$. Hence, $\binom{n}{k} \cdot k! = n!/(n-k)!$. Consequently, $\binom{n}{k} = n!/[k!(n-k)!]$. We summarize with

Theorem 2.10 (The Number of Combinations). *In any set S with $|S| = n$, there*

are exactly

$$C(n,k) = \binom{n}{k} = \frac{n!}{k!(n-k)!} \tag{2:3}$$

subsets of cardinality k, where $0 \le k \le n$.

One often refers to the number $C(n,k)$ as *the number of combinations of n objects taken k at a time*. Also, $\binom{n}{k}$ is referred to as a *binomial coefficient*.

Hopefully, the reader can recall the binomial theorem. For any numbers a and b, and any positive integer n, we have

$$(a+b)^n = \sum_{k=0}^{n} \binom{n}{k} a^{n-k} b^k,$$

so with $a = b = 1$, we get

$$2^n = (1+1)^n = \sum_{k=0}^{n} \binom{n}{k} = \binom{n}{0} + \binom{n}{1} + \cdots + \binom{n}{n-1} + \binom{n}{n}.$$

So we have, using Theorem 2.10, the following theorem.

Theorem 2.11. *A set S of n elements has exactly 2^n distinct subsets.*

Another proof. Let $\mathscr{P}(S) = \{A \mid A \subseteq S\}$, where $S = \{s_1, s_2, \ldots, s_n\}$, and let

$$\mathbb{Z}_2^n = \{(b_1, \ldots, b_n) \mid b_i \in \mathbb{Z}_2,\ 1 \le i \le n\}.$$

Define the function $b : \mathscr{P}(S) \to \mathbb{Z}_2^n$ by $b(A) = (b_1, \ldots, b_n)$ where

$$b_i = \begin{cases} 0, & \text{if } s_i \notin A; \\ 1, & \text{if } s_i \in A. \end{cases}$$

Therefore, $b_i = 1$ if and only if $s_i \in A$. So, if $A \ne B$, then there is an element in one set and not the other. Suppose $s_i \in A \setminus B$. Then $b(A) = (a_1, \ldots, a_n)$ where $a_i = 1$, whereas $b(B) = (b_1, \ldots, b_n)$ where $b_i = 0$. Hence, $A \ne B$ implies that $b(A) \ne b(B)$. So b is injective.

Take $d = (d_1, \ldots, d_n) \in \mathbb{Z}_2^n$ and from d construct the set $D = \{s_i \in S \mid d_i = 1\}$. Then if $b(D) = (e_1, \ldots, e_n)$, then $e_i = 1$ if and only if $s_i \in S$, which is true if and only if $d_i = 1$. Hence, $(e_1, \ldots, e_n) = (d_1, \ldots, d_n)$. Now we have that b is surjective, since $b(D) = d$, and therefore that b is bijective. Hence $|\mathscr{P}(S)| = |Z_2^n| = 2^n$.

From a set S of n objects, if we choose a subset A of cardinality k, then we also have chosen a subset $S \setminus A$ of cardinality $n-k$. We then have a partition of S into two disjoint subsets. Hence, $C(n,k)$ is the number of partitions of S into two disjoint subsets, one subset having $m_1 = k$ elements, and the other having $m_2 = n-k$ elements, and $m_1 + m_2 = n$. (We also obtain the equation $C(n,k) = C(n,n-k)$.)

Suppose we want to count the number of ways to partition S into t pairwise disjoint subsets $A_1, A_2, \ldots, A_t$, where $S = A_1 \cup A_2 \cup \cdots \cup A_t$, where each $|A_i| = m_i$, $i = 1, 2, \ldots, t$, and so $m_1 + m_2 + \cdots + m_t = n$. We obtain such a partition by applying the General Principle **2.2** and at the same time we get the desired number of ways.

First, select a subset A_1 of S with exactly m_1 elements. This can be done in $n_1 = C(n, m_1)$ ways. This leaves us a set $S \setminus A_1$ of $n - m_1$ elements from which we make our second choice, a subset A_2 of $S \setminus A_1$ of m_2 elements. This can be done in $n_2 = C(n - m_1, m_2)$ ways, and leaves a set $S \setminus (A_1 \cup A_2)$ of $n - (m_1 + m_2)$ elements. We continue in this way, but let us be exact.

Suppose we have selected the pairwise disjoint subsets A_1, A_2, ..., A_k of S, with $m_i = |A_i|$, $1 \leq i \leq k$, in such a way that for each $j \in \{1, 2, \ldots, k\}$, A_j was selected as a subset of cardinality m_j from the set $S \setminus (A_1 \cup \cdots \cup A_{j-1})$ of cardinality $s_j = n - (m_1 + \cdots + m_{j-1})$. So A_j was one of $n_j = C(s_j, m_j)$ possible choices. We next select A_{k+1}, a subset of $S \setminus (A_1 \cup \cdots \cup A_k)$, where

$$|A_{k+1}| = m_{k+1}, \quad |S \setminus (A_1 \cup \cdots \cup A_k)| = s_{k+1} = n - (m_1 + \cdots + m_k),$$

and so A_{k+1} is one of $n_{k+1} = C(s_{k+1}, m_{k+1})$ possible choices. We continue until we get the set A_{t-1}. In $S \setminus (A_1 \cup \cdots \cup A_{t-1})$ there are $s_t = n - (m_1 + \cdots + m_{t-1}) = m_t$

elements remaining. Let $A_t = S \setminus (A_1 \cup \cdots \cup A_{t-1})$. We have $n_t = C(s_t, m_t) = 1$ choice for A_t. So we have made a sequence of t choices.

There were $n_1 = C(n, m_1)$ ways to make our choice for A_1. There were $n_2 = C(s_2, m_2)$ ways to make our choice for A_2. And in general, there were

$$n_j = C(s_j), m_j)$$

ways to make our choice for A_j. Our result is a partition $\mathscr{A} = \{A_1, A_2, \ldots, A_t\}$, where each $|A_i| = m_i$, and $m_1 + m_2 + \cdots + m_t = n$. The number of such partitions, being the number of such sequence of choices is

$$\begin{aligned} n_1 n_2 \cdots n_t &= C(n, m_1) C(s_2, m_2) \cdots C(s_t, m_t) \\ &= C(n, m_1) C(n - m_1, m_2) \cdots C(n - (m_1 + \cdots + m_{t-1}), m_t). \end{aligned}$$

Another way of writing this number is

$$\frac{n!}{m_1!(n-m_1)!} \cdot \frac{(n-m_1)!}{m_2![n-(m_1+m_2)]!} \cdot \cdots \cdot \frac{[n-(m_1+\cdots+m_{t-1})]!}{m_t![n-(m_1+\cdots+m_t)]!},$$

which can be rewritten simply as

$$\frac{n!}{m_1! m_2! \cdots m_t!}.$$

This important result is formalized.

Theorem 2.12. *For a finite set S with $|S| = n$, suppose $m_1, m_2, \ldots, m_t$ are positive integers such that $m_1 + m_2 + \cdots + m_t = n$. Then the number of partitions $\mathscr{A} = \{A_1, A_2, \ldots, A_t\}$ of S, where each $|A_i| = m_i$, is*

$$\binom{n}{m_1, m_2, \ldots, m_t} = \frac{n!}{m_1! m_2! \cdots m_t!}. \tag{2:4}$$

The number $\binom{n}{m_1, m_2, \ldots, m_t}$ is often called a *multinomial coefficient*.

We are now in a position to solve the problem stated in Problem 1.39. This is an example of combinatorics at its best. The solution we provide is extremely elegant.

Suppose we have n golf balls, numbered temporarily 1, 2, ..., n, and denoted by $b_1, b_2, \ldots, b_n$. We put these n balls uniformly on a line in sequential order, and in such a way to have a space between each adjacent pair of balls. There are $n-1$ such spaces which we denote by s_1, s_2, ..., s_{n-1}. Such an arrangement can be expressed symbolically by the $(2n-1)$-tuple

$$(b_1, s_1, b_2, s_2, \ldots, b_{n-1}, s_{n-1}, b_n). \tag{2:5}$$

This $(2n-1)$-tuple is of the form $(x_1, x_2, \ldots, x_{2n-1})$ where the ith odd position has b_i and the jth even position has s_j as entries.

We will first determine the number of solutions $(n_1, \ldots, n_k)$ to (1:1) among the positive integers. When we have this problem solved, we will use our result to determining the number of solutions $(n_1, \ldots, n_k)$ to (1:1) among the nonnegative integers.

Suppose $(n_1, \ldots, n_k)$ is a solution to (1:1) among the positive integers. This represents taking the first n_1 balls, namely $b_1, \ldots, b_{n_1}$ and putting them in basket 1. Then take the next n_2 balls, namely,

$$b_{n_1+1}, b_{n_1+2}, \ldots, b_{n_1+n_2},$$

and putting them in basket 2. In general, put the balls

$$b_{m_i+1}, b_{m_i+2}, \ldots, b_{m_i+n_i},$$

where $m_1 = 0$ and for $i > 1$, $m_i = n_1 + \cdots + n_{i-1}$, in basket i.

Another way to say the same thing is as follows. Take the $u_1 = n_1$ space s_{u_1} and put in basket 1 all the balls b_i where $1 \leq i \leq u_1$. So basket 1 gets the first n_1 balls. Next, take the $u_2 = n_1 + n_2$ space s_{u_2} and put in basket 2 all the balls b_i where $u_1 < i \leq u_2$. In general, for $2 \leq j \leq k-1$, take the $u_j = n_1 + \cdots + n_j$ space s_{u_j} and put in basket j the balls b_i where $u_{j-1} < i \leq u_j$. Finally, put in basket k the balls b_i where $u_{k-1} < i \leq n$. The important thing to learn from this is that a solution $(n_1, \ldots, n_k)$ to (1:1) among the positive integers determines

$k-1$ positions $s_{u_1},\ldots,s_{u_{k-1}}$, with $u_1 < u_2 < \cdots < u_{k-1}$, which in turn, amounts to selecting $k-1$ integers $u_1,u_2,\ldots,u_{k-1}$ from the set $\{1,2,\ldots,n-1\}$, where $u_1 < u_2 < \cdots < u_{k-1}$.

Conversely, if $s_{u_1},\ldots,s_{u_{k-1}}$, with $u_1 < u_2 < \cdots < u_{k-1}$, is a selection of $k-1$ of the $n-1$ positions, one gets a solution $(n_1,\ldots,n_k)$ to (1:1) among the positive integers. Put balls b_i, $1 \le i \le u_1$ in basket 1. Put balls b_i, $u_1 < i \le u_2$, in basket 2. For $2 \le j \le k-1$, put balls b_i, $u_{j-1} < i \le u_j$, in basket j. Finally, put balls b_i, $u_{k-1} < i \le n$, in basket k. Since there is at least one ball between two different positions s_i and s_j, we have that each basket has at least one ball. In fact, basket 1 has $n_1 = u_1$ balls, basket 2 has $n_2 = u_2 - u_1$ balls, and for $2 \le j \le k-1$, basket j has $n_j = u_j - u_{j-1}$ balls. Finally, basket k has $n_k = n - u_{k-1}$ balls. Now

$$\begin{aligned}
&n_1 + n_2 + \cdots + n_{k-1} + n_k \\
&\quad = u_1 + (u_2 - u_1) + \cdots + (u_{k-1} - u_{k-2}) + (n - u_{k-1}) \\
&\quad = n.
\end{aligned}$$

Hence, $(n_1,\ldots,n_k)$ is a solution to (1:1) among the positive integers.

So, a solution $(n_1,\ldots,n_k)$ to (1:1) among the positive integers is equivalent to selecting $k-1$ of the $n-1$ spaces $s_1,\ldots,s_{n-1}$. From Theorem 2.10, we know there are exactly $\binom{n-1}{k-1}$ such selections. This completes the proof of

Theorem 2.13. *The number of solutions* $(n_1,\ldots,n_k)$ *to* (1:1) *among the positive integers is*

$$\binom{n-1}{k-1}.$$

For each selection $\{u_1,\ldots,u_{k-1}\} \subseteq \{1,2,\ldots,n-1\}$, *where* $u_1 < u_2 < \cdots < u_{k-1}$, *we have the solution* $(n_1,\cdots,n_k)$ *to* (1:1) *among the positive integers given by* $n_1 = u_1$, $n_k = n - u_{k-1}$, *and for* $2 \le j \le k-1$, $n_j = u_j - u_{j-1}$.

We will use Theorem 2.13 to solve (1:1) among the nonnegative integers. Make sure that the n golf balls to be distributed among the k baskets are all white. Add to this collection exactly k red golf balls. Now consider placing the $n+k$ golf balls into the k baskets in such a way that each basket has at least one golf ball. There are, of course, by Theorem 2.13, exactly $\binom{n+k-1}{k-1}$ ways to do this. (Note that $\binom{n+k-1}{k-1} = \binom{n+k-1}{n}$.) For any way in which this is done, there may or may not be a basket without a red ball. Each basket has at least one red ball if and only if each basket has exactly one red ball. So, some basket has no red ball if and only if some other basket has more than one red ball. Take a basket without a red ball. It has at least one white ball, and there is another basket with 2 or more red balls. Exchange a white ball for a red ball so that the basket which had no red ball now has a red ball. The number of balls in each basket has not changed, and one more basket has a red ball. Continue this until each basket has a red ball. (This red ball may be the only ball in that basket.) Now remove the red ball from each basket.

Suppose $(m_1, \ldots, m_k)$ is a solution to

$$y_1 + y_2 + \cdots + y_k = n + k, \tag{2:6}$$

and that each m_i is a positive integer. This represents each basket i having m_i balls. After removing the red ball from each of the k baskets, we now have a total of n balls in the k baskets, and there are $n_i = m_i - 1$ balls in basket i. Now $n_i \geq 0$, and

$$\begin{aligned} n_1 + n_2 + \cdots + n_k &= (m_1 - 1) + (m_2 - 1) + \cdots + (m_k - 1) \\ &= (m_1 + m_2 + \cdots + m_k) - k = n. \end{aligned}$$

Hence, $(n_1, \ldots, n_k)$ is a solution to (1:1) among the nonnegative integers.

Conversely, if $(n_1, \ldots, n_k)$ is a solution to (1:1) among the nonnegative integers, and if $m_i = n_i + 1$, then $(m_1, \ldots, m_k)$ is a solution to (2:6) among the positive integers. So we have

Theorem 2.14. *The number of solutions $(n_1, \ldots, n_k)$ to (1:1) among the nonnegative*

integers is

$$\binom{n+k-1}{k-1} = \binom{n+k-1}{n}.$$

To get a solution to (2:6)*, find a solution* $(m_1,\ldots,m_k)$ *among the positive integers, using Theorem 2.13. Then* $(m_1-1,\ldots,m_k-1)$ *is a solution to* (1:1) *among the nonnegative integers.*

In Problem 1.40, you were asked to formulate conjectures about relationships among the parameters $v, b, r, k,$ and λ of a BIBD. The desired relationships, which we can now derive, are given in the next theorem.

Theorem 2.15. *Let* $(V,\mathscr{B})$ *be a BIBD with parameters*

$$v = |V|,\, b = |\mathscr{B}|,\ r,\ k,\ \text{and}\ \lambda.$$

Then

$$vr = bk \quad \text{and} \quad \lambda(v-1) = r(k-1).$$

Proof. Choose a point $x \in V$. There are v such choices. Then choose a block $B \in \mathscr{B}$ with $x \in B$. There are r such choices. This gives us vr choices of (x,B) with $x \in B$. On the other hand, choose a block $B \in \mathscr{B}$. This can be done in b ways. Now choose an $x \in B$. This gives us bk choices of (B,x) with $x \in B$. The map $(x,B) \mapsto (B,x)$, where $x \in B$, is a bijection between $\{(x,B) \mid B \subset \mathscr{B},\ x \in B\}$ and $\{(B,x) \mid x \in B,\ B \in \mathscr{B}\}$. Hence, $vr = bk$.

Next, choose a pair $\{x,y\}$ of distinct points $x,y \in V$. This can be done in $\binom{v}{2} = v(v-1)/2$ ways. With an $\{x,y\}$ choose a block $B \in \mathscr{B}$ with $x,y \in B$. This can be done in λ ways, and so we have $\lambda v(v-1)/2$ such choices $(\{x,y\},B)$ with $x,y \in B$. On the other hand, choose a block $B \in \mathscr{B}$. This can be done in b ways. For the block B, choose a pair $\{x,y\}$ of distinct points $x,y \in B$. This can be done in $\binom{k}{2} = k(k-1)/2$ ways. So, there are $bk(k-1)/2$ such sequences of choices $(B,\{x,y\})$ with $x,y \in B$. As before, $(\{x,y\},B) \mapsto (B,\{x,y\})$, with $x,y \in B$,

$x \neq y$, is a bijection. Hence,

$$\frac{\lambda v(v-1)}{2} = \frac{bk(k-1)}{2}.$$

But $bk = vr$, so we have $\lambda v(v-1) = vr(k-1)$, and finally $\lambda(v-1) = r(k-1)$. We now have finished the proof of Theorem 2.15. □

There is another nice application of our new counting skills.

Theorem 2.16. *Let $(V, \mathscr{B})$ be a circular BIBD with parameters*

$$v = |V|, \ b = |\mathscr{B}|, \ r, \ k, \ \textit{and} \ \lambda.$$

Then $\lambda(k-2) \leq v-2$.

Proof. Each of the b circles in $\mathscr{B}$ has $\binom{k}{3} = k(k-1)(k-2)/6$ triples x, y, z of distinct points belonging to it. There are $bk(k-1)(k-2)/6$ such triples belonging to the circles of $\mathscr{B}$. But each triple $a, b, c \in V$ of distinct points belongs to at most one circle, and there are $\binom{v}{3} = v(v-1)(v-2)/6$ such triples. So, we cannot have the inequality

$$\frac{bk(k-1)(k-2)}{6} > \frac{v(v-1)(v-2)}{6}.$$

Since $bk = vr$, we cannot have the inequality

$$r(k-1)(k-2) > (v-1)(v-2).$$

Since $r(k-1) = \lambda(v-1)$, we cannot have the inequality

$$\lambda(k-2) > v-2.$$

So, we must have $\lambda(k-2) \leq v-2$. □

Corollary 2.17. *Let $(V, \mathscr{B})$ be a circular BIBD with parameters v, b, r, k, and λ. Suppose every three distinct points $x, y, z \in V$ belong to exactly one circle $B \in \mathscr{B}$. Then $\lambda(k-2) = v-2$.*

Proof. Use the proof of Theorem 2.16 as a model. Eliminate the word "cannot," and in place of the word "inequality," use the word "equality." Then replace $>$ with $=$. □

Corollary 2.18. *Let $(V, \mathscr{B})$ be a circular BIBD with parameters v, b, r, k, and λ. If $\lambda = k-1$, then $k \leq (3+\sqrt{4v-7})/2$.*

Proof. From Theorem 2.16, we have $\lambda(k-2) \leq v-2$, and with $\lambda = k-1$, we get $(k-1)(k-2) \leq v-2$. With $k^2-3k+2 \leq v-2$, we get $4k^2-12k+9 \leq 4v-7$, and then $(2k-3)^2 \leq 4v-7$, from which the inequality $k \leq (3+\sqrt{4v-7})/2$ follows. □

We now apply our new counting skills to some basic algebraic ideas. A "multiplication table" for a binary operation on a set $S = \{a_1, a_2, \ldots, a_n\}$ has the structure

$\cdot$	a_1	a_2	$\cdots$	a_j	$\cdots$	a_n
a_1	a_1a_1	a_1a_2	$\cdots$	a_1a_i	$\cdots$	a_1a_n
a_2	a_2a_1	a_2a_2	$\cdots$	a_2a_i	$\cdots$	a_2a_n
$\vdots$	$\vdots$	$\vdots$		$\vdots$		$\vdots$
a_i	a_ia_1	a_ia_2	$\cdots$	a_ia_j	$\cdots$	a_ia_n
$\vdots$	$\vdots$	$\vdots$		$\vdots$		$\vdots$
a_n	a_na_1	a_na_2	$\cdots$	a_na_j	$\cdots$	a_na_n

For example, if $S = \mathbb{Z}_3 = \{0, 1, 2\}$, two such binary operations are

$+$	0	1	2
0	0	1	2
1	1	2	0
2	2	0	1

and

$\cdot$	0	1	2
0	0	0	0
1	0	1	2
2	0	2	1

These are exactly the binary operations of the field $(\mathbb{Z}_3, +, \cdot)$. But there are many more binary operations on $\mathbb{Z}_3$. If we illustrate $*$ of (g) from Examples A3.2, then we get the multiplication table

$$\begin{array}{c|ccc} + & 0 & 1 & 2 \\ \hline 0 & 0 & 1 & 2 \\ 1 & 0 & 1 & 2 \\ 2 & 0 & 1 & 2 \end{array}.$$

Just how many binary operations are there for $\mathbb{Z}_3$? How many are commutative? How many are associative? One of these questions is difficult to answer. Which one is it?

Actually, it is difficult enough to answer these questions even for the set $\mathbb{Z}_2 = \{0,1\}$. But, they can be answered.

Exercise 2.19. How many binary operations can be defined on $\mathbb{Z}_2 = \{0,1\}$. How many of these are commutative? How many are associative? How many are both commutative and associative? How many make $\mathbb{Z}_2$ into a group?

After doing Exercise 2.19, then you should be able to do the next exercise.

Exercise 2.20. Let $S = \{a_1, a_2, \ldots, a_n\}$. For example, $S = \{0, 1, \ldots, n-1\}$. How many binary operations can be defined on S? How many of these binary operations are commutative? (I dare not ask you to find the number of associative binary operations on S. But, if you can program a computer, you might try to answer this for small values of n, say $n = 3, 4, 5$. If you do this, check to see what percent of the binary operations are associative.)

Exercise 2.21. Think about a multiplication table for a group $(G, \cdot)$. Think about the two groups $(\mathbb{Z}_6, +)$ and $(\mathscr{U}(M(Y)), \circ)$ of Exercise A3.21. Now construct two Latin squares of order 6. (See Exercise 1.36.) Hopefully, your two Latin squares are essentially different. From one of your Latin squares construct many more. How many Latin squares do you think there are? (Do not try to answer this, but you should be able to construct 1440 of them.) Do you believe there are two orthogonal Latin squares of order 6? (There does not exist a pair of orthogonal Latin squares of order 6.)

For a finite field F with $|F| = q$, exactly how many irreducible polynomials $a \in F[x]$ are there with deg $a = 2$? What percent of the polynomials $a \in F[x]$ with deg $a = 2$ are irreducible? We can now answer these questions.

Theorem 2.22. *Let F be a finite field of order q. Then there are exactly $q(q-1)^2/2$ irreducible polynomials $p \in F[x]$ with* deg $p = 2$. *So $50 - \frac{50}{q}$ percent of the polynomials of degree 2 in $F[x]$ are irreducible.*

Proof. If $ax^2 + bx + c \in F[x]$, $a \neq 0$, is irreducible, then so is $x^2 + (a^{-1}b)x + (a^{-1}c)$, and conversely. In fact, for every irreducible $f = x^2 + dx + e \in F[x]$, there are $q - 1 = |F^*|$ irreducible $\alpha f = \alpha x^2 + \alpha dx + \alpha e \in F[x]$, one for each $\alpha \in F^*$. See Lemma A2.37. So, we will count the irreducible $x^2 + dx + c \in F[x]$.

There are q^2 polynomials $x^2 + dx + e \in F[x]$. If there are R reducible polynomials $x^2 + dx + e \in F[x]$, then there are $q^2 - R$ irreducible $x^2 + dx + e \in F[x]$. We will calculate R.

Using Lemma A2.39, the reducible $x^2 + dx + e \in F[x]$ are the $(x-\alpha)(x-\beta)$, with $\alpha, \beta \in F$. But $(x-\alpha)(x-\beta) = (x-\beta)(x-\alpha)$. Now there are q of the form $(x-\alpha)(x-\alpha)$, and there are $\binom{q}{2} = q(q-1)/2$ of the form $(x-\alpha)(x-\beta)$ with $\alpha \neq \beta$ Hence, $R = q + q(q-1)/2 = q(q+1)/2$.

So we have $q^2 - q(q+1)/2 = q(q-1)/2$ irreducible $x^2 + dx + e \in F[x]$, and therefore there are $q(q-1)^2/2$ irreducible $ax^2 + bx + c \in F[x]$ with $a \neq 0$.

Altogether, there are $(q-1)q^2$ polynomials $ax^2 + bx + c \in F[x]$ with $a \neq 0$. Hence,

$$100\left(\frac{q(q-1)^2}{2(q-1)q^2}\right) = 50 - \frac{50}{q}$$

percent of the polynomials of degree 2 are irreducible. □

Corollary 2.23. *For any finite field F, there is an irreducible polynomial $a \in F[x]$ with* deg $a = 2$. *If $|F| = q$, then there is a field K with $|K| = q^2$.*

Proof. Since $0, 1 \in F$ and $0 \neq 1$, then $q \geq 2$. Hence, at least 25% of the polynomials of degree 2 are irreducible. For such an irreducible $a \in F[x]$, deg $a = 2$, we know that $F[x]/(a)$ is a field and that $|F[x]/(a)| = q^2$. See Theorem A2.32. So let $K = F[x]/(a)$. □

Exercise 2.24. Using Theorem 2.22 and Corollary 2.23 as models, investigate the irreducible polynomials $a \in F[x]$ with $\deg a = 3$ and with $|F| = q < \infty$. How many are there? What percent are irreducible? You may wish to execute the following steps.

(a) Prove that the polynomial $x^3 + bx^2 + cx + d \in F[x]$ is irreducible if and only if $\alpha x^3 + \alpha bx^2 + \alpha cx + \alpha d \in F[x]$ is irreducible for every $\alpha \in F^*$.

(b) Show that $x^3 + bx^2 + cx + d$ is reducible if and only if

$$x^3 + bx^2 + cx + d = (x^2 + \alpha x + \beta)(x + \gamma)$$

or

$$x^3 + bx^2 + cx + d = (x + \alpha)(x + \beta)(x + \gamma)$$

for some $\alpha, \beta, \gamma \in F$.

(c) Calculate the number R of reducible $a \in F[x]$ with $\deg a = 3$.

The counting techniques we have developed in this section are very powerful. There is a rich theory built upon these in the applications of basic probability theory. It takes a fair amount of practice in order to handle the nuances which arise. For that reason, we include here just a small sample of exercises to illustrate the power of these tools. We would recommend anyone to develop facility in using these ideas, and would refer those interested to the references [R] and [B & F] for further reading and excellent selections of exercises.

Exercise 2.25. (a) Prove that

$$\binom{m+n}{k} = \binom{m}{0}\binom{n}{k} + \binom{m}{1}\binom{n}{k-1} + \cdots + \binom{m}{k}\binom{n}{0}$$

for positive integers k, m, and n, with $k \leq m$ and $k \leq n$.

(b) How many ways can n golf balls be placed in k baskets, labelled basket 1, basket 2, $\cdots$, basket k, in such a way that basket i contains at least s_i golf balls? (Of course, $\sum_{i=1}^{k} s_i \leq n$.)

(c) Prove that

$$\binom{2n}{n} = \binom{n}{0}^2 + \binom{n}{1}^2 + \cdots + \binom{n}{n}^2.$$

(d) How many solutions are there to the equation

$$x_1 + x_2 + \cdots + x_k = n$$

among the nonnegative integers for which exactly s of the x_i's are equal to 0?

(e) There are 20 men and 20 women to participate in a chess tournament. These 40 people are to be paired in 20 sets of two people for their initial match. If pairing is made at random, how many ways can this be done? Of these, how many ways result in a pairing where there is no male-female pairs? How many ways result in a pairing with exactly $2i$ male-female pairs, $1 \leq i \leq 10$.

(f) How many integers are there with the following properties?

(1) The integer must be greater than 64,000.
(2) No digit can be used twice.
(3) The digits 5 and 6 cannot be used.

Exercise 2.26. Let $(N, \mathscr{B})$ be a BIBD with $\lambda = k$. Suppose also that every triple $x, y, z \in N$ of distinct points belong to exactly one $B \in \mathscr{B}$. Prove that $(N, \mathscr{B})$ is a Möbius plane. (Is this really true?)

3 Basic field theory applications

In this section, we want to (1) construct an affine plane from any field F, (2) construct a projective plane from any field F, (3) construct a Möbius plane from any field F, and (4) construct a family of $|F|-1$ mutually orthogonal Latin squares from any finite field F. The latter three are all closely related to affine planes, so that is where we begin.

3.1 Affine planes constructed from fields

Exactly when did you learn to associate a real number with a point on a line, and conversely? Perhaps it is not clear when you learned to do this. The one-to-one correspondence between the real numbers and the points on a line permeate our mathematical education these days. But this was not always the case. Today, we speak of "the real line" to mean a line with a fixed point called 0. This point 0 cuts the real line into two parts, the part on one side, usually the right side, corresponding to the positive real numbers, and the points on the other side correspond to the negative real numbers. A point corresponding to the positive number 1 is chosen, and then all other numbers correspond to points on the line in relationship to our chosen 0 and 1.

With the elements $(a,b) \in \mathbb{R} \times \mathbb{R}$, we associate a point in the plane, and conversely. Our classical euclidean geometry is now seen in terms of the coordinates $(a,b) \in \mathbb{R} \times \mathbb{R}$. With this, we describe lines by equations like $y = mx + b$ or $x = a$, and circles by equations like $(x-a)^2 + (y-b)^2 = r^2$. Actually, the line defined by $y = mx + b$ is

$$L(m,b) = \{(x, mx+b) \mid x \in \mathbb{R}\},$$

the line defined by $x = a$ is

$$L(\infty, a) = \{(a,y) \mid y \in \mathbb{R}\},$$

and the circle defined by $(x-a)^2 + (y-b)^2 = r^2$ is

$$C\big((a,b);r\big) = \{(x,y) \in \mathbb{R} \times \mathbb{R} \mid (x-a)^2 + (y-b)^2 = r^2\}.$$

If $\mathscr{L} = \{L(m,b),\ L(\infty,a) \mid m,b,a \in \mathbb{R}\}$, then $(\mathbb{R} \times \mathbb{R}, \mathscr{L})$ is a model of classical geometry. It is a very nice model, since we are really quite skilled with working with real numbers. For example, we can easily find the point in common to the line $L(3,4)$ and the line $L(-1,2)$, where $L(3,4)$ is the line defined by $y = 3x+4$ and the line $L(-1,2)$ is the line defined by $y = -x+2$. Hence, to get the point in common, we set

$$3x+4 = -x+2$$

and solve for x, giving $x = -1/2$. With $x = -1/2$, we can use either $y = 3x+4$ or $y = -x+2$ to compute $y = 5/2$. So $(-1/2, 5/2)$ is the point in common to $L(3,4)$ and $L(-1,2)$.

For our purposes, it is enough now to realize that $(\mathbb{R} \times \mathbb{R}, \mathscr{L})$ is an affine plane. Let's state the definition.

Definitions 3.1. An *affine plane* is a pair $(P, \mathscr{L})$, where P is a nonempty set of *points* and $\mathscr{L}$ is a nonempty set of subsets of P. The members $L \in \mathscr{L}$ are called *lines*. The points and the lines must have the following properties.

(a) If $x, y \in P$ and $x \neq y$, then there is exactly one line $[x, y] \in \mathscr{L}$ containing both x and y.

(b) If $L \in \mathscr{L}$, $x \in P$, and $x \notin L$, then there is exactly one line $M \in \mathscr{L}$ such that $x \in M$ and $L \cap M = \varnothing$. We say that L and M are *parallel* and we write $L \parallel M$.

(c) If $L \in \mathscr{L}$, then $|L| \geq 2$.

(d) If $L \in \mathscr{L}$, then there is a point $x \in P$ such that $x \notin L$.

Exercise 3.2. Prove that $(\mathbb{R} \times \mathbb{R}, \mathscr{L})$ defined above is an affine plane.

The simplest affine plane consists of four points, the points of $\mathbb{Z}_2 \times \mathbb{Z}_2$ are

$(0,0)$, $(0,1)$, $(1,0)$, and $(1,1)$. There are six lines in $\mathscr{L}_2$, namely

$$\begin{aligned} L_1 &= \{(0,0),\ (0,1)\}, & L_2 &= \{(1,0),\ (1,1)\}, \\ M_1 &= \{(0,0),\ (1,1)\}, & M_2 &= \{(0,1),\ (1,0)\}, \\ N_1 &= \{(0,0),\ (1,0)\}, & N_3 &= \{(0,1),\ (1,1)\}. \end{aligned}$$

Exercise 3.3. Prove that $(\mathbb{Z}_2 \times \mathbb{Z}_2, \mathscr{L}_2)$ is an affine plane.

Our example $(\mathbb{Z}_2 \times \mathbb{Z}_2, \mathscr{L}_2)$ really is quite simple, but this example, together with our example $(\mathbb{R} \times \mathbb{R}, \mathscr{L})$, illustrate a major application of fields. For every field $(F, +, \cdot)$, we get an affine plane $(F \times F, \mathscr{L}_F)$. So, let's construct $\mathscr{L}_F$.

We've essentially revealed the main points. For $m, b, a \in F$, we define

$$L(m,b) = \{(x, mx+b) \mid x \in F\} \tag{3:7}$$

and

$$L(\infty,a) = \{(a,y) \mid y \in F\}. \tag{3:8}$$

For $m \in F$, let

$$\mathscr{L}(m) = \{L(m,b) \mid b \in F\} \tag{3:9}$$

and define

$$\mathscr{L}(\infty) = \{L(\infty,a) \mid a \in F\}. \tag{3:10}$$

Then each $\mathscr{L}(\alpha)$ will be a family of parallel lines. Of course, we let

$$\mathscr{L}_F = \mathscr{L} = \{L(m,b),\ L(\infty,a) \mid m,b,a \in F\}, \tag{3:11}$$

that is,

$$\mathscr{L}_F = \mathscr{L} = \mathscr{L}(\infty) \cup [\cup_{m \in F} \mathscr{L}(m)].$$

(We often use $\mathscr{L}$ for $\mathscr{L}_F$ to make the notation simpler.)

Our goal is to prove

Theorem 3.4. *For any field F, $(F \times F, \mathscr{L}_F)$ is an affine plane.*

Proof. Let's take care of the easy requirements first. Since $|F| \geq 2$, we have $|L| \geq 2$ for each $L \in \mathscr{L} = \mathscr{L}_F$. For if $x_1 \neq x_2$, then $(x_1, mx_1 + b) \neq (x_2, mx_2 + b)$, and $(a, x_1) \neq (a, x_2)$. This takes care of (c) of Definitions 3.1. Requirement (d) is just as easy. Take any $L(m,b)$. For any $a \in F$, $(a, ma+b) \in L(m,b)$. Let $c \in F \setminus \{a\}$. Then $(c, ma+b) \in F \times F$ and $(c, ma+b) \notin L(m,b)$. Similarly, for any $L(\infty, a)$, we have $(a,d) \in L(\infty, a)$ for any $d \in F$, but if $c \neq a$, then $(c,d) \in F \times F$, but $(c,d) \notin L(\infty, a)$. This takes care of (d) of Definitions 3.1.

Let (a,b) and (c,d) be two distinct points in $F \times F$. We want to show the existence of a line $L \in \mathscr{L}$ with $(a,b), (c,d) \in L$. There are two cases. Either $a = c$ or $a \neq c$. If $a = c$, then $(c,d) = (a,d)$ and so $b \neq d$. We have $(a,b), (a,d) \in L(\infty, a)$. Suppose $a \neq c$. Let

$$m = (d-b)(c-a)^{-1}, \quad \text{and} \quad e = -(d-b)(c-a)^{-1}a + b.$$

The line $L(m,e)$ contains each $(x, mx+e)$ for $x \in F$. In particular, for $x = a$, $L(m,e)$ contains

$$(a, ma+e) = (a, (d-b)(c-a)^{-1}a - (d-b)(c-a)^{-1}a + b) = (a,b),$$

and for $x = c$, $L(m,e)$ contains

$$\begin{aligned}(c, mc+e) &= (c, (d-b)(c-a)^{-1}c - (d-b)(c-a)^{-1}a + b) \\ &= (c, (d-b)(c-a)^{-1}(c-a) + b) = (c, (d-b) + b) = (c,d).\end{aligned}$$

Hence, $(a,b), (c,d) \in L(m,e)$. We summarize this paragraph by noting that two distinct points $(a,b), (c,d) \in F \times F$ belong to a line $L \in \mathscr{L}$. We now turn to the question of uniqueness.

Again, suppose $a = c$. We have $(a,b), (a,d) \in L(\infty, a)$. If $(a,b), (a,d) \in L(\infty, e)$, then $(a,b) = (e,y)$ for some $y \in F$. Hence, $a = e$, and so $L(\infty, e) = L(\infty, a)$. If $(a,b), (c,d) \in L(m,e)$ for some $m \in F$, then $(a,b) = (a, ma+e)$ and $(a,d) = (a, ma+e)$. Hence, $b = ma+e$ and $d = ma+e$, which means that $b = d$, and so

$(a,b) = (a,d)$, or $(a,b) = (c,d)$, a contradiction. Hence, if $(a,b),(a,d) \in F \times F$ are two distinct points, they belong only to the line $L(\infty,a)$.

Next let $a \neq c$. We have $(a,b),(c,d) \in L(m,e)$ above. If $(a,b),(c,d) \in L(\infty,u)$ for some $u \in F$, then $(a,b) = (u,y)$ for some $y \in F$ and $(c,d) = (u,y')$ for some $y' \in F$. Hence, $a = u = c$, which cannot be. Thus, there is no $L(\infty,u) \in \mathscr{L}(\infty)$ which contains both (a,b) and (c,d).

Suppose $(a,b),(c,d) \in L(m',e')$. Then there are $x,x' \in F$ such that

$$(a,b) = (x,m'x+e') \text{ and } (c,d) = (x',m'x'+e').$$

Hence, $a = x$ and $c = x'$. Consequently,

$$(a,b) = (a,m'a+e') \text{ and } (c,d) = (c,m'c+e').$$

With $b = m'a+e'$ and $d = m'c+e'$, we have $b - m'a = e' = d - m'c$, which yields

$$d - b = m'c - m'a = m'(c-a).$$

Consequently,

$$m' = (d-b)(c-a)^{-1} = m.$$

So $L(m',e') = L(m,e')$.

We have $(a,b),(c,d) \in L(m,e) \cap L(m,e')$. So $(a,b) = (a,ma+e) = (a,ma+e')$. Hence, $e = e'$, which in turn implies that $L(m,e') = L(m,e)$. So $L(m',e') = L(m,e)$, and $L(m,e)$ is the only line containing (a,b) and (c,d) if $a \neq c$. We now have (a) of Definitions 3.1.

Turning our attention to requirement (b) of Definitions 3.1, we take a line L. First, let $L \in \mathscr{L}(\infty)$, so $L = L(\infty,a)$ for some $a \in F$. If $(c,d) \notin L$, then $c \neq a$. Now $(c,d) \in L(\infty,c)$, and $L(\infty,a) \cap L(\infty,c) = \varnothing$. For if $(s,t) \in L(\infty,a) \cap L(\infty,c)$, then $(s,t) = (a,t) = (c,t)$, and so $a = c$. We again have a line $L(\infty,c)$ with (c,d) in $L(\infty,c)$ and $L(\infty,c) \parallel L(\infty,a)$. But is it the only one? Another would belong to some $\mathscr{L}(m)$, with $m \in F$. Consider an $L(m,b) \in \mathscr{L}(m)$. Then the intersection

$L(m,b) \cap L(\infty,a)$ is nonempty since $(a, ma+b) \in L(m,b) \cap L(\infty,a)$. This gives the uniqueness of the line $L(\infty,c)$. Hence (b) is satisfied for $L \in \mathscr{L}(\infty)$.

Continuing toward satisfying (b), we let $L \in \mathscr{L}(m)$, $m \in F$. Let's assume that $L = L(m,b)$. Take $(c,d) \notin L(m,b)$. Now $(c, mc+b) \in L(m,b)$ is the only point in $L(m,b)$ with first coordinate c, so if $(c,d) \notin L(m,b)$, then $d \neq mc+b$. But $d = mc + (-mc+d)$, and so $(c,d) \in L(m, -mc+d)$. If (s,t) is a member of $L(m,b) \cap L(m, -mc+d)$, then $(s,t) = (s, ms+b) = (s, ms-mc+d)$, which implies that $b = -mc+d$, or $mc+b = d$. But we have $d \neq mc+b$, and so $L(m,b) \cap L(m,-mc+d) = \varnothing$, which means that $L(m,b) \parallel L(m,-mc+d)$. So, we have a line $L(m,-mc+d)$ on (c,d) and parallel to $L(m,b)$. We must be assured that it is the only one.

If $(c,d) \in L(m,b')$, then $(c,d) = (c, mc+b')$. Hence $b' = -mc+d$ and the line $L(m,-mc+d)$ is the only one in $\mathscr{L}(m)$ containing the point (c,d). We have seen earlier in this proof that if $L \in \mathscr{L}(\infty)$ and $M \in \mathscr{L}(m)$, $m \in F$, then $L \cap M \neq \varnothing$. So, if there is another line $M \in \mathscr{L}$ on (c,d) and parallel to $L(m,b)$, then $M \in \mathscr{L}(m')$ for some $m' \in F$, $m' \neq m$. Let $M = L(m',b')$ with $m' \neq m$. We will see that $L(m',b') \cap L(m,b) \neq \varnothing$. Just as we did at the beginning of our discussion for $(\mathbb{R} \times \mathbb{R}, \mathscr{L})$, take $(x,y) \in L(m',b') \cap L(m,b)$. Then $(x,y) = (x, m'x+b') = (x, mx+b)$. So $m'x+b' = mx+b$. (Note how we have an equation of the form $ax = bx+c$ here!) Solving for x, we get $x = (m'-m)^{-1}(b-b')$. (Note the importance of the condition $m' \neq m$.) With $x = (m'-m)^{-1}(b-b')$, then

$$m'x+b' = m'(m'-m)^{-1}(b-b')+b'$$

and

$$mx+b = m(m'-m)^{-1}(b-b')+b.$$

Hence,

$$\begin{aligned}(m'x+b')-(mx+b) &= (b'-b)+(m'-m)^{-1}(b-b')(m'-m) \\ &= (b'-b)+(b-b') = 0.\end{aligned}$$

Consequently,

$$((m'-m)^{-1}(b-b'),\ m'(m'-m)^{-1}(b-b')+b') \in L(m',b') \cap L(m,b),$$

and so $L(m',b')$ is not parallel to $L(m,b)$. From this, we can now infer that $L(m,-mc+d)$ is the only line on (c,d) parallel to $L(m,b)$. This completes our proof that requirement (b) of Definitions 3.1 is met, and also completes our proof of the theorem. □

Corollary 3.5. *For a field F and any $\alpha \in F \cup \{\infty\}$, we have that $\mathscr{L}(\alpha)$ is a partition of $F \times F$.*

Proof. By (a) of Definitions 3.1 and the proof of Theorem 3.4, we saw that $\cup\mathscr{L}(\alpha) = F \times F$. By (b) of Definitions 3.1 and the proof of Theorem 3.4, we saw that for $L, M \in \mathscr{L}(\alpha)$, if $L \neq M$, then $L \cap M = \varnothing$. That is all there is to it! □

With Theorem 3.4 and a field F, we can construct an affine plane $(F \times F, \mathscr{L}_F)$ where each line L has exactly $|F|$ points.

Definition 3.6. If $(P, \mathscr{L})$ is an affine plane with P a finite set, and if $|L| = |M|$ whenever $L, M \in \mathscr{L}$, then the number $|L|$ is called the *order* of the affine plane $(P, \mathscr{L})$.

Exercise 3.7. Construct an affine plane of order 3. Then construct one of order 4.

Exercise 3.8. Prove that there exists an affine plane of order n for each n a member in $\{2, 3, 4, 5, 7, 8, 9\}$. Do you have any idea why 6 was not included as a possibility for n? What about 10?

3.2 Projective planes constructed from fields

As one looks into the distance at railroad tracks, the two tracks appear to meet at a point. This is the basic idea of a projective plane. All parallel lines appear to meet at the same point. In that case, there would be no such thing as a pair of lines L and M such that $L \cap M = \varnothing$. However, if one has two pairs of parallel lines, say L, L' and M, M', with $L \parallel L'$, $M \parallel M'$, but with L not parallel to M, then it would appear that L and L' meet at a point a, that M and M' meet at a point b, and that $a \neq b$. We express this idea by saying that parallel lines "meet at infinity." But if L and L' meet at a and M and M' meet at b, and $a \neq b$, then a and b are two different "points at infinity." There should be a unique line on both a and b, so one thinks of the "line at infinity" consisting of all the "points at infinity." We will begin with the definition of a projective plane, and then use the ideas above to construct a projective plane from any field, using the field's affine plane.

Definition 3.9. A *projective plane* is an ordered pair $(P, \mathscr{L})$, where $P \neq \varnothing$ is a set of *points*, and $\mathscr{L} \neq \varnothing$ is a family of subsets of P. The members $L \in \mathscr{L}$ are called *lines*, and the points and lines must satisfy the following properties.

(a) If $x, y \in P$ and $x \neq y$, then there is a unique line $L \in \mathscr{L}$ with $x, y \in L$.

(b) If $L, M \in \mathscr{L}$ and $L \neq M$, then there is a unique point $x \in P$ with $x \in L \cap M$.

(c) If $L \in \mathscr{L}$, then $|L| \geq 3$.

(d) If $L \in \mathscr{L}$, then there is a point $x \in P$ such that $x \notin L$.

Look at the affine plane we constructed in Example 1.14. One can identify the parallel lines immediately. Then look at the projective plane we displayed in Example 1.19. See how the "points at infinity" were added, a to the K_i, b to the L_i, c to the M_i, and d to the N_i, and finally, the "line at infinity" $I = \{a, b, c, d\}$ was added. We can do the same thing for any affine plane. We illustrate this with the affine planes $(F \times F, \mathscr{L}_F)$ of Theorem 3.4.

For a field F, we have the family of parallel lines $\mathscr{L}(\alpha)$, for $\alpha \in F \cup \{\infty\}$, as described in (3:7) through (3:10). To each $L \in \mathscr{L}(\alpha)$, add the "point at infinity" α. So

$$L' = L \cup \{\alpha\}.$$

For example,

$$L'(m,b) = \{(x, mx+b) \mid x \in F\} \cup \{m\},$$

and

$$L'(\infty, a) = \{(a,y) \mid y \in F\} \cup \{\infty\}.$$

Let $\mathscr{L}'(\alpha) = \{L' \mid L \in \mathscr{L}(\alpha)\}$ and define $I = F \cup \{\infty\}$. With

$$\mathscr{L}' = \{I\} \cup \left[\cup_{\alpha \in I} \mathscr{L}'(\alpha)\right],$$

and with

$$P' = (F \times F) \cup I,$$

we want to show that $(P', \mathscr{L}')$ is a projective plane.

Theorem 3.10. *Let F be any field. From the affine plane of Theorem 3.4, let $(P', \mathscr{L}')$ be as described above. Then $(P', \mathscr{L}')$ is a projective plane.*

Proof. Again, we will handle the easy requirements first. Take $L' \in \mathscr{L}'$. If $L' = I$, then $|I| = |F| + 1 \geq 3$. If $L' \neq I$, then $L' = L \cup \{\alpha\}$, and since $|L| \geq 2$ and $\alpha \notin L$, we have $|L'| \geq 3$. This gives us (c) of Definition 3.9. For $L' \in \mathscr{L}'$, $L' = L \cup \{\alpha\}$ for some α. There is a point $(a,b) \in F \times F$ where $(a,b) \notin L$, and certainly $(a,b) \neq \alpha$, so $(a,b) \notin L'$. Hence (d) of Definition 3.9 is satisfied.

Next, we will show that (a) of Definition 3.9 is satisfied. Let $p, p' \in P'$ be two distinct points. There are three cases to consider. First, there is the case that $p, p' \in F \times F$; second, there is the case that $p \in F \times F$ and $p' \notin F \times F$; and third, there is the case that $p, p' \notin F \times F$. When $p, p' \in F \times F$, then by Theorem 3.4, there is a unique line $L \in \mathscr{L}$ such that $p, p' \in L$. So $p, p' \in L' = L \cup \{\alpha\}$ for some α. It cannot be that $p, p' \in M'$ if $M' \in \mathscr{L}'$ and $M' \neq L'$, since $\{p, p'\} \cap I = \varnothing$. Hence, $p, p' \in L'$ and $p, p' \notin M'$ if $M' \in \mathscr{L}'$ and $M' \neq L'$. If $p \in F \times F$ and $p' \notin F \times F$, then $p' \in I$. So $p' = \alpha$ for some $\alpha \in I$, and p belongs to exactly one $L \in \mathscr{L}(\alpha)$. This means that $p, p' \in L' \in \mathscr{L}'(\alpha)$ and $p, p' \notin M'$ if $M' \in \mathscr{L}'(\alpha)$ and $M' \neq L'$. Finally, if $p, p' \notin F \times F$, then $p, p' \in I \in \mathscr{L}'$. If $p, p' \in L' \in \mathscr{L}'(\alpha)$, then

$p = \alpha = p'$. Since $p \neq p'$, we cannot have $p, p' \in L' \in \mathscr{L}'(\alpha)$ for any $\alpha \in I$, and so I is the only line in $\mathscr{L}'$ containing both p and p'. In all three cases, p and p' belong to a unique line $L' \in \mathscr{L}'$. Hence, (a) of Definition 3.9 is satisfied.

There remains to see that (b) of Definition 3.9 is fulfilled. Take two distinct $L', M' \in \mathscr{L}'$. There are two cases to consider. Either $I \in \{L', M'\}$ or $I \notin \{L', M'\}$. If $I = L'$, say, take $M' \in \mathscr{L}'(\alpha)$ for some $\alpha \in I$. Hence, $M' = M \cup \{\alpha\}$, and so $L' \cap M' = I \cap M' = \{\alpha\}$. If $I \notin \{L', M'\}$, there are two subcases to consider. Either there is an $\alpha \in I$ such that $L', M' \in \mathscr{L}'(\alpha)$, or there is not. If there is an $\alpha \in I$ such that $L', M' \in \mathscr{L}'(\alpha)$, then $L' = L \cup \{\alpha\}$ and $M' = M \cup \{\alpha\}$ with $L, M \in \mathscr{L}(\alpha)$. Hence, $L \cap M = \varnothing$, and so $L' \cap M' = \{a\}$. Finally, if $L' \in \mathscr{L}'(\alpha)$ and $M' \in \mathscr{L}'(\mu)$, with $\alpha \neq \mu$, then $L' = L \cup \{\alpha\}$ and $M' = M \cup \{\mu\}$. Hence, $L' \cap M' = L \cap M$. Since $L \in \mathscr{L}(\alpha)$ and $M \in \mathscr{L}(\mu)$, with $\alpha \neq \mu$, we have $|L \cap M| = 1$. Hence, $|L' \cap M'| = 1$. This completes the verification of (b) of Definition 3.9, and also the proof of Theorem 3.10. □

Theorem 3.11. *For any finite field F with $|F| = q$, there is a BIBD $(P', \mathscr{L}')$ with parameters $v = |P'| = q^2 + q + 1$, $b = |\mathscr{L}'| = q^2 + q + 1$, $r = q + 1$, $k = q + 1$, and $\lambda = 1$.*

Proof. Of course, $(P', \mathscr{L}')$ is the projective plane of Theorem 3.10. By (a) of Definition 3.9, we have $\lambda = 1$. If $\alpha \in I$, then α belongs to each $L' \in \mathscr{L}'(\alpha)$ and $\alpha \in I$. There are q lines in $\mathscr{L}'(\alpha)$, so α belongs to $q + 1$ lines in $\mathscr{L}'$. If $p \notin I$, then p belongs to exactly one $L \in \mathscr{L}(\alpha)$ for each $\alpha \in I$. Since $I = F \cup \{\infty\}$, we have p belonging to exactly $q + 1$ lines in $\mathscr{L}$, and hence $q + 1$ lines in $\mathscr{L}'$. Now we have $r = q + 1$.

With $L' \in \mathscr{L}'$, then $|L'| = q + 1$ if $L' = L \cup \{\alpha\}$ for some $\alpha \in I$, and $|L'| = q + 1$ if $L' = I$, since $I = F \cup \{\infty\}$. Hence, $k = q + 1$, also. Now $P' = (F \times F) \cup I$, and $(F \times F) \cap I = \varnothing$, so $|P| = q^2 + (q + 1)$, and so $v = q^2 + q + 1$. Finally, for each $\alpha \in I$, there are q lines in $\mathscr{L}'(\alpha)$, and $|I| = q + 1$, so this accounts for $(q + 1)q$ lines. But then there is the line I. Altogether, $b = q^2 + q + 1$. Now we are finished with the proof. □

Exercise 3.12. Prove that for every field F with $|F| = q$, there is a BIBD $(P, \mathscr{L})$ with parameters $v = |P| = q^2$, $b = |\mathscr{L}| = q^2 + q$, $r = q + 1$, $k = q$, and $\lambda = 1$.

Exercise 3.13. Construct a BIBD with parameters $v = 7$, $b = 7$, $r = 3$, $k = 3$, and $\lambda = 1$.

3.3 Möbius planes constructed from fields

In Theorem 3.4, we took an arbitrary field F and constructed an affine plane $(F \times F, \mathscr{L}_F)$. In Theorem 3.10, we took the affine plane $(F \times F, \mathscr{L}_F)$ and constructed a projective plane $(P', \mathscr{L}')$. In Theorem A2.40 and Definition A2.42, we learned that if there is a quadratic extension $K = F[x]/(p)$ of F, then we can make $F \times F$ into a field, which we also call a quadratic extension. So, let F be a field with a quadratic extension K.

In constructing the projective plane $(P', \mathscr{L}')$ we thought of different families of parallel lines in $(F \times F, \mathscr{L}_F)$ meeting at different points at infinity. We added all these points to $F \times F$ and made a line at infinity. But, for Möbius planes, we want all lines in the affine plane $(F \times F, \mathscr{L}_F)$ to meet at the same point ∞, and we just add ∞ to $F \times F$. But we also want to make a quadratic extension of F. So, let K be a quadratic extension of F. (As we saw in Theorem A2.40, we can identify K with $F \times F$.) Remember, that for any finite field F, there is a quadratic extension K of F. (Can you think of a field without a quadratic extension?) We add ∞ to K, but when we think of F as a subfield of K, then we also think of ∞ as being added to F. So

$$F' = F \cup \{\infty\} \quad \text{and} \quad K' = K \cup \{\infty\},$$

where ∞ is some symbol not in $K \cup F$, but ∞ is chosen for its suggestive meaning. We also think of F as being a subfield of the field K.

For $a,b,c,d \in K$, with $\{c,d\} \neq \{0\}$, define a function $[a,b,c,d] : K' \to K'$ as follows. First, if $c = 0$, then $d \neq 0$, and so we define

$$[a,b,c,d](x) = \begin{cases} (a/d)x + (b/d), & \text{if } x \neq \infty; \\ \infty, & \text{if } x = \infty. \end{cases}$$

If $c \neq 0$, define

$$[a,b,c,d](x) = \begin{cases} (ax+b)/(cx+d), & \text{if } x \notin \{-d/c, \infty\}; \\ \infty, & \text{if } x = -d/c; \\ a/c, & \text{if } x = \infty. \end{cases}$$

We want $[a,b,c,d]$ to be a bijection. This will be provided by the following theorem.

Theorem 3.14. *For a quadratic extension field K of a field F, the mapping $[a,b,c,d]$ is a bijection if and only if $ad - bc \neq 0$.*

Proof. Suppose that $[a,b,c,d]$ is a bijection. Consider first the case where $c = 0$. Then $d \neq 0$, and $[a,b,c,d](x) = (a/d)x + (b/d)$, if $x \neq \infty$. If $a = 0$, then $[a,b,c,d](x) = b/d$ for every $x \in K$. Since $|K| \geq 4$, we no longer have a bijection. Hence, $a \neq 0$. This means $0 \neq ad = ad - 0 = ad - bc$.

Next, suppose $c \neq 0$. If $ad - bc = 0$, then let λ be the common value $d/b = c/a$. So $d = \lambda b$ and $c = \lambda a$, hence $(c,d) = (\lambda a, \lambda b)$. For $u \in K \setminus \{-d/c, \infty\}$, we have

$$[a,b,c,d](u) = \frac{au+b}{cu+d} = \frac{au+b}{\lambda(au+b)} = \frac{1}{\lambda}.$$

Again, $|K| \geq 4$, and so $[a,b,c,d]$ is not a bijection. If we insist that $[a,b,c,d]$ is a bijection, then we must accept that $ad - bc \neq 0$.

The converse requires more detail. Suppose $ad - bc \neq 0$. If $c = 0$, then $ad \neq 0$, and so $a/d \neq 0$. Certainly, $[a,b,c,d]$ restricted to K defines a bijection on K, and since $[a,b,c,d](\infty) = \infty$, we have that $[a,b,c,d]$ is a bijection.

Next, suppose that $c \neq 0$. Our first step is to show that $[a,b,c,d]$ is injective, and so, for $u, v \in K'$, we want to show that $[a,b,c,d](u) = [a,b,c,d](v)$ implies $u = v$. We will do this in four steps. First, consider the case where u and v are in $K' \setminus \{-d/c, \infty\}$; second, consider $u = -d/c$; third, consider $u = \infty$; and fourth, consider $\{u, v\} = \{-d/c, \infty\}$.

If $u, v \in K' \setminus \{-d/c, \infty\}$ and $[a,b,c,d](u) = [a,b,c,d](v)$, then

$$\frac{au+b}{cu+d} = \frac{av+b}{cv+d},$$

which implies that $[ad - bc]u = [ad - bc]v$. Since $ad - bc \neq 0$, we then conclude that $u = v$.

If $u = -d/c$, then $[a,b,c,d](u) = \infty$. If $[a,b,c,d](v) = \infty$, then $v = -d/c$, also. Hence, $u = v$. If $u = \infty$, then $[a,b,c,d](u) = a/c$. If $[a,b,c,d](v) = a/c$, then $v \neq -d/c$. If $v \notin \{-d/c, \infty\}$, then we have

$$\frac{av+b}{cv+d} = \frac{a}{c},$$

from which we conclude that $acv + ad = acv + bc$, and so $ad - bc = 0$. We are left only with the possibility that $v = \infty$. Hence, $u = v$.

Finally, if $\{u, v\} = \{-d/c, \infty\}$, then

$$\{[a,b,c,d](u),\ [a,b,c,d](v)\} = \{a/c, \infty\},$$

and so $[a,b,c,d](u) \neq [a,b,c,d](v)$.

In all cases, $[a,b,c,d](u) = [a,b,c,d](v)$ implies $u = v$. Hence, $[a,b,c,d]$ is an injection when $ad - bc \neq 0$. There remains to show that $[a,b,c,d]$ is surjective when $ad - bc \neq 0$ and $c \neq 0$.

We have $[a,b,c,d](-d/c) = \infty$ and $[a,b,c,d](\infty) = a/c$. So, take an element $s \in K' \setminus \{a/c, \infty\}$. Since $s \neq a/c$, we have $(a - cs)^{-1} \in K$. Let $t = (a - cs)^{-1}(sd - b)$.

Now

$$[a,b,c,d](t) = \frac{a(a-cs)^{-1}(sd-b)+b}{c(a-cs)^{-1}(sd-b)+d}$$
$$= \frac{a(sd-b)+b(a-cs)}{c(sd-b)+d(a-cs)} = \frac{s(ad-bc)}{ad-bc} = s,$$

since $ad-bc \neq 0$. This makes $[a,b,c,d]$ surjective and injective, and, therefore, $[a,b,c,d]$ is a bijection. □

With $a,b,c,d \in F$ such that $ad-bc \neq 0$, we cannot help but think of $\left(\begin{smallmatrix} a & b \\ c & d \end{smallmatrix}\right)$ as an element in $GL(2,K)$. (See Examples A4.2.) With $ad-bc \neq 0$, we have that $[a,b,c,d] : K' \to K'$ is a bijection on K'. Hence, $[a,b,c,d] \in \mathscr{U}(M(K'))$. This defines a mapping $\Gamma : GL(2,K) \to \mathscr{U}(M(K'))$ defined by

$$\Gamma\left[\begin{pmatrix} a & b \\ c & d \end{pmatrix}\right] = [a,b,c,d].$$

It is often convenient and suggestive to write

$$[a,b,c,d](x) = \frac{ax+b}{cx+d}.$$

Suppose we have $f,g \in \mathscr{U}(M(K'))$ where

$$f(x) = \frac{ax+b}{cx+d} \qquad \text{and} \qquad g(x) = \frac{a'x+b'}{c'x+d'}.$$

Then

$$(f \circ g)(x) = f(g(x)) = \frac{ag(x)+b}{cg(x)+d} = \frac{(aa'x+ab')/(c'x+d')+b}{(ca'x+cb')/(c'x+d')+d}$$
$$= \frac{aa'x+ab'+bc'x+bd'}{ca'x+cb'+dc'x+dd'} = \frac{(aa'+bc')x+(ab'+bd')}{(ca'+dc')x+(cb'+dd')}$$
$$= \Gamma\left[\begin{pmatrix} aa'+bc' & ab'+bd' \\ ca'+dc' & cb'+dd' \end{pmatrix}\right](x)$$
$$= \Gamma\left[\begin{pmatrix} a & b \\ c & d \end{pmatrix}\begin{pmatrix} a' & b' \\ c' & d' \end{pmatrix}\right](x).$$

This is exactly what we need to have

$$\Gamma\left[\begin{pmatrix} a & b \\ c & d \end{pmatrix}\begin{pmatrix} a' & b' \\ c' & d' \end{pmatrix}\right] = \Gamma\left[\begin{pmatrix} a & b \\ c & d \end{pmatrix}\right]\Gamma\left[\begin{pmatrix} a' & b' \\ c' & d' \end{pmatrix}\right],$$

which means that $\Gamma : GL(2,K) \to \mathscr{U}(M(K'))$ is a group homomorphism.

Is Γ a monomorphism? Well,

$$\Gamma\left[\begin{pmatrix} a & 0 \\ 0 & a \end{pmatrix}\right] = [a,0,0,a]$$

and

$$[a,0,0,a](x) = \frac{ax}{a} = x$$

for every $a \in K^* = K \setminus \{0\}$. So Γ is not a monomorphism. In any event, all $[a,b,c,d]$ with $ad - bc \neq 0$ form a subgroup of $\mathscr{U}(M(K'))$, called the *group of linear fractional transformations*, and is denoted by $LF(K)$. A mapping $[a,b,c,d] \in LF(K)$ is a *linear fractional transformation.*

We should summarize some of these observations.

Theorem 3.15. *Let K be a quadratic extension of the field F, and define the function $\Gamma : GL(2,K) \to LF(K)$ by*

$$\Gamma\left[\begin{pmatrix} a & b \\ c & d \end{pmatrix}\right] = [a,b,c,d]$$

where

$$[a,b,c,d](x) = \frac{ax+b}{cx+d}.$$

Then Γ is a group epimorphism.

The group $LF(K)$ will help us construct our family of Möbius planes. We mentioned earlier that we wanted to add a point ∞ to every line $L \in \mathscr{L}_F$ of the affine plane $(F \times F, \mathscr{L}_F)$. We want to do more. In fact, we want to think of each $L \cup \{\infty\}$, $L \in \mathscr{L}_F$, not as a "line," but as a "circle" with "infinite radius" in $F \times F \cup \{\infty\} = K \cup \{\infty\} = K'$. The best place to see this is when $F = \mathbb{R}$ and $K = \mathbb{C}$, which we will do shortly. But first, let's make a few observations about $LF(K)$.

Lemma 3.16. *For $[a,b,c,d] \in LF(K)$, $[a,b,c,d] \neq 1_{K'}$, there are at most two solutions to the equation*

$$[a,b,c,d](z) = z.$$

Proof. If $[a,b,c,d](z) = z$, then

$$\frac{az+b}{cz+d} = z,$$

and so $cz^2 + dz = az + b$. Hence, z is a root for $cx^2 + (d-a)x - b \in K[x]$. There are at most two such solutions in K. □

These next two results are very significant in mathematics.

Theorem 3.17. *For three distinct points $z_2, z_3, z_4 \in K'$, there is a unique $S \in LF(K')$ such that $S(z_2) = 1$, $S(z_3) = 0$, and $S(z_4) = \infty$.*

Proof. Suppose $S, T \in LF(K)$ and $S(z_i) = T(z_i)$, $2 \leq i \leq 4$. Then 1, 0, ∞ are three distinct solutions to the equation $S \circ T^{-1}(z) = z$. By the result of Lemma 3.16, we must conclude that $S \circ T^{-1} = 1_{K'}$, which gives us $S = T$. This demonstrates uniqueness, but we must still show existence. This will be done with the participation of the reader.

If $\infty \in \{z_2, z_3, z_4\}$, define

$$S(z) = \begin{cases} (z-z_3)/(z-z_4), & \text{if } z_2 = \infty; \\ (z_2-z_4)/(z-z_4), & \text{if } z_3 = \infty; \\ (z-z_3)/(z_2-z_3), & \text{if } z_4 = \infty. \end{cases}$$

One should now verify that $S(z_2) = 1$, $S(z_3) = 0$, and $S(z_4) = \infty$ in each of these three cases.

If $\infty \notin \{z_2, z_3, z_4\}$, let

$$S(z) = \left(\frac{z - z_3}{z - z_4}\right) \Big/ \left(\frac{z_2 - z_3}{z_2 - z_4}\right).$$

Certainly, one should verify that $S(z_2) = 1$, $S(z_3) = 0$, and $S(z_4) = \infty$ in this case, also. If one does this required checking, then one can be assured of the existence of the S promised in the statement of Theorem 3.17. □

We are now in a position to prove the companion result to Theorem 3.17.

Theorem 3.18. *Let z_2, z_3, z_4 be three distinct points in K'. Also, let w_2, w_3, w_4 be three distinct points in K'. Then there is a unique $M \in LF(K)$ such that $M(z_2) = w_2$, $M(z_3) = w_3$, and $M(z_4) = w_4$.*

Proof. For z_2, z_3, and z_4, we let $S \in LF(K)$ be as promised by Theorem 3.17. For w_2, w_3, and w_4, we let $T \in LF(K)$ be as promised by Theorem 3.17. Then

$$S(z_2) = 1 = T(w_2), \;\; S(z_3) = 0 = T(z_3), \;\text{ and }\; S(z_4) = \infty = T(w_4).$$

Consequently,

$$T^{-1} \circ S(z_2) = w_2, \;\; T^{-1} \circ S(z_3) = w_3, \;\text{ and }\; T^{-1} \circ S(z_4) = w_4.$$

With $M = T^{-1} \circ S \in LF(K)$, we have the existence. If $M' \in LF(K)$ and $M'(z_i) = w_i$, $2 \le i \le 4$, then $M^{-1} \circ M'(z_i) = z_i$, $2 \le i \le 4$. Hence, the equation $M^{-1} \circ M'(z) = z$ has three distinct solutions. By Lemma 3.16, this makes $M^{-1} \circ M' = 1_{K'}$, and therefore, $M = M'$. Now we have the required uniqueness. □

Definition 3.19. For three distinct points $z_2, z_3, z_4 \in K'$, let S be the unique linear fractional transformation in $LF(K)$ such that $S(z_2) = 1$, $S(z_3) = 0$, and $S(z_4) = \infty$. For $z_1 \in K'$, the *cross ratio* of z_1, z_2, z_3, z_4 is $S(z_1)$, but is denoted by (z_1, z_2, z_3, z_4). So

$$(*, z_2, z_3, z_4) : K' \to K'$$

is a mapping defined by

$$(*, z_2, z_3, z_4)(z) = (z, z_2, z_3, z_4).$$

Theorem Theorem 3.18 is a prototype of some very significant applications of field theory to the subject of sharply transitive permutation groups. We are not in a position here to develop an adequate appreciation of this result to group theory, but we will at least give a few definitions.

For a nonempty set X, subgroups of $\mathscr{U}(M(X))$ have been studied heavily for their transitivity properties. A subgroup $\Gamma < \mathscr{U}(M(X))$ is *k-transitive on X* if for every pair of k-tuples $(x_1, \ldots, x_k)$, $(y_1, \ldots, y_k) \in X^k$, with

$$|\{x_1, \ldots, x_k\}| = |\{y_1, \ldots, y_k\}| = k,$$

there is a $T \in \Gamma$ such that $T(x_i) = y_i$, for each i, $1 \leq i \leq k$. If such a unique $T \in \Gamma$ exists for each pair $(x_1, \ldots, x_k)$ and $(y_1, \ldots, y_k)$, then Γ is *sharply k-transitive*. The mathematical community still does not know for sure if it has all the sharply 2-transitive groups. All sharply 3-transitive groups are modeled after the $LF(K)$, and Theorem 3.18 tells us that each $LF(K)$ is sharply 3-transitive.

It will be helpful to study $LF(\mathbb{C})$. Actually, this group alone has been heavily studied, and there is a rich body of literature on the many properties of $LF(\mathbb{C})$. There are numerous applications to engineering and to geometry, and a serious study of $LF(\mathbb{C})$ is highly recommended. But here, we will use some properties of $LF(\mathbb{C})$ for motivation to get Möbius planes from any quadratic extension K.

Perhaps we should collect together some basic properties and notation of the complex numbers. For $x, y \in \mathbb{R}$, a complex number $z \in \mathbb{C}$ has the form $z = x + iy$, where $i^2 = -1$. $\mathscr{R}(z) = x$ is the *real part* of z, and $\mathscr{I}(z) = y$ is the *imaginary part* of z. There is a very important field automorphism $c : \mathbb{C} \to \mathbb{C}$ defined by $c(z) = \bar{z}$, where $\bar{z} = x - iy$ if $z = x + iy$. The number $\bar{z}$ is the *(complex) conjugate* of z. Certainly $c \circ c(z) = z$, and so $c^2 = 1_{\mathbb{C}}$, making $(\{1_{\mathbb{C}}, c\}, \circ)$ a group isomorphic to $(\mathbb{Z}_2,+)$. Note that

$$z\bar{z}=x^2+y^2=|z|^2, \tag{3:12}$$

where $|z|=\sqrt{x^2+y^2}$ is the *absolute value* or *length* of $z=x+iy$. Always remember to identify $z=x+iy$ with the point (x,y). Sometimes (x,y) is used for $x+iy$, and we will even write $z=x+iy=(x,y)$. If $z=x+iy$ and $c=a+ib$, with $x,y,a,b\in\mathbb{R}$, then

$$|z-c|=\sqrt{(x-a)^2+(y-b)^2},$$

and so $|z-c|$ is the distance between z and c. Hence,

$$|z-c|=r, \tag{3:13}$$

for $r>0$, is the equation of the circle of radius r and center c. The solutions to the equation (3:13) make up the points of this circle. Also, notice that $z+\bar{z}=2x$ if $z=x+iy$, so

$$\mathscr{R}(z)=\frac{1}{2}(z+\bar{z}). \tag{3:14}$$

Similarly, $z-\bar{z}=2iy$, and so

$$\mathscr{I}(z)=\frac{1}{2i}(z-\bar{z}). \tag{3:15}$$

When we add ∞ to $\mathbb{C}$, we will be set to construct a Möbius plane from $\mathbb{C}$, and it is on this model that our construction will be built. So, $\mathbb{C}'=\mathbb{C}\cup\{\infty\}$ and $\mathbb{R}'=\mathbb{R}\cup\{\infty\}$. The elements of $\mathbb{C}'$ will be our *points*. But we do not want a family of "lines." Rather, we want a family of "circles," but we want to enlarge our idea for a circle in a Möbius plane. Our circles will include the traditional circles,

$$|z-a|=r,$$

for $r>0$ and $a\in\mathbb{C}$. But our circles will also include all traditional lines L with ∞ tacked on. So, if $L\in\mathscr{L}_{\mathbb{R}}$, for the affine plane $(\mathbb{R}\times\mathbb{R},\mathscr{L}_{\mathbb{R}})$, then

$$L'=L\cup\{\infty\},$$

and L' is a "circle" with infinite radius. If

$$C(a;r) = \{z \in \mathbb{C} \mid |z-a| = r\},$$

for $a \in \mathbb{C}$ and $r > 0$, and if

$$\mathscr{C} = \{C(a;r) \mid a \in \mathbb{C},\ r \in \mathbb{R}_+\} \cup \{L' \mid L \in \mathscr{L}_{\mathbb{R}}\},$$

then $(\mathbb{C}', \mathscr{C})$ will be a Möbius plane. But just what is a Möbius plane?

Definitions 3.20. An ordered pair $(P, \mathscr{C})$ is a *Möbius plane*, or *inversive plane*, if P is a set of *points* and $\mathscr{C}$ is a family of subsets of P. The members $C \in \mathscr{C}$ are called *circles*. The points and circles must have the following properties.

(a) $|P| \geq 4$.

(b) If $A \in \mathscr{C}$, then $|A| \geq 3$.

(c) There is an $A \in \mathscr{C}$ and an $x \in P$ with $x \notin A$.

(d) If $a, b, c \in P$ are three distinct points, then there is a unique circle $A \in \mathscr{C}$ with $a, b, c \in A$.

(e) If $x, y \in P$, $A \in \mathscr{C}$, $x \in A$, and $y \notin A$, then there is a unique circle $B \in \mathscr{C}$ such that $x, y \in B$ and $A \cap B = \{x\}$. So A and B are *tangent*.

At this point, the reader is encouraged to think about the $(\mathbb{C}', \mathscr{C})$ which we constructed above. Can you believe that $(\mathbb{C}', \mathscr{C})$ is a Möbius plane?

Of course, we want to prove that $(\mathbb{C}', \mathscr{C})$ is a Möbius plane, but right now we are more concerned about a way to get all the circles $C \in \mathscr{C}$ using $LF(\mathbb{C})$. This will provide the motivation for constructing our Möbius plane from any quadratic extension K. For now, we refer to the elements of $\mathbb{C}'$ as *points* and the elements $C \in \mathscr{C}$ as *circles*.

Perhaps the following lemma will be reassuring.

Lemma 3.21. *Let z_2, z_3, z_4 be three distinct points in $\mathbb{C}'$. Then there is a unique circle $C \in \mathscr{C}$ with $z_2, z_3, z_4 \in C$.*

Proof. If $\infty \notin \{z_2, z_3, z_4\}$, then $z_2, z_3, z_4 \in \mathbb{C}$, and since $\mathbb{C}$ is the classical euclidean plane, there is a unique classical circle C containing z_2, z_3, z_4, unless z_2, z_3, z_4 are collinear in a unique line $L \in \mathscr{L}_{\mathbb{R}}$. In that case, z_2, z_3, z_4 belong to the circle $L' = L \cup \{\infty\}$. In any event, z_2, z_3, z_4 belong to a unique circle in $\mathscr{C}$.

If $\infty \in \{z_2, z_3, z_4\}$, then two of the points belong to a unique line $M \in \mathscr{L}_{\mathbb{R}}$, and so $z_2, z_3, z_4 \in M' = M \cup \{\infty\}$, a circle in $\mathscr{C}$. □

Unless you have the courage for a lot of technical detail, which you should develop sometime if you want to be seriously active in mathematics, we would suggest you accept Lemmas 3.22 and 3.23 on faith, and start again with Theorem 3.24. Perhaps Lemmas 3.22 and 3.23 are more suitable for a second or third reading.

Lemma 3.22. *Let $S = (*, z_2, z_3, z_4) = [a, b, c, d]$ be the transformation of Definition 3.19. Suppose $a\bar{c} \in \mathbb{R}$, and let $C \in \mathscr{C}$ be the unique circle with $z_2, z_3, z_4 \in C$. Then $C = L' = L \cup \{\infty\}$ for some $L \in \mathscr{L}_{\mathbb{R}}$, and $S(C) = \mathbb{R}' = \mathbb{R} \cup \{\infty\}$. Consequently, $S^{-1}(\mathbb{R}') = C$.*

Proof. Let $B = S^{-1}(\mathbb{R})$. (We know that $z_4 = S^{-1}(\infty)$, and $z_4 \in C$.) If $z = x + iy \in B$, then $S(z) \in \mathbb{R}$. With the condition that $a\bar{c} \in \mathbb{R}$, which we have, we will develop 8 equivalent equations and prove their equivalence as we go along. The first one will be true if $S(z) \in \mathbb{R}$. The last one will be true if $z = x + iy = (x, y) \in L$. As a consequence of $S(z) \in \mathbb{R}$, we have

$$S(z) = \overline{S(z)}; \tag{1}$$

$$\frac{az+b}{cz+d} = \frac{\bar{a}\bar{z}+\bar{b}}{\bar{c}\bar{z}+\bar{d}}; \tag{2}$$

and

$$(a\bar{c} - \bar{a}c)|z|^2 + (a\bar{d} - \bar{b}c)z + (b\bar{c} - d\bar{a})\bar{z} + (b\bar{d} - \bar{b}d) = 0. \tag{3}$$

Certainly, (2) implies (3), and (3) implies (2) if $cz+d \neq 0$. If $cz+d=0$, then $z=-d/c$ and so $S(z)=\infty$. Hence, $z=z_4$ and $z \notin B$. So, we are safe in having $cz+d \neq 0$.

As long as $a\bar{c} \in \mathbb{R}$, then $\overline{a\bar{c}} = \bar{a}c \in \mathbb{R}$, and so $a\bar{c} - \bar{a}c = 0$. Hence, (3) is equivalent to

$$(4) \qquad (a\bar{d} - \bar{b}c)z + (b\bar{c} - d\bar{a})\bar{z} + (b\bar{d} - \bar{b}d) = 0.$$

Let $\alpha/2 = a\bar{d} - \bar{b}c$ and let

$$\beta = i(b\bar{d} - \bar{b}d) = i[b\bar{d} - \overline{b\bar{d}}] = i[2i\mathscr{I}(b\bar{d})] = -2\mathscr{I}(b\bar{d}),$$

so $\beta \in \mathbb{R}$. Hence, (4) is equivalent to

$$(5) \qquad (\alpha z - \overline{\alpha z})/2 + \beta/i = 0,$$

which is equivalent to

$$(6) \qquad \frac{i}{2}[2i\mathscr{I}(\alpha z)] + \beta = 0,$$

and to

$$(7) \qquad \mathscr{I}(\alpha z) - \beta = 0.$$

If $\alpha = \alpha_1 + i\alpha_2$, $z = x+iy$, $\alpha_1, \alpha_2, x, y \in \mathbb{R}$, then $\alpha z = (\alpha_1 x - \alpha_2 y) + (\alpha_1 y + \alpha_2 x)i$, and so (7) is equivalent to

$$(8) \qquad \alpha_2 x + \alpha_1 y = \beta.$$

We've already noted that $\beta \in \mathbb{R}$, and so (8) is the equation of some line $L \in \mathscr{L}_{\mathbb{R}}$. Hence, $(x,y) = x+iy \in L$, and so $z \in L' = L \cup \{\infty\}$. Among other things, we have the existence of L' defined from (8), as a consequence of $S(z) \in \mathbb{R}$. Hence, $z \in B = S^{-1}(\mathbb{R})$ implies $z \in L'$.

Conversely, if $z = x+iy = (x,y) \in L$, then (8) is true, and so (1) is true. With $S(z) = \overline{S(z)}$, we have $S(z) \in \mathbb{R}$. So, $L \subseteq B$.

We have, since $S(z_4) = \infty$, that $L' = S^{-1}(\mathbb{R}')$, and that $z_2, z_3, z_4 \in L'$. But C is the unique circle with $z_2, z_3, z_4 \in C$. Hence, $C = L'$. □

Lemma 3.23. *Let $S = (*, z_2, z_3, z_4) = [a,b,c,d]$ be the transformation of Definition 3.19. Suppose $a\bar{c} \notin \mathbb{R}$, and let $C \in \mathscr{C}$ be the unique circle with $z_2, z_3, z_4 \in C$. Then $C = C(\alpha; r)$ for some classical circle $C(\alpha; r)$ in the Euclidean plane $\mathbb{C}$, and $S(C) = \mathbb{R}' = \mathbb{R} \cup \{\infty\}$. Consequently, $S^{-1}(\mathbb{R}') = C$.*

Proof. Let $B = S^{-1}(\mathbb{R})$. We certainly have $z_2, z_3 \in B$, and $S(z_4) = \infty$. With the condition that $a\bar{c} \notin \mathbb{R}$, which we have, we will develop 9 equivalent equations and prove their equivalence as we go along. The first one will be true if $S(z) \in \mathbb{R}$. The last will be true if $z \in C(\alpha; r)$, $z \neq z_4$. As a consequence of $S(z) \in \mathbb{R}$, we have

$$S(z) = \overline{S(z)}; \tag{1}$$

$$\frac{az+b}{cz+d} = \frac{\bar{a}\bar{z}+\bar{b}}{\bar{c}\bar{z}+\bar{d}}; \tag{2}$$

$$(a\bar{c} - \bar{a}c)|z|^2 + (a\bar{d} - \bar{b}c)z + (b\bar{c} - d\bar{a})\bar{z} + (b\bar{d} - \bar{b}d) = 0. \tag{3}$$

As before, (2) implies (3), and (3) implies (2) if $cz + d \neq 0$. If $cz + d = 0$, then $z = z_4$ and $S(z) = \infty$, so $z \notin B$. Again, we're safe in having $cz + d \neq 0$.

With $a\bar{c} \notin \mathbb{R}$, we have $a\bar{c} - \bar{a}c \neq 0$. So (3) is equivalent to

$$|z|^2 + \left[\frac{a\bar{d} - \bar{b}c}{a\bar{c} - \bar{a}c}\right] z + \left[\frac{b\bar{c} - d\bar{a}}{a\bar{c} - \bar{a}c}\right] \bar{z} + \left[\frac{b\bar{d} - \bar{b}d}{a\bar{c} - \bar{a}c}\right] = 0. \tag{4}$$

Let $\gamma = (b\bar{c} - d\bar{a})/(a\bar{c} - \bar{a}c)$ and $-\delta = (b\bar{d} - \bar{b}d)/(a\bar{c} - \bar{a}c)$, and so (4) is equivalent to

$$|z|^2 + \bar{\gamma}z + \gamma\bar{z} - \delta = 0 \tag{5}$$

and to

(6) $$|z|^2+\overline{\gamma}z+\gamma\overline{z}+|\gamma|^2=|\gamma|^2+\delta.$$

From (3:12), we get

(7) $$|z+\gamma|^2=|\gamma|^2+\delta.$$

Now

$$\begin{aligned}|\gamma|^2+\delta&=\left[\frac{b\overline{c}-d\overline{a}}{a\overline{c}-\overline{a}c}\right]\left[\frac{\overline{b}c-\overline{d}a}{\overline{a}c-a\overline{c}}\right]+\frac{\overline{b}d-b\overline{d}}{a\overline{c}-\overline{a}c}\\&=\frac{(b\overline{c}-d\overline{a})(\overline{b}c-\overline{d}a)+(\overline{b}d-b\overline{d})(\overline{a}c-a\overline{c})}{|a\overline{c}-\overline{a}c|^2}\\&=\frac{a\overline{a}d\overline{d}-a\overline{b}\overline{c}d-\overline{a}bc\overline{d}+b\overline{b}c\overline{c}}{|a\overline{c}-\overline{a}c|^2}=\frac{(ad-bc)(\overline{a}\overline{d}-\overline{b}\overline{c})}{|a\overline{c}-\overline{a}c|^2}\\&=\frac{|ad-bc|^2}{|a\overline{c}-\overline{a}c|^2}>0,\end{aligned}$$

since $ad-bc\neq 0$ and $a\overline{c}-\overline{a}c\neq 0$. Hence, (7) is equivalent to

(8) $$|z+\gamma|=\sqrt{|\gamma|^2+\delta}=\lambda,$$

with $\lambda>0$. Certainly, (8) is equivalent to

(9) $$|z-(-\gamma)|=\lambda,$$

which says that $z\in C(-\gamma;\lambda)$, and $z\neq z_4$. We now have, as a consequence of $S(z)\in\mathbb{R}$ and $a\overline{c}\notin\mathbb{R}$, the existence of a circle $C(-\gamma;\lambda)$, which depends only upon a,b,c,d, hence upon S, and $z\in B=S^{-1}(\mathbb{R})$ implies $z\in C(-\gamma;\lambda)$, $z\neq z_4$.

Conversely, if $z=x+iy=(x,y)\in C(-\gamma;\lambda)$, $z\neq z_4$, then (9) is true, and so (1) is true. With $S(z)=\overline{S(z)}$, we know that $S(z)\in\mathbb{R}$. Hence, $C(-\gamma;\lambda)\setminus\{z_4\}\subseteq B$.

Since $S(z_4)=\infty$, we have that $C(-\gamma;\lambda)=S^{-1}(\mathbb{R}')$ and $z_2,z_3,z_4\in C(-\gamma;\lambda)$. But C is the unique circle with $z_2,z_3,z_4\in C$, hence $C=C(-\gamma;\lambda)$. □

An important point is made in the following theorem.

Theorem 3.24. *For* $(\mathbb{C}', \mathscr{C})$, *we have*

$$\mathscr{C} = \{L(\mathbb{R}') \mid L \in LF(\mathbb{C})\}.$$

Proof. Let $C = C(\alpha; \lambda)$ be a classical circle in the euclidean plane $\mathbb{C}$. Then it has three distinct points $z_2, z_3, z_4 \in C$, and with them, we construct $S = (*, z_2, z_3, z_4)$ of Definition 3.19. From Lemma 3.23, we have $S(C) = \mathbb{R}'$, and so $C = S^{-1}(\mathbb{R}')$.

Let $L' = L \cup \{\infty\}$ be a circle from a line $L \in \mathscr{L}_{\mathbb{R}}$. The line L also has three distinct points $z_2, z_3, z_4 \in L$ from which we construct $S = (*, z_2, z_3, z_4)$ of Definition 3.19. From Lemma 3.22, we know that $S(L') = \mathbb{R}'$, and so $L' = S^{-1}(\mathbb{R}')$.

We now have that $\mathscr{C} \subseteq \{L(\mathbb{R}') \mid L \in LF(\mathbb{C})\}$. So, take some $L(\mathbb{R}')$ with $L \in LF(\mathbb{C})$. Certainly, $L(1) = z_2$, $L(0) = z_3$, and $L(\infty) = z_4$, and $z_2, z_3, z_4 \in L(\mathbb{R}')$. If $M = (*, z_2, z_3, z_4)$, then $M \circ L(1) = 1$, $M \circ L(0) = 0$, and $M \circ L(\infty) = \infty$. Hence,

$$M \circ L(z) = z$$

has three distinct solutions, namely, z_2, z_3, and z_4. From Lemma 3.16, we learn that $M \circ L = 1_{\mathbb{C}'}$, and so $M = L^{-1}$. But $M^{-1} = L$, and $M^{-1}(\mathbb{R}')$ is the unique circle containing z_2, z_3, and z_4. Since $M^{-1}(\mathbb{R}') = L(\mathbb{R}')$, we have that $\{L(\mathbb{R}') \mid L \in LF(\mathbb{C})\} \subseteq \mathscr{C}$. This completes the proof of Theorem 3.24. □

So, with a quadratic extension K of F, we can make a set

$$\mathscr{C} = \{L(F') \mid L \in LF(K)\},$$

with $F' = F \cup \{\infty\}$, of course. Motivated by Theorem 3.24, we call the sets $L(F') \in \mathscr{C}$ *circles*. The main goal of this section is the next theorem.

Theorem 3.25. *Let* K *be a quadratic extension field of a field* F. *Let* $K' = K \cup \{\infty\}$ *and* $F' = F \cup \{\infty\}$, *as above. Define*

$$\mathscr{C} = \{L(F') \mid L \in LF(K)\}.$$

Then $(K', \mathscr{C})$ *is a Möbius plane.*

Proof. Certainly $|K'| \geq 4$, and $|A| \geq 3$ for every $A \in \mathscr{C}$. Since $LF(K) \subseteq \mathscr{U}(M(K'))$, and since $F' \subset K'$, we have an $A \in \mathscr{C}$ and an $x \in K'$ with $x \notin A$. From Theorem 3.18, we can be assured that if $a, b, c \in K'$ are three distinct points, then there is an $A \in \mathscr{C}$ with $a, b, c \in A$. We must show uniqueness, and also property (e) of Definitions 3.20.

If $a, b, c \in B$ and $B \in \mathscr{C}$, then $B = M(F')$ for some $M \in LF(K)$. Also, $A = L(F')$ for some $L \in LF(K)$. There is a unique $N \in LF(K)$ for which $N(a) = 1, N(b) = 0$, and $N(c) = \infty$, and so $0, 1, \infty \in N(A) = N \circ L(F')$, and $0, 1, \infty \in N(B) = N \circ M(F')$. Hence, $N(A) = N(B)$, and since $N \in LF(K)$, we have that $A = B$. This gives us the desired uniqueness.

For property (e), we will first show existence. Showing uniqueness will be rather involved. So our immediate goal is to prove the following statement.

($*$) *If $x, y \in K'$, $A \in \mathscr{C}$, $x \in A$, and $y \notin A$, then there is a circle $B \in \mathscr{C}$ such that $x, y \in B$ and $A \cap B = \{x\}$.*

First, let $A = F'$, $x = 0$, and $a = y$. Consider $L = [a, 0, 1, a]$. So $L(z) = az/(z+a)$. Then $L(0) = 0$ and $L(\infty) = a$. Let $B = L(F')$. So $0, a \in B$, which is to say that $x, y \in B$. Now $\infty \notin L(F') = B$, since $y = a \notin A = F'$. Let $u \in F$ and $t = L(u)$. Then $t = au/(u+a)$, and so $t(u+a) = au$. Since $y = a \in K \setminus F$, and since K is a quadratic extension of F, then $a = a_1 + a_2 w$ where $a_1, a_2 \in F$ and $w \notin F$, and $a_2 \neq 0$. From $t(u + a_1 + a_2 w) = u(a_1 + a_2 w)$, we may conclude that $tu + ta_1 = ua_1$ and $ta_2 w = ua_2 w$. From $ta_2 w = ua_2 w$ and $a_2 \neq 0$, we get $t = u = 0$. Therefore, $\infty \notin B$, and if $t, u \in A \setminus \{\infty\}$ such that $t = L(u) \in L(F') = B$, then $t = u = 0$. Hence, $A \cap B = \{0\} = \{x\}$.

Now, we remove the extra conditions imposed on A and x. From Theorem 3.17 and Lemma 3.21, there is an $S \in LF(K)$ such that $S(x) = 0$ and $S(A) = F'$. With $y \notin A$, then $S(y) \notin F'$. By the previous paragraph, we have the existence of a $C \in \mathscr{C}$ such that $S(y) \in C$ and $F' \cap C = \{0\}$. So $S^{-1}(C) \in \mathscr{C}$,

$y \in S^{-1}(C)$, and $\{x\} = \{S^{-1}(0)\} = S^{-1}(F') \cap S^{-1}(C) = A \cap S^{-1}(C)$. Let $B = S^{-1}(C)$. Hence, $x, y \in B$ and $A \cap B = \{x\}$.

We now have the existence requirement of (e) of Definitions 3.20. The uniqueness requires three technical steps in preparation.

(1) *Let $a \in K \setminus F$, $S(x) = ax/(x+a)$, and $B = S(F')$. Suppose $L \in LF(K)$, $L(0) = 0$, $L(\infty) = a$, and $L(F') = B$. Then $L(x) = ax/(x+\lambda a)$ for some $\lambda \in F$, and conversely.*

Proof of **(1)**. From $L(0) = 0$ and $L(\infty) = a$, we can write $L(x) = ax/(x+d)$ for some d. Since $a/(1+a) = S(1) \in B$, we have $a/(1+a) = a\lambda/(\lambda + d)$ for some $\lambda \in F$. Hence, $a(\lambda + d) = a\lambda(1+a)$. As a consequence, $\lambda a^2 = ad$, and with $a \neq 0$, we have $d = \lambda a$. Now we have $L(x) = ax/(x+\lambda a)$, as desired. Conversely, if $L(x) = ax/(x+\lambda a)$ for some $\lambda \in F$, then

$$L(\lambda) = a\lambda/(\lambda + \lambda a) = a/(1+a).$$

With $0, a, a/(1+a) \in L(F') \cap B$, we are assured that $L(F') = B$. □

(2) *If $a \in K \setminus F$ and $L(x) = ax/(x+\lambda)$, $\lambda \in F$, then $0, \infty \in F' \cap L(F')$.*

Proof of **(2)**. Certainly, $0, \infty \in F'$, and $L(0) = 0$ and $L(-\lambda) = \infty$. □

(3) *For $a \in K \setminus F$, $L(x) = ax/(x+d)$, $d \in K \setminus F$, and $d \notin Fa = \{\lambda a \mid \lambda \in F\}$, we have $|F' \cap L(F')| = 2$.*

Proof of **(3)**. Certainly, $0 \in F' \cap L(F')$, since $L(0) = 0$. Let $a = a_1 + a_2 w$ and $d = d_1 + d_2 w$, with $a_1, a_2, d_1, d_2 \in F$ and $w \in K \setminus F$. Since $a, d \in K \setminus F$, we have $0 \notin \{a_2, d_2\}$. Now $a_1 d_2 - a_2 d_1 \neq 0$. For if $a_1 d_2 - a_2 d_1 = 0$, then set $\mu = a_1/a_2 = d_1/d_2$, and so $a_1 = \mu a_2$, $d_1 = \mu d_2$, $a = a_1 + a_2 w = \mu a_2 + a_2 w = a_2(\mu + w)$, and

$d = d_2(\mu + w)$. Hence, $d = d_2(\mu + w) = d_2 a_2^{-1} a_2(\mu + w) = d_2 a_2^{-1} a \in Fa$. So $t = (a_1 d_2 - a_2 d_1)/d_2 \neq 0$, and $u = (a_1 d_2 - a_2 d_1)/a_2 \neq 0$. Now

$$\begin{aligned} L(u) &= \frac{au}{u+d} = \frac{(a_1 + a_2 w)(a_1 d_2 - a_2 d_1)/a_2}{(a_1 d_2 - a_2 d_1)/a_2 + (d_1 + d_2 w)} \\ &= \frac{(a_1 + a_2 w)(a_1 d_2 - a_2 d_1)}{(a_1 d_2 - a_2 d_1) + a_2(d_1 + d_2 w)} = \frac{(a_1 + a_2 w)(a_1 d_2 - a_2 d_1)}{d_2(a_1 + a_2 w)} \\ &= \frac{a_1 d_2 - a_2 d_1}{d_2} = t. \end{aligned}$$

Now $t \neq 0$ and $t \in L(F')$. So $0, t \in F' \cap L(F')$ and so $|F' \cap L(F')| \geq 2$. But $L(\infty) = a \notin F'$, and therefore $L(F') \neq F'$. Since (d) of Definitions 3.20 has been established, we know that $|F' \cap L(F')| = 2$. □

We are now ready to demonstrate uniqueness of B in $(*)$ above. As in the proof of $(*)$, we will start with $A = F'$, $x = 0$, and $a = y \notin A = F'$. So $a \in K \setminus F'$. Also, $L(z) = az/(z+a)$, $L(0) = 0$, and $L(\infty) = a$, and $B = L(F')$. Therefore, we have $0, a, a/(1+a) \in B$, and $A \cap B = \{0\}$.

Suppose $x, y \in C$, that is, $0, a \in C$. Now $C = M(F')$ for some $M \in LF(K)$, and so there are $x_1, x_2 \in F'$ such that $M(x_1) = 0$ and $M(x_2) = a$. Also, there is an $N \in LF(K)$ such that $N(0) = x_1$ and $N(\infty) = x_2$, and certainly $N(F') = F'$. So $M \circ N(0) = 0$, $M \circ N(\infty) = a$, and $M \circ N(F') = C$. Hence, $M \circ N = [a, 0, 1, d]$ for some $d \in K$. Either $d \in F$ or $d \notin F$. If $d \in F$, then, by (2), we also have that 0 and ∞ are contained in $F' \cap M \circ N(F')$, that is, $0, \infty \subset A \cap C$. As such, $C \neq B$, since $A \cap B = \{0\}$. If $d \notin F$, then either $d \notin Fa$ or $d \in Fa$. If $d \notin Fa$, then by (3), $|F' \cap M \circ N(F')| = 2$, that is, $|A \cap C| = 2$. As such, $C \neq B$, since $A \cap B = \{0\}$. Finally, for $d \notin F$ and $d \in Fa$, we have $d = \lambda a$ for some $\lambda \in F$. We have $a \in K \setminus F$, $L(z) = az/(z+a)$, $B = L(F')$, and $M \circ N(z) = az/(z + \lambda a)$ for some $\lambda \in F$. Applying (1), we get $M \circ N(F') = B$. Since $M \circ N(F') = C$, we have $C = B$. This gives us the uniqueness of B when the extra conditions are placed upon A and x.

We can now remove the conditions that $A = F'$ and $x = 0$. Take $A \in \mathscr{C}$,

$x, y \in K'$, $x \in A$, and $y \notin A$. Let $B \in \mathscr{C}$ with $x, y \in B$ and $A \cap B = \{x\}$. If $D \in \mathscr{C}$, $x, y \in D$, and $A \cap D = \{x\}$, then there is an $L \in LF(K)$ such that $L(x) = 0$ and $L(A) = F'$. As such, $L(B), L(D) \in \mathscr{C}$ and $L(y) \in L(B) \cap L(D)$, $F' \cap L(B) = \{0\}$, and $F' \cap L(D) = \{0\}$. The preceding paragraph assures us that $L(B) = L(D)$. Since L is a bijection of K', we have $B = D$. This completes the proof of the uniqueness of the B in $(*)$, and also the proof that property (e) of Definitions 3.20 is satisfied. And, finally, we have an end to the proof of Theorem 3.25, and so we have that $(K', \mathscr{C})$ is a Möbius plane. □

Theorem 3.26. *Let K be a finite quadratic extension of F where $|F| = q$. Then the Möbius plane $(K', \mathscr{C})$ is a BIBD with parameters $v = q^2 + 1$, $b = q(q^2 + 1)$, $r = q(q + 1)$, and $k = \lambda = q + 1$.*

Proof. With $|F| = q$, we certainly have $|K'| = q^2 + 1$. With $|F'| = q + 1$, we have $|C| = q + 1$ for each $C \in \mathscr{C}$, since each $L \in LF(K)$ is a bijection on K'. So, $k = q + 1$. If 0 belongs to exactly r distinct blocks $B_1 = L_1(F'), \cdots, B_r = L_r(F')$, each $L_i \in LF(K)$, and if $M(0) = x$ with $M \in LF(K)$, then x belongs to the r distinct blocks

$$M(B_1) = M \circ L_1(F'),\ M(B_2) = M \circ L_2(F'),\ \ldots,\ M(B_r) = M \circ L_r(F'),$$

and these r distinct blocks belong to $\mathscr{C}$. Conversely, if x belongs to exactly r' distinct blocks

$$B'_1 = L'_1(F'),\ B'_2 = L'_2(F'),\ \ldots,\ B'_{r'} = L'(F'),$$

and if $M'(x) = 0$ for some $M' \in LF(K)$, then $r \le r'$ and 0 belongs to the r' distinct blocks

$$M' \circ L'_1(F'),\ M' \circ L'_2(F'),\ \ldots,\ M' \circ L'_{r'}(F')$$

in $\mathscr{C}$, and so $r' \le r$. Hence, $r = r'$, and an arbitrary $x \in K'$ belongs to exactly r distinct blocks of $\mathscr{C}$. We will calculate r later.

Now, take $a,b \in K'$, $a \neq b$. For any $x \in K' \setminus \{a,b\}$, a,b,x belong to exactly one circle, or block

$$B = \{a,b,x_{11},x_{12},\ldots,x_{1,q-1}\}.$$

Let λ be the number of blocks containing a and b. For each such block containing a and b, we have exactly $q-1$ elements of $K' \setminus \{a,b\}$, and a,b are the only points in common with any two of them. Hence, $q^2-1 = \lambda(q-1)$, and so $\lambda = q+1 = k$.

At this point, we are assured that $(K',\mathscr{C})$ is a BIBD, but we do not have the value for the parameters b and r. From $\lambda(v-1) = r(k-1)$ we have $r = q(q+1)$. Also, $vr = bk$ gives us $b = q(q^2+1)$. This finishes our proof of the theorem. □

How big is $LF(K)$? Each $L \in LF(K)$ is $(*,z_2,z_3,z_4)$ for some unique elements $z_2,z_3,z_4 \in K'$. We choose z_2 in q^2+1 ways, z_3 in q^2 ways, and z_4 in q^2-1 ways. Hence, $|LF(K)| = (q^2+1)q^2(q^2-1)$.

But $b = q(q^2+1)$. This suggests that there may be $q(q^2-1)$ different $L \in LF(K)$ defining the same $B \in \mathscr{C}$. In fact, if $M \in LF(F)$, and $L(F') = B$, then $L \circ M(F') = L(F') = B$. So, each $L \circ M$ defines the same block as L does. There are, as one can easily show, exactly $(q+1)q(q-1) = q(q^2-1)$ elements in $LF(F)$.

So we have a lot of redundancy in computing a Möbius plane. It would be nice to have a better way.

Exercise 3.27. Construct a Möbius plane from the field $\mathbb{Z}_3$.

In Definitions 1.24, we defined what it meant for a BIBD to be a Möbius plane. In Definitions 3.20, we defined a Möbius plane independently of being a BIBD. This would be in very poor taste were it not for the fact that the two definitions are equivalent.

Theorem 3.28. *Suppose a BIBD $(P,\mathscr{C})$ is a Möbius plane according to Definitions 1.24. Then $(P,\mathscr{C})$ is a Möbius plane according to Definitions 3.20. Conversely,* *if $(P,\mathscr{C})$ is a Möbius plane according to Definitions 3.20, and if P is finite, then $(P,\mathscr{C})$ is a BIBD with parameters $(v,b,r,k,\lambda) = (n^2+1, n(n^2+1), n(n+1), n+1, n+1)$ for some n, and $(P,\mathscr{C})$ is a Möbius plane according to Definitions 1.24.*

Proof. Assume Definitions 1.24. Then

$$(v,b,r,k,\lambda) = (n^2+1, n(n^2+1), n(n+1), n+1, n+1), \text{ with } n \geq 2.$$

So, $v \geq 4$ and $|A| > 1$ for each $A \in \mathscr{C}$. Since $|A| = n+1 < n^2+1 = |P|$, there is an $x \in P$ with $x \notin A$. From Definitions 1.24, each triple of distinct points belong to exactly one block $B \in \mathscr{C}$. Hence, (a), (b), (c), and (d) of Definitions 3.20 are satisfied.

Take $x,y \in P, A \in \mathscr{C}, x \in A, y \notin A$. Now, if $A = \{x, x_1, \ldots, x_n\}$, then y, x, x_i belong to exactly one block $B_i \in \mathscr{C}$, $1 \leq i \leq n$, and if $B_i = B_j$, then $x, x_i, x_j \in B_i \cap B_j$, and so $A = B_i = B_j$. So the B_i, $1 \leq i \leq n$, are distinct. These are n of the $n+1$ blocks containing both x and y. There is another one containing both x and y, call it B. If some $x_i \in B$, then $y, x, x_i \in B \cap B_i$, and so $B = B_i$, a contradiction. This means that $A \cap B = \{x\}$, or B is tangent to A. Hence, $(P,\mathscr{C})$ is a Möbius plane according to Definitions 3.20.

For the converse, we have $|P| < \infty$, and a Möbius plane according to Definitions 3.20. For three distinct points $x,y,z \in P$, there is a unique $C \in \mathscr{C}$ with $x,y,z \in C$. There are at least 4 points in P, and a point x not in some block $C \in \mathscr{C}$. We will first show that $(P,\mathscr{C})$ is a BIBD.

Given $A \in \mathscr{C}, x \in A, y \in P, y \notin A$, then there is a unique $B \in \mathscr{C}$ with $B \cap A = \{x\}$ and with $x,y \in B$. Let $|A| = k$ with $A = \{x, x_2, \ldots, x_k\}$. For each i, $2 \leq i \leq k$, there is a unique $B_i \in \mathscr{C}$ with $y, x, x_i \in B_i$. If $B_i = B_j$, then $y, x, x_i, x_j \in B_i \cap B_j$, and so $B_i = A = B_j$. This makes the B_i distinct, and there are $k-1$ of them. Hence, $B, B_2, \cdots, B_k$ are the distinct blocks containing both x and y.

Now, if $x,y \in P, x \neq y$, then there is a block $A \in \mathscr{C}$ with $x \in A$ and $y \notin A$. Why is this? With $|P| \geq 4$, there are three distinct points $x,a,b \in P$, and a unique C with $x,a,b \in C$. If $y \notin C$, we can let $C = A$. If $y \in C$, then there is a point $z \in P$

with $z \notin C$. Otherwise, $C = P$. With $C = P$, there is an $A' \in \mathscr{C}$ and an $x' \in P$ with $x' \notin A'$, and $|A'| \geq 3$. So, there are three distinct points $u, v, w \in A' \cap C$, since $C = P$ and $|A'| \geq 3$. Since $A' \neq C$, we have two distinct blocks, A' and C, each containing the three distinct points u, v, and w. This contradiction, along with $y \in C$, assures us that there is a point $z \in P$ with $z \notin C$. So, x, a, z are three distinct points belonging to a unique block A, and $y \notin A$, for if $y \in A$, then $x, a, y \in A \cap C$, and so $A = C$, which, in turn, implies that $z \in C$.

So, with an arbitrary $x, y \in P$, $x \neq y$, there is an $A \in \mathscr{C}$ with $x \in A$ and $y \notin A$. The number of blocks containing both x and y is $|A|$. The same argument works for $y, x \in P$, $y \neq x$, and so there is a $B \in \mathscr{C}$ with $y \in B$ and $x \notin B$. The number of blocks containing both y and x is $|B|$. Hence, $|A| = |B|$. In fact, this is true for any $z \in P$, $z \neq x$. Hence, the number of blocks containing both x and y is $|A| = k$, and also $|A| = |B| = k$ if $A, B \in \mathscr{C}$. This makes $\lambda = k = |A|$ for any $A \in \mathscr{C}$.

There remains to show the existence of r, where r is the number of blocks containing any point $x \in P$. Take $x \in P$. There are $v - 1 = |P| - 1$ other elements $y \in P$. Each of these has $\lambda = k$ blocks $B_{y1}, \ldots, B_{yk}$ with $x, y \in B_{yi}$, $1 \leq i \leq k$. So, we have

$$(x,y) \mapsto \mathscr{B}_y = \{B_{y1}, B_{y2}, \ldots, B_{yk}\}$$

and $P = \cup_{i=1}^{k} B_{yi}$, since, for any $z \in P \setminus \{x, y\}$, x, y, z belong to exactly one block B_{yi}.

We have these $v - 1$ families $\mathscr{B}_{y_1}, \cdots, \mathscr{B}_{y_{v-1}}$, and each has k blocks, so there are $(v-1)k$ blocks contained in these families $\mathscr{B}_{y_i}$, with blocks being counted several times. If

$$B_{yi} = \{x, y, y_2, \ldots, y_{k-1}\} \in \mathscr{B}_y,$$

then $B_{yi} \in \mathscr{B}_{y_j}$ for each j, $2 \leq j \leq k-2$. That is, each B_{yi} is duplicated exactly $k - 1$ times. This means, there are exactly $(v-1)k/(k-1)$ blocks containing x. So, $r = (v-1)k/(k-1)$, and we have a BIBD $(P, \mathscr{C})$.

But, what are the parameters of such a BIBD? We know that $k = \lambda$ and $r = (v-1)k/(k-1)$. From $vr = bk$, we get $b = vr/k = v(v-1)/(k-1)$. From Corollary 2.17, we have $v - 2 = k(k-2)$, since $k = \lambda$. So

$$v = (k-1)^2 + 1.$$

Let $n = k - 1$. Then $v = n^2 + 1$, $k = \lambda = n + 1$, $r = n(n+1)$, and $b = n(n^2 + 1)$. This, altogether, means that $(P, \mathscr{C})$ is a BIBD and a Möbius plane according to Definitions 1.24, and completes our proof. □

3.4 Mutually orthogonal Latin squares from fields

Look at Exercise 1.37 again. Euler's problem involving 36 officers from 6 regiments and 6 ranks has fascinated mathematicians, both amateurs and professionals, up to today. Hopefully, no reader spent too many nice weekends believing that a solution to Exercise 1.37 was near.

If the reader had success with Exercise 1.35, perhaps he could extend his good fortune. Arrange 16 officers from 4 regiments and 4 ranks into a square so that each row and each column contains exactly one officer of each rank and one officer from each regiment. Then arrange 25 officers from 5 regiments and 5 ranks into a square so that each row and each column contains exactly one officer of each rank and one officer from each regiment.

Can you state the problem in general? Arrange n^2 officers from n regiments and n ranks into a square so that each row and each column contains exactly one officer of each rank and one officer of each regiment. When you finish this subsection, you should have no trouble doing this for $n \in \{3,4,5,7,8,9\}$, and we will show you how close you may come for $n = 6$. Actually, with every finite field F with $|F| = q$, we will be able to construct a family of $q - 1$ mutually orthogonal Latin squares. Our method will be extended to all finite planar nearrings later.

Hopefully, the reader notices in Exercise 1.37 that a solution to our general problem for any given n is equivalent to constructing a pair of orthogonal Latin squares of order n.

As with everything in this section, our construction method starts with a field, and in this case, we need our field F to be finite. Also, every construction in this section has been related to the affine plane $(F \times F, \mathscr{L}_F)$ constructed from this field F. We will continue in this pattern, and will hope that the reader is developing an appreciation for fields, both finite and infinite, and for affine planes.

We refer back to our constructions of (3:7)–(3:11). Suppose our field F has elements $0 = n_1, n_2, ..., n_q$. Then $F \times F = \{(n_i, n_j) \mid i, j \in \{1,2,\ldots,q\}\}$ and $|F \times F| = q^2$. Next, we construct a $(q+1)$ by q^2 matrix M. Label the $q+1$ rows with $\mathscr{L}(0)$, $\mathscr{L}(n_2), \ldots, \mathscr{L}(n_q)$, $\mathscr{L}(\infty)$. Label the q^2 columns with

$$\begin{array}{llll} (n_1,n_1),(n_1,n_2),\ldots,(n_1,n_q), & & & (n_2,n_1),(n_2,n_2),\ldots,(n_2,n_q), \\ \ldots & \ldots & \ldots & (n_q,n_1),(n_q,n_2),\ldots,(n_q,n_q). \end{array}$$

We speak of the $(\mathscr{L}(n_h),\ (n_s,n_t))$-entry of M as the element of F which we put in the intersection of the row labeled with $\mathscr{L}(n_h)$ and the column labeled with (n_s,n_t).

Now $\mathscr{L}(n_h) = \{L(n_h,b) \mid b \in F\}$ is a partition of $F \times F$, and so there is a unique $b \in F$ for which $(n_s,n_t) \in L(n_h,b)$. This $b \in F$ is the element we put in the $(\mathscr{L}(n_h),(n_s,n_t))$-position of M. That is, $b = b^h_{st} = -n_h n_s + n_t$. We really could use a good example at this point.

Example 3.29. Let $F = \mathbb{Z}_3$, so

$$F \times F = \{(0,0),(0,1),(0,2),(1,0),(1,1),(1,2),(2,0),(2,1),(2,2)\};$$

$$L(0,0) = \{(0,0),(1,0),(2,0)\}, \quad L(0,1) = \{(0,1),(1,1),(2,1)\},$$
$$L(0,2) = \{(0,2),(1,2),(2,2)\},$$

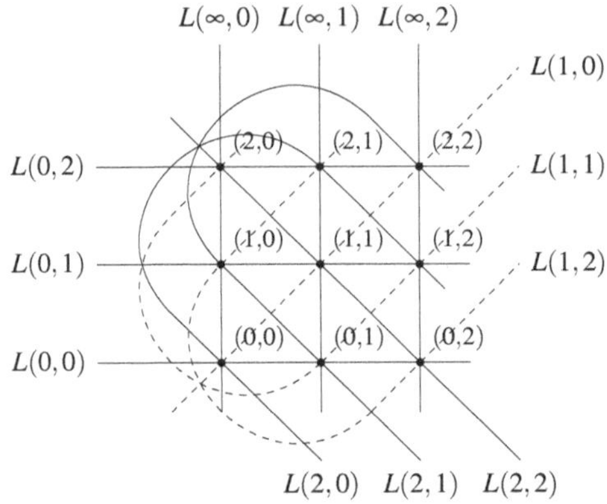

Figure 1: The affine plane $(F \times F, \mathscr{L}_F)$.

$$L(1,0) = \{(0,0),(1,1),(2,2)\}, \quad L(1,1) = \{(0,1),(1,2),(2,0)\},$$
$$L(1,2) = \{(0,2),(1,0),(2,1)\},$$

$$L(2,0) = \{(0,0),(1,2),(2,1)\}, \quad L(2,1) = \{(0,1),(1,0),(2,2)\},$$
$$L(2,2) = \{(0,2),(1,1),(2,0)\},$$

$$L(\infty,0) = \{(0,0),(0,1),(0,2)\}, \quad L(\infty,1) = \{(1,0),(1,1),(1,2)\},$$
$$L(\infty,2) = \{(2,0),(2,1),(2,2)\},$$

and

$$\mathscr{L}(0) = \{L(0,0),L(0,1),L(0,2)\}; \quad \mathscr{L}(1) = \{L(1,0),L(1,1),L(1,2)\};$$
$$\mathscr{L}(2) = \{L(2,0),L(2,1),L(2,2)\}; \quad \mathscr{L}(\infty) = \{L(\infty,0),L(\infty,1),L(\infty,2)\}.$$

The affine plane $(F \times F, \mathscr{L}_F)$ is illustrated in Figure 1.

The matrix M is illustrated in the following array.

M	(0,0)	(0,1)	(0,2)	(1,0)	(1,1)	(1,2)	(2,0)	(2,1)	(2,2)
$\mathscr{L}(0)$	0	1	2	0	1	2	0	1	2
$\mathscr{L}(1)$	0	1	2	2	0	1	1	2	0
$\mathscr{L}(2)$	0	1	2	1	2	0	2	0	1
$\mathscr{L}(\infty)$	0	0	0	1	1	1	2	2	2

In this array for M, we take the $\mathscr{L}(1)$-row and arrange its 3^2 entries into the matrix

$$A_1 = \begin{pmatrix} 0 & 1 & 2 \\ 2 & 0 & 1 \\ 1 & 2 & 0 \end{pmatrix},$$

and take the $\mathscr{L}(2)$-row and arrange its 3^2 entries into the matrix

$$A_2 = \begin{pmatrix} 0 & 1 & 2 \\ 1 & 2 & 0 \\ 2 & 0 & 1 \end{pmatrix}.$$

Referring to Problem 1.30, we compute $A_1 * A_2$ and get

$$A_1 * A_2 = \begin{pmatrix} (0,0) & (1,1) & (2,2) \\ (2,1) & (0,2) & (1,0) \\ (1,2) & (2,0) & (0,1) \end{pmatrix}.$$

One can quickly see that A_1 and A_2 are orthogonal Latin squares. This was not an accident.

In the general situation, for an arbitrary finite field F, take any row labelled by $\mathscr{L}(n_s)$, $2 \le s \le q$, of the matrix M. Suppose the $\mathscr{L}(n_s)$-row is

$$(b^s_{11} \quad b^s_{12} \quad \cdots \quad b^s_{1q} \quad b^s_{21} \quad b^s_{22} \quad \cdots \quad b^s_{2q} \quad \cdots \quad b^s_{q1} \quad b^s_{q2} \quad \cdots \quad b^s_{qq}).$$

(Here, the superscript s is not an exponent. It is only a bookkeeping device.) The row has q^2 entries, from which we construct the q by q square matrix

$$A_s = \begin{pmatrix} b^s_{11} & b^s_{12} & \cdots & b^s_{1q} \\ b^s_{21} & b^s_{22} & \cdots & b^s_{2q} \\ \vdots & \vdots & \ddots & \vdots \\ b^s_{q1} & b^s_{q2} & \cdots & b^s_{qq} \end{pmatrix}.$$

This gives us a family $\mathscr{A} = \{A_2, A_3, \ldots, A_q\}$ of $q-1$ square $q \times q$ matrices.

We have built the foundation for a very nice theorem.

Theorem 3.30. *For a finite field F with $|F| = q$, the $q-1$ matrices in the family $\mathscr{A} = \{A_2, A_3, \ldots, A_q\}$ are mutually orthogonal Latin squares.*

Proof. Our first step will be to show that each A_s is a Latin square. To do this, it may help to make some observations about how A_s was constructed from M.

First, the rows of M are labeled with $\mathscr{L}(n_s)$, $n_s \in F' = F \cup \{\infty\}$. But, in constructing $A_2, \ldots, A_q$, we do not use the rows labeled with $\mathscr{L}(0)$ and $\mathscr{L}(\infty)$. Exactly what does $\mathscr{L}(n_s)$ suggest? With $L(m,b)$ suggesting a line $y = mx + b$, we think of m as the "slope" and b as the "y-intercept." So, $\mathscr{L}(n_s)$ consists of all the lines with slope n_s. We are using only the "nonvertical" and "nonhorizontal" lines, since $n_s \notin \{0, \infty\}$.

The columns of M are labeled with (n_i, n_j), but notice, for $1 \le i \le q$, we have (n_i, n_1), (n_i, n_2), …, (n_i, n_q) as consecutive labels. These, then, are the points on the "vertical" line $x = n_i$. As such, each line $x = n_i$ will intersect each $L(n_s, b)$ exactly once, if $n_s \neq \infty$. Hence, $b^s_{ij} = -n_s n_i + n_j$.

The matrix A_s is constructed from the $\mathscr{L}(n_s)$-row. The entries in the ith row

$$(b^s_{i1} \quad b^s_{i2} \quad \cdots \quad b^s_{iq})$$

are in the columns labeled (n_i, n_1), …, (n_i, n_q). As such, $(n_i, n_j) = (n_i, n_s n_i + b^s_{ij})$. So, if $b^s_{ij} = b^s_{ik}$, then $-n_s n_i + n_j = -n_s n_i + n_k$, and so $(n_i, n_j) = (n_i, n_k)$. This assures us that $|\{b^s_{i1},\ldots,b^s_{iq}\}| = q$. Hence, the ith row of A_s consists of q distinct elements of F, since each $b^s_{ij} \in F$. This is one of our requirements for A_s to be a Latin square.

Consider now the jth column of A_s. It consists of $\{b^s_{1j}, b^s_{2j}, \ldots, b^s_{qj}\}$ with each $b^s_{ij} \in F$, and b^s_{ij} is in the column labeled (n_i, n_j). Hence we have $(n_i, n_j) = (n_i, n_s n_i + b^s_{ij})$. If $b^s_{ij} = b^s_{kj}$, then $-n_s n_i + n_j = -n_s n_k + n_j$, and so $n_i = n_k$, since $n_s \notin \{0, \infty\}$. This means that $(n_i, n_j) = (n_k, n_j)$, and so $|\{b^s_{1j}, b^s_{2j}, \ldots, b^s_{qj}\}| = q$. This is the other requirement for A_s to be a Latin square. Hence, for $2 \leq s \leq q$, A_s is a Latin square.

Now consider two distinct Latin squares $A_s, A_t \in \mathscr{A}$. Let $A_s = (b^s_{ij})$ and $A_t = (b^t_{ij})$. The matrix A_s was constructed from the $\mathscr{L}(n_s)$-row of M, namely

$$(b^s_{11} \quad b^s_{12} \quad \ldots \quad b^s_{1q} \quad b^s_{21} \quad b^s_{22} \quad \ldots \quad b^s_{2q} \quad \ldots \quad b^s_{q1} \quad b^s_{q2} \quad \ldots \quad b^s_{qq})$$

and A_t was constructed from the $\mathscr{L}(n_t)$-row of M, namely

$$(b^t_{11} \quad b^t_{12} \quad \ldots \quad b^t_{1q} \quad b^t_{21} \quad b^t_{22} \quad \ldots \quad b^t_{2q} \quad \ldots \quad b^t_{q1} \quad b^t_{q2} \quad \ldots \quad b^t_{qq}).$$

To see that A_s and A_t are orthogonal, we must construct the matrix $A_s * A_t$ of Definitions 1.31, and here

$$A_s * A_t = \begin{pmatrix} (b^s_{11}, b^t_{11}) & (b^s_{12}, b^t_{12}) & \cdots & (b^s_{1q}, b^t_{1q}) \\ (b^s_{21}, b^t_{21}) & (b^s_{22}, b^t_{22}) & \cdots & (b^s_{2q}, b^t_{2q}) \\ \vdots & \vdots & \ddots & \vdots \\ (b^s_{q1}, b^t_{q1}) & (b^s_{q2}, b^t_{q2}) & \cdots & (b^s_{qq}, b^t_{qq}) \end{pmatrix}.$$

Then we must be assured that

$$\{(b^s_{ij}, b^t_{ij}) \mid i, j \in \{1, 2, \ldots, q\}\} = \{(x, y) \mid x, y \in F\} = F \times F.$$

Suppose $(b^s_{ij}, b^t_{ij}) = (b^s_{kl}, b^t_{kl})$. What does it mean for $b^s_{ij} = b^s_{kl}$?

Now b^s_{ij} is in the $\mathscr{L}(n_s)$-row and in the (n_i, n_j)-column of M, and $(n_i, n_j) \in L(n_s, b^s_{ij})$. Similarly, b^s_{kl} is in the $\mathscr{L}(n_s)$-row and in the (n_k, n_l)-column of M, and $(n_k, n_l) \in L(n_s, b^s_{kl})$. With $b^s_{ij} = b^s_{kl}$ we have $L(n_s, b^s_{ij}) = L(n_s, b^s_{kl})$; therefore, $(n_i, n_j), (n_k, n_l) \in L(n_s, b^s_{ij})$.

Next, what does it mean for $b^t_{ij} = b^t_{kl}$? As in the preceding paragraph, it means that $(n_i, n_j), (n_k, n_l) \in L(n_t, b^t_{ij})$. Since $n_s \neq n_t$ follows from $A_s \neq A_t$, we have that the two distinct lines $L(n_s, b^s_{ij})$ and $L(n_t, b^t_{ij})$ have in common the points (n_i, n_j) and (n_k, n_l). Since $(F \times F, \mathscr{L}_F)$ is an affine plane, we have no choice but to conclude that $(n_i, n_j) = (n_k, n_l)$ and that $(i, j) = (k, l)$. This means that the mapping $\phi : F \times F \to \{(b^s_{ij}, b^t_{ij}) \mid i, j \in \{1, 2, \ldots, q\}\}$, defined by $\phi(n_i, n_j) = (b^s_{ij}, b^t_{ij})$ is a bijection, which in turn completes the proof that A_i and A_j are orthogonal, and also completes the proof of the theorem. □

Exercise 3.31. Construct the four mutually orthogonal Latin squares given in Example 1.32. Then construct a family of three mutually orthogonal Latin squares of order 4.

By now, it is hoped that every reader has forgiven the author for the cruel joke of Exercise 1.37. Let me state now very clearly that no solution of Exercise 1.37 exists, and, moreover, no solution will ever exist. That is, there do not exist two orthogonal Latin squares of order 6, but to get a proof, the reader will have to look beyond these pages.

Often, we come close to a goal in mathematics, only to be rejected by one small point or two. We want to show now how close one can come to an impossibility. We will almost get two orthogonal Latin squares of order 6.

Exercise 3.32. Consider the matrix

$$A = \begin{pmatrix} 1 & 2 & 4 & 0 & 3 & 5 \\ 3 & 1 & 0 & 5 & 2 & 4 \\ 2 & 0 & 1 & 4 & 5 & 3 \\ 0 & 3 & 5 & 1 & 4 & 2 \\ 4 & 5 & 2 & 3 & 1 & 0 \\ 5 & 4 & 3 & 2 & a & b \end{pmatrix}.$$

Choose a and b so that A is a Latin square of order 6. How many choices do you have?

Exercise 3.33. Consider the matrix

$$B = \begin{pmatrix} 4 & 0 & 1 & 5 & 2 & 3 \\ 0 & 5 & 4 & 1 & 3 & 2 \\ 1 & 3 & 2 & 0 & 4 & 5 \\ 2 & 1 & 0 & 3 & 5 & 4 \\ 3 & 2 & 5 & 4 & 0 & 1 \\ 5 & 4 & 3 & 2 & x & y \end{pmatrix}.$$

Choose x and y so that B is a Latin square of order 6. How many choices do you have?

Exercise 3.34. Consider the matrix $A * B$ for A and B from Exercise 3.32 and Exercise 3.33, respectively. So

$$A * B = \begin{pmatrix} (1,4) & (2,0) & (4,1) & (0,5) & (3,2) & (5,3) \\ (3,0) & (1,5) & (0,4) & (5,1) & (2,3) & (4,2) \\ (2,1) & (0,3) & (1,2) & (4,0) & (5,4) & (3,5) \\ (0,2) & (3,1) & (5,0) & (1,3) & (4,5) & (2,4) \\ (4,3) & (5,2) & (2,5) & (3,4) & (1,0) & (0,1) \\ (5,5) & (4,4) & (3,3) & (2,2) & (a,x) & (b,y) \end{pmatrix}.$$

For $\mathbb{Z}_6 = \{0,1,2,3,4,5\}$, how many of the 36 elements of $\mathbb{Z}_6 \times \mathbb{Z}_6$ are listed in $A * B$? What are the missing ones? Can you put in the missing ones in place of (a,x) and (b,y) to make orthogonal Latin squares? Can you put in the missing ones in place of (a,x) and (b,y) to make something interesting?

Just how many Latin squares can there be in a family $\mathscr{A} = \{A_1,\ldots,A_m\}$ of mutually orthogonal Latin squares of order n? This next result shows us that $m \leq n-1$.

Theorem 3.35. *Let $\mathscr{A} = \{A_1, \ldots, A_m\}$ be a family of distinct mutually orthogonal Latin squares of order n. Then $m \le n-1$.*

Proof. Denote $\mathbb{Z}_n = \{0, 1, \ldots, n-1\}$. We assume each Latin square $A_k = (a^k_{ij})$ has each $a^k_{ij} \in \mathbb{Z}_n$ as in Definitions 1.31. So,

$$(a^k_{11}, a^k_{12}, \ldots, a^k_{1n}) = (\mu_k(0), \mu_k(1), \ldots, \mu_k(n-1))$$

defines a bijective function $\mu_k : \mathbb{Z}_n \to \mathbb{Z}_n$. Let $\pi_k = \mu_k^{-1}$. We have $\pi_k(a^k_{ij}) = \pi_k \circ \mu_k(j-1) = j-1$ for each j, $1 \le j \le n$. Define

$$B_k = \Pi_k(A_k) = (b^k_{ij}) = (\pi_k(a^k_{ij})).$$

Let $\mathscr{B} = \{B_1, B_2, \ldots, B_m\}$. We will show that $\mathscr{B}$ is also a family of mutually orthogonal Latin squares.

First, take any $B_s \in \mathscr{B}$. With $B_s = (b^s_{ij}) = (\pi_s(a^s_{ij}))$, we infer from $b^s_{iu} = b^s_{iv}$ that $\pi_s(a^s_{iu}) = \pi_s(a^s_{iv})$, and so $a^s_{iu} = a^s_{iv}$. This forces $u = v$, since A_s is a Latin square. Similarly, we infer from $b^k_{uj} = b^k_{vj}$ that $u = v$. Hence, each B_s is a Latin square.

Notice that each $B_s = (b^s_{ij})$ has first row

$$(b^s_{11}, b^s_{12}, \ldots, b^s_{1n}) = (0, 1, \ldots, n-1).$$

To see that $\mathscr{B} = \{B_1, B_2, \ldots, B_m\}$ is a mutually orthogonal family, take $B_s, B_t \in \mathscr{B}$, $s \ne t$, and consider

$$B_s * B_t = \begin{pmatrix} (b^s_{11}, b^t_{11}) & \cdots & (b^s_{1n}, b^t_{1n}) \\ \vdots & \ddots & \vdots \\ (b^1_{n1}, b^t_{n1}) & \cdots & (b^s_{nn}, b^t_{nn}) \end{pmatrix} = \left((b^s_{ij}, b^t_{ij})\right).$$

Suppose $\left((b^s_{ij}, b^t_{ij})\right) = \left((b^s_{uv}, b^t_{uv})\right)$. Then

$$\left(\pi_s(a^s_{ij}), \pi_t(a^t_{ij})\right) = \left(\pi_s(a^s_{uv}), \pi_t(a^t_{uv})\right),$$

and so $(a^s_{ij}, a^t_{ij}) = (a^s_{uv}, a^t_{uv})$, since π_s and π_t are bijective. Since $\mathscr{A}$ is a mutually orthogonal family of Latin squares, A_s and A_t are orthogonal, and so $(i, j) = (u, v)$. Now we are assured that $\mathscr{B}$ is a mutually orthogonal family of Latin squares.

If $b^s_{21} = b^t_{21} = k-1$, then in $B_s * B_t$, we have

$$(b^s_{21}, b^t_{21}) = (k-1, k-1) = (b^s_{1k}, b^t_{1k}),$$

and consequently, $s = t$. Hence, the set $\{b^1_{21}, b^2_{21}, \ldots, b^m_{21}\}$ consists of m distinct values in $\mathbb{Z}_n$, none of which can be $b^1_{11} = 0$, hence they are m distinct values in $\{1, 2, \ldots, n-1\}$. So, $m \leq n-1$, as promised. □

Exercise 3.36. For a finite field $F = \{0 = n_1, n_2, \ldots, n_q\}$, with $|F| = q$, and for each h, $2 \leq h \leq q$, define $A_h = (n_h n_i + n_j)$. Show that $\mathscr{A} = \{A_2, A_3, \ldots, A_q\}$ is a family of $q-1$ mutually orthogonal Latin squares of order q.

With Exercise 3.36, one should wonder why we made our development here to get Theorem 3.30. It was so the reader would have some insight into doing

Exercise 3.37. For a family $\mathscr{A} = \{A_2, A_3, \ldots, A_q\}$ of $q-1$ distinct mutually orthogonal Latin squares of order q, construct a BIBD with parameters

$$(v, b, r, k, \lambda) = (n^2, n(n+1), n+1, n, 1)$$

for some appropriate n.

Chapter II

The equation $ax = bx + c$

In §3, we explored four basic applications of fields: (1) constructing affine planes; (2) constructing projective planes; (3) constructing Möbius planes; and (4) constructing families of mutually orthogonal Latin squares. The latter three applications depended upon the first, namely, the construction of affine planes.

A very basic idea of plane geometry is that nonparallel lines should intersect at exactly one point. There are variations of this idea. Two nonparallel lines could intersect in at most one point, or two nonparallel lines could intersect at more than one point. Each of these options leads to interesting mathematics, but we will stay focused on the idea that two nonparallel lines will intersect at exactly one point.

For the affine plane $(F \times F, \mathscr{L}_F)$ constructed from a field F in §3, we constructed and identified "nonvertical" lines $L(m,b)$, $m \neq \infty$, with the equation $y = mx + b$. If $L(m,b)$ and $L(m',b')$ are two nonparallel lines in $\mathscr{L}_F$, with $\infty \notin \{m,m'\}$, then $m \neq m'$. To find the unique point (x_0,y_0) in common to these two lines, we took advantage of the facts that $(x_0,y_0) \in L(m,b) \cap L(m',b')$. So, x_0 was the unique solution to the equation

$$mx + b = m'x + b',$$

where $m \neq m'$, $m, m' \in F$. In the process of solving for x, we might write the equation

$$mx = m'x + (b' - b).$$

Here, $m, m', b' - b \in F$ are constants, and $m \neq m'$. With a for m, b for m', and c for $(b' - b)$, this last equation becomes

$$ax = bx + c,$$

with $a \neq b$. We want this equation to have a unique solution. Usually, we would not even stop to notice this equation, for we would quickly write

$$(a - b)x = c$$

and so $x = (a - b)^{-1}c$. Exactly what gave us the right to take these steps? First, we have $ax - bx = (a - b)x$, which is a consequence of the fact that multiplication $\cdot$ is right distributive over addition $+$ in a field. Next, with $a \neq b$, we have $a - b \neq 0$, and so $(a - b)^{-1}$ exists.

In this chapter, we focus on the equation $ax = bx + c$. We are willing to relax other properties of a field, but we will want $ax = bx + c$ to have a unique solution as much as possible. After all, this has historically led us to some very important geometric consequences.

From what we have just seen, we should be able to relax commutativity of multiplication with no problem at all. Relaxing the right distributivity of multiplication $\cdot$ over addition $+$ will cause some problem. Finally, we will relax the property of the existence of $(a - b)^{-1}$ if $a \neq b$. This last stage, surprisingly enough, offers the most diversity.

4 Division rings, or skew fields

Division rings are very close to being fields. That is why they are sometimes called skew fields. The only requirement for a division ring to be a field is commutativity of multiplication. We present here the requirements listed in §A2 for a field.

(A1) $+$ is commutative.

(A2) $+$ is associative.

(A3) 0 is the identity with respect to $+$.

(A4) Each a has an inverse $-a$ with respect to $+$.

(M1) $\cdot$ is commutative.

(M2) $\cdot$ is associative.

(M3) 1 is the identity with respect to $\cdot$.

(M4) Each $a \neq 0$ has an inverse a^{-1} with respect to $\cdot$.

(DL) $\cdot$ is left distributive over $+$.

(DR) $\cdot$ is right distributive over $+$.

So, we state formally the definition of a division ring.

Definition 4.1. Let D be a set having at least two distinct elements, denoted by 0 and 1. Also, suppose that $+$ and $\cdot$ are two binary operations on D. Then the algebraic structure $(D,+,\cdot)$ is a *division ring*, or *skew field*, if the binary operations have properties (A1)–(A4), (M2)–(M4), (DL), and (DR), as defined above.

So, every field is a division ring. Of course, we want to show that there is a division ring which is not a field. We will find one inside $M_2(\mathbb{C})$, the ring of two by two matrices over the field of complex numbers $\mathbb{C}$. See Corollary A2.50.

Let $Q = \left\{ \begin{pmatrix} a & b \\ -\bar{b} & \bar{a} \end{pmatrix} \;\middle|\; a, b \in \mathbb{C} \right\}$. Then

Theorem 4.2. *Q is a subring of $M_2(\mathbb{C})$. It is also a division ring.*

Proof. Let $\left(\begin{smallmatrix} a & b \\ -\overline{b} & \overline{a}\end{smallmatrix}\right), \left(\begin{smallmatrix} x & y \\ -\overline{y} & \overline{x}\end{smallmatrix}\right) \in Q$. Then

$$\begin{pmatrix} a & b \\ -\overline{b} & \overline{a}\end{pmatrix} + \begin{pmatrix} x & y \\ -\overline{y} & \overline{x}\end{pmatrix} = \begin{pmatrix} a+x & b+y \\ -\overline{b}-\overline{y} & \overline{a}+\overline{x}\end{pmatrix} = \begin{pmatrix} a+x & b+y \\ -\overline{(b+y)} & \overline{a+x}\end{pmatrix},$$

$$-\begin{pmatrix} a & b \\ -\overline{b} & \overline{a}\end{pmatrix} = \begin{pmatrix} -a & -b \\ \overline{b} & -\overline{a}\end{pmatrix} = \begin{pmatrix} -a & -b \\ -\overline{(-b)} & \overline{-a}\end{pmatrix},$$

and

$$\begin{aligned}\begin{pmatrix} a & b \\ -\overline{b} & \overline{a}\end{pmatrix} \circ \begin{pmatrix} x & y \\ -\overline{y} & \overline{x}\end{pmatrix} &= \begin{pmatrix} ax - b\overline{y} & ay + b\overline{x} \\ -\overline{b}x - \overline{a}\overline{y} & -\overline{b}y + \overline{a}\overline{x}\end{pmatrix} \\ &= \begin{pmatrix} ax - b\overline{y} & ay + b\overline{x} \\ -\overline{a}\overline{y} - \overline{b}x & \overline{a}\overline{x} - \overline{b}y\end{pmatrix} = \begin{pmatrix} ax - b\overline{y} & ay + b\overline{x} \\ -\overline{(ay + b\overline{x})} & \overline{ax - b\overline{y}}\end{pmatrix}.\end{aligned}$$

Hence, Q is a subring of $M_2(\mathbb{C})$.

Certainly, $\left(\begin{smallmatrix} 1 & 0 \\ 0 & 1\end{smallmatrix}\right) \in Q$. Take any $A = \left(\begin{smallmatrix} a & b \\ -\overline{b} & \overline{a}\end{smallmatrix}\right) \in Q$, $\left(\begin{smallmatrix} a & b \\ -\overline{b} & \overline{a}\end{smallmatrix}\right) \neq \left(\begin{smallmatrix} 0 & 0 \\ 0 & 0\end{smallmatrix}\right)$. Then $\Delta(A) = |a|^2 + |b|^2 \neq 0$. Let $A' = \left(\begin{smallmatrix} \overline{a}/\Delta(A) & -b/\Delta(A) \\ \overline{b}/\Delta(A) & a/\Delta(A)\end{smallmatrix}\right)$. Then

$$\begin{aligned} AA' &= \begin{pmatrix} a & b \\ -\overline{b} & \overline{a}\end{pmatrix} \circ \begin{pmatrix} \overline{a}/\Delta(A) & -b/\Delta(A) \\ \overline{b}/\Delta(A) & a/\Delta(A)\end{pmatrix} \\ &= \begin{pmatrix} (|a|^2+|b|^2)/\Delta(A) & 0 \\ 0 & (|a|^2+|b|^2)/\Delta(A)\end{pmatrix} = \begin{pmatrix} 1 & 0 \\ 0 & 1\end{pmatrix},\end{aligned}$$

and

$$\begin{aligned} A'A &= \begin{pmatrix} \overline{a}/\Delta(A) & -b/\Delta(A) \\ \overline{b}/\Delta(A) & a/\Delta(A)\end{pmatrix} \circ \begin{pmatrix} a & b \\ -\overline{b} & \overline{a}\end{pmatrix} \\ &= \begin{pmatrix} (|a|^2+|b|^2)/\Delta(A) & 0 \\ 0 & (|a|^2+|b|^2)/\Delta(A)\end{pmatrix} = \begin{pmatrix} 1 & 0 \\ 0 & 1\end{pmatrix}.\end{aligned}$$

Hence,

$$A^{-1} = \begin{pmatrix} a & b \\ -\overline{b} & \overline{a}\end{pmatrix}^{-1} = \begin{pmatrix} \overline{a}/\Delta(A) & -b/\Delta(A) \\ \overline{b}/\Delta(A) & a/\Delta(A)\end{pmatrix}$$

for any $A \in Q, A \neq \left(\begin{smallmatrix} 0 & 0 \\ 0 & 0 \end{smallmatrix}\right)$.

But

$$A^{-1} = \begin{pmatrix} \overline{a}/\Delta(A) & -b/\Delta(A) \\ -\overline{[-b/\Delta(A)]} & \overline{(\overline{a}/\Delta(A))} \end{pmatrix},$$

so $A^{-1} \in Q$, also. Hence, $(Q, +, \cdot)$ is a division ring.

But what about commutativity of $\circ$? Note that $\iota = \left(\begin{smallmatrix} i & 0 \\ 0 & -i \end{smallmatrix}\right)$ and $\kappa = \left(\begin{smallmatrix} 0 & 1 \\ -1 & 0 \end{smallmatrix}\right)$ are in Q, that $\iota\kappa = \left(\begin{smallmatrix} 0 & i \\ i & 0 \end{smallmatrix}\right)$, and $\kappa\iota = \left(\begin{smallmatrix} 0 & -i \\ -i & 0 \end{smallmatrix}\right)$. Hence, multiplication is not commutative, and so $(Q, +, \cdot)$ is not a field. □

The division ring Q is often called the division ring of *real quaternions*, but many authors do not introduce Q as we have done so here. If the reader will permit liberties with mathematical notation, we will put our introduction to the real quaternions in traditional form. Make the following identifications.

$$1 = \begin{pmatrix} 1 & 0 \\ 0 & 1 \end{pmatrix}, \quad i = \begin{pmatrix} i & 0 \\ 0 & -i \end{pmatrix}, \quad j = \begin{pmatrix} 0 & 1 \\ -1 & 0 \end{pmatrix}, \quad k = \begin{pmatrix} 0 & i \\ i & 0 \end{pmatrix}.$$

If we multiply a matrix $A \in M_2(R)$ by an element $\lambda \in R$ as

$$\lambda A = \lambda \begin{pmatrix} a & b \\ c & d \end{pmatrix} = \begin{pmatrix} \lambda a & \lambda b \\ \lambda c & \lambda d \end{pmatrix},$$

then

$$\begin{aligned} a + bi + cj + dk &= a \begin{pmatrix} 1 & 0 \\ 0 & 1 \end{pmatrix} + b \begin{pmatrix} i & 0 \\ 0 & -i \end{pmatrix} + c \begin{pmatrix} 0 & 1 \\ -1 & 0 \end{pmatrix} + d \begin{pmatrix} 0 & i \\ i & 0 \end{pmatrix} \\ &= \begin{pmatrix} a + bi & c + di \\ -c + di & a - bi \end{pmatrix} = \begin{pmatrix} a + bi & c + di \\ -\overline{(c + di)} & \overline{a + bi} \end{pmatrix}, \end{aligned}$$

an element of Q. One traditionally uses the notation $a + bi + cj + dk$ in place of

$$\begin{pmatrix} a + bi & c + di \\ -\overline{(c + di)} & \overline{a + bi} \end{pmatrix}.$$

The reader is strongly encouraged to do the following exercise before continuing.

Exercise 4.3. With $1, i, j$, and k as defined above, prove the following properties.

(a) $i^2 = j^2 = k^2 = -1$.

(b) $ij = k$, $jk = i$, $ki = j$.

(c) $ji = -k$, $kj = -i$, $ik = -j$.

Exercise 4.4. Prove that

$$(a+bi+cj+dk)+(x+yi+zj+uk) = (a+x)+(b+y)i+(c+z)j+(d+u)k$$

and that

$$\begin{aligned}(a+bi+cj+dk)\cdot(x+yi+zj+uk) &= (ax-by-cz-du)\\ &\quad +(ay+bx+cu-dz)i+(az-bu+cx+dy)j+(au+bz-cy+dx)k.\end{aligned}$$

Exercise 4.5. Let $Q = \{a+bi+cj+dk \mid a,b,c,d \in \mathbb{R}\}$ and let $+$ and $\cdot$ be as in Exercise 4.4. Prove that $(Q,+,\cdot)$ is a division ring without the use of Theorem 4.2. After you have done this, you can decide for yourself if you prefer your proof that $(Q,+,\cdot)$ is a division ring to that of Theorem 4.2.

Exercise 4.6. The division ring $(Q,+,\cdot)$ has three subfields each isomorphic to $(\mathbb{C},+,\cdot)$. It also has a subfield isomorphic to $(\mathbb{R},+,\cdot)$. Prove that the subfields $F_i = \{a+bi \mid a,b \in \mathbb{R}\}$, $F_j = \{a+cj \mid a,c \in \mathbb{R}\}$, and $F_k = \{a+dk \mid a,d \in \mathbb{R}\}$ are each isomorphic to $(\mathbb{C},+,\cdot)$, and that the subfield

$$\mathbb{R}' = \left\{ \begin{pmatrix} a & 0 \\ 0 & a \end{pmatrix} \,\middle|\, a \in \mathbb{R} \right\}$$

is isomorphic to $(\mathbb{R},+,\cdot)$.

For $q = a+bi+cj+dk \in Q$, define the *conjugate* of q by $\overline{q} = a-bi-cj-dk$.

Proposition 4.7. *For $q = a + bi + cj + dk \in Q$, we have*

$$\overline{q}q = q\overline{q} = a^2 + b^2 + c^2 + d^2,$$

and if $q \neq 0$, then

$$q^{-1} = (q\overline{q})^{-1}\overline{q}.$$

Proof. First

$$\begin{aligned}\overline{q}q &= (a - bi - cj - dk)(a + bi + cj + dk) \\ &= a^2 + abi + acj + adk - abi - b^2i^2 - bcij - bdik \\ &\qquad -acj - bcji - c^2j^2 - cdjk - adk - bdki - cdkj - d^2k^2 \\ &= a^2 + b^2 + c^2 + d^2 + (ab - ab)i + (ac - ac)j + (ad - ad)k \\ &\qquad -(bc - bc)ij - (bd - bd)ik - (cd - cd)jk = a^2 + b^2 + c^2 + d^2.\end{aligned}$$

Similarly, $q\overline{q} = a^2 + b^2 + c^2 + d^2$.

From

$$q[(q\overline{q})^{-1}\overline{q}] = (q\overline{q})^{-1}(q\overline{q}) = 1$$

and

$$[(q\overline{q})^{-1}\overline{q}]q = (q\overline{q})^{-1}(\overline{q}q) = (q\overline{q})^{-1}(q\overline{q}) = 1,$$

we have that $q^{-1} = (q\overline{q})^{-1}\overline{q}$. □

The reader hopefully realizes that Q is an extension of the idea of complex numbers. One almost gets a bigger field Q containing $\mathbb{C}$, but one loses commutativity of multiplication. As with complex numbers, we define the *norm*, or *length* or *absolute value* of $q = a + bi + cj + dk \in Q$ by

$$|q| = (a^2 + b^2 + c^2 + d^2)^{1/2}.$$

Within the complex numbers, $T = \{z \in \mathbb{C} \mid |z| = 1\}$ plays an important role. The same will be true for $U = \{q \in Q \mid |q| = 1\}$.

II The equation $ax = bx + c$

Let $q = a + bi + cj + dk = \begin{pmatrix} a+bi & c+di \\ -\overline{(c+di)} & \overline{a+bi} \end{pmatrix} = A$. Then

$$|q|^2 = a^2 + b^2 + c^2 + d^2 = \Delta(A).$$

The function Δ is a special case of a very important function for matrices.

Take $A = \left(\begin{smallmatrix} a & b \\ c & d \end{smallmatrix}\right) \in M_2(R)$, where R is a ring with identity. Define $\Delta(A) = ad - cb$. This is the *determinant* of A. It has the following property.

Theorem 4.8. *The determinant function $\Delta : M_2(R) \to R$, R a commutative ring with identity, has the property that*

$$\Delta(AB) = \Delta(A)\Delta(B)$$

for all $A, B \in M_2(R)$.

Proof. Let $A = \left(\begin{smallmatrix} a & b \\ c & d \end{smallmatrix}\right)$ and $B = \left(\begin{smallmatrix} x & y \\ z & u \end{smallmatrix}\right)$. Then

$$AB = \begin{pmatrix} ax + bz & ay + bu \\ cx + dz & cy + du \end{pmatrix},$$

so

$$\begin{aligned} \Delta(AB) &= (ax + bz)(cy + du) - (cx + dz)(ay + bu) \\ &= ad(xu - yz) - bc(xu - yz) \\ &= (ad - cb)(xu - zy) = \Delta(A)\Delta(B). \end{aligned}$$

□

We have already noted that any $A \in Q$ has $\Delta(A) \neq 0$ if $A \neq \left(\begin{smallmatrix} 0 & 0 \\ 0 & 0 \end{smallmatrix}\right)$. Note that $\Delta : Q^* \to \mathbb{R}_+$, and so by Theorem 4.8, Δ is a group epimorphism from $Q^* = Q \setminus \{0\}$ onto $\mathbb{R}_+$, the multiplicative group of positive numbers. From Theorem A4.27, we know that $\ker \Delta = \{A \in Q^* \mid \Delta(A) = 1\}$ is a subgroup of Q^*.

What does it mean for A to be in $\ker \Delta$? As we have noted earlier, if $A = a + bi + cj + dk = q$, then $\Delta(A) = |q|^2$. Hence, $\Delta(A) = 1$ if and only if $|A| = |q| = 1$. So, U is a subgroup of Q^*. We summarize our observations.

Theorem 4.9. *Let $\Delta : Q^* \to \mathbb{R}_+$ be the determinant function restricted to $Q^* = Q \setminus \{0\}$. Then Δ is a group epimorphism and*

$$\ker \Delta = U = \{q \in Q \mid |q| = 1\},$$

so U is a subgroup of Q^*.

There is another subgroup of Q^* that will be important for us. Let K be the set $K = \left\{ \left(\begin{smallmatrix} r & 0 \\ 0 & r \end{smallmatrix} \right) \,\middle|\, r \in \mathbb{R}_+ \right\}$. One easily sees that K is a normal subgroup of Q^* that is isomorphic to $\mathbb{R}_+$, the multiplicative group of positive real numbers. For $A \in Q^*$, we have $\Delta(A) = |A|^2$, and $\left(\begin{smallmatrix} |A| & 0 \\ 0 & |A| \end{smallmatrix} \right) \in K$. Also, $|A|^{-1}A$ has the property that $\Delta(|A|^{-1}A) = 1$, and so $|A|^{-1}A \in U$. Now why is $\Delta(|A|^{-1}A) = 1$? If $A = \left(\begin{smallmatrix} a & b \\ -\bar{b} & \bar{a} \end{smallmatrix} \right)$, then $|A|^{-1}A = \left(\begin{smallmatrix} a/|A| & b/|A| \\ -\bar{b}/|A| & \bar{a}/|A| \end{smallmatrix} \right)$, and so

$$\Delta(|A|^{-1}A) = (a\bar{a}/|A|^2) + (b\bar{b}/|A|^2) = (|a|^2 + |b|^2)/|A|^2 = |A|^2/|A|^2 = 1.$$

Next, notice that

$$A = \left(|A|^{-1}A\right) \begin{pmatrix} |A| & 0 \\ 0 & |A| \end{pmatrix} = (|A|^{-1}A)(|A|I)$$

where $I = \left(\begin{smallmatrix} 1 & 0 \\ 0 & 1 \end{smallmatrix} \right)$. Now $|A|^{-1}A \in U$ and $|A|I \in K$, so A has a factorization $A = A_1A_2$ where $A_1 \in U$ and $A_2 \in K$. This factorization is unique, for if $A = B_1B_2$ with $B_1 \in U$ and $B_2 \in K$, then $A_1A_2 = B_1B_2$, and so $B_1^{-1}A_1 = B_2A_2^{-1} \in U \cap K = \{I\}$. Hence, $A_1 = B_1$ and $A_2 = B_2$. Summarizing, we have

Theorem 4.10. *Every $A \in Q^*$ has a unique factorization $A = A_1A_2$ where*

$$A_1 = |A|^{-1}A \in U \ \text{ and } \ A_2 = |A|I \in K.$$

In addition, Q^* is the direct product of the subgroups U and K.

Exercise 4.11. Let R be a commutative ring with identity. Then, for any $r \in R$ and any $A = \left(\begin{smallmatrix} a & b \\ c & d \end{smallmatrix} \right) \in M_2(R)$, we have

$$\begin{pmatrix} r & 0 \\ 0 & r \end{pmatrix} \begin{pmatrix} a & b \\ c & d \end{pmatrix} = \begin{pmatrix} a & b \\ c & d \end{pmatrix} \begin{pmatrix} r & 0 \\ 0 & r \end{pmatrix}.$$

That is, $(rI)A = A(rI)$.

An equation $ax = bx + c$, $a, b, c \in D$, a division ring, with $a \neq b$, has a unique solution, and the proof is exactly as for $a, b, c \in F$, a field. We could, if we were interested, construct affine planes $(D \times D, \mathscr{L}_D)$ from a division ring exactly as was done for fields in §3. Division rings have been heavily studied, and it may surprise the reader to learn that there are no finite division rings that are not fields. A proof of this classical result can be found in [H].

5 Nearfields

At the beginning of §4, the requirements of a field were listed. If (M1), commutativity of multiplication, is eliminated, we have the requirements for a division ring. If either (DL) or (DR) is also eliminated, we have the requirements for a nearfield. To be in harmony with the best reference on the subject, we will eliminate (DR). To obtain an understanding of the many fascinating secrets of nearfields, one should learn German and study Wähling's labor of love [Wä]. Let us begin with a formal definition.

Definition 5.1. Let K be a set having at least two distinct elements, denoted by 0 and 1. Also, suppose that $+$ and $\cdot$ are two binary operations on K. Then the algebraic structure $(K, +, \cdot)$ is a *(left) nearfield* if the binary operations have properties (A1)–(A4), (M2)–(M4), and (DL), as listed at the beginning of §4.

Of course, every field and every division ring is a nearfield, but the reader should expect more from an author than such examples. We can almost provide another example quite easily.

Example 5.2. From the field $(\mathbb{Z}_2, +, \cdot)$, replace $\cdot$ by $*$ where $*$ is defined on $\mathbb{Z}_2$ by $a * b = b$ for all $a, b \in \mathbb{Z}_2 = \{0, 1\}$. Then $(\mathbb{Z}_2, +, *)$ is almost a nearfield. Here we have $0 * 1 = 1$ and not $0 * 1 = 0$, as required by (M3). But $(\mathbb{Z}_2, +, *)$ satisfies all the other requirements for a nearfield. Note that 1 is a "left identity" for $*$, so (M3) is half satisfied.

The main goals of this section are (1) to introduce the concept of a nearfield, (2) to develop a method to construct a family of examples of nearfields which are not division rings, (3) to point out that the equation $ax = bx + c$ is particularly significant for nearfields, and (4) to illustrate that nearfields are particularly significant for geometry.

All of our nearfields will be constructed from fields, and there are only seven finite nearfields which cannot be constructed from fields. To construct examples of nearfields from fields, we need to develop a few ideas which are traditionally introduced along with fields. But these ideas are equally valid for nearfields and for division rings, and so we present them here where we really need them.

From a nearfield $(K,+,\cdot)$, we have two groups, namely the additive group $(K,+)$ and the multiplicative group $(K^*,\cdot)$, where $K^* = K \setminus \{0\}$. Now $1 \in K$ is the identity for $(K^*,\cdot)$, and, as an element of K, its order in the group $(K,+)$ is not 1. With $2 \cdot 1 = 1+1$, and $n \cdot 1 = (n-1) \cdot 1 + 1$, and $(-1) \cdot 1 = -1$, and $(-n) \cdot 1 = -(n \cdot 1)$, we have meaning for $m \cdot 1$ for every $m \in \mathbb{Z}$. If there does not exist a positive integer m for which $m \cdot 1 = 0$, then the additive order of 1 is $|1| = \infty$, and we say that K has *characteristic* 0. The other alternative is for the existence of a positive $m \in Z$ for which $m \cdot 1 = 0$, in which case, there is a smallest positive integer p for which $p \cdot 1 = 0$. In this case the additive order of 1 is $|1| = p$. Just what kind of a number is p? Well, we will see now that it is a prime. If it were not a prime, then $p = st$ where $s,t \in \{2,3,\ldots,p-1\}$. From $0 = p \cdot 1 = (st) \cdot 1 = s \cdot (t \cdot 1) = (s \cdot 1) \cdot (t \cdot 1)$, we have $s \cdot 1 \neq 0$ and $t \cdot 1 \neq 0$, so $s \cdot 1, t \cdot 1 \in K^*$, a group, and therefore $(s \cdot 1) \cdot (t \cdot 1) \in K^*$. But $0 \notin K^*$. Hence, p is a prime.

If one doesn't look too carefully at the proof that $(st) \cdot 1 = (s \cdot 1) \cdot (t \cdot 1)$, one might be caught sleeping. Just what justifies us to write

$$(st) \cdot 1 = s \cdot (t \cdot 1) = (s \cdot 1) \cdot (t \cdot 1)?$$

Well,

$$(st) \cdot 1 = 1 + 1 + \cdots + 1,$$

with st 1's in the summation. Hence, there are s blocks of 1's with t 1's in each block. That is,

$$(st) \cdot 1 = (1 + \cdots + 1) + \cdots + (1 + \cdots + 1),$$

where there are s blocks $(1 + \cdots + 1)$ and each such block $(1 + \cdots + 1)$ has t 1's. Likewise,

$$\begin{aligned}(s \cdot 1) \cdot (t \cdot 1) &= (s \cdot 1) \cdot (1 + \cdots + 1) \\ &= (s \cdot 1) \cdot 1 + \cdots + (s \cdot 1) \cdot 1 = (s \cdot 1) + \cdots + (s \cdot 1),\end{aligned}$$

where there are t 1's in the first $(1 + \cdots + 1)$. Each of the t $(s \cdot 1)$'s is really $1 + \cdots + 1$, where there are s 1's, and so

$$(s \cdot 1) + \cdots + (s \cdot 1) = (1 + \cdots + 1) + \cdots + (1 + \cdots + 1) = (ts) \cdot 1 = (st) \cdot 1.$$

These types of details should not slip by the careful reader.

So, if $|1| \neq \infty$, then $|1| = p$, a prime, and we say that K has *characteristic* p. What we have seen about 1 is true for any $a \in K^*$. If m is a positive integer such that $m \cdot a = 0$, where $2 \cdot a = a + a$, and $m \cdot a = (m-1) \cdot a + a$, then $0 = a + \cdots + a = a \cdot (1 + \cdots + 1)$. The only way for this to happen is for $1 + \cdots + 1 = 0$, and we have discussed that case above. It might be a good idea to formalize these observations.

Proposition 5.3. *Let $(K, +, \cdot)$ be a nearfield. If K has characteristic $p > 0$, then $p \cdot a = 0$ for every $a \in K$, and if $m \cdot a = 0$ for some positive integer m and some $a \in K^*$, then $m = pn$ for some integer n. If K has characteristic 0, then for any $m \in \mathbb{Z}$, $m \neq 0$ and any $a \in K$, $a \neq 0$, then $m \cdot a \neq 0$.*

Exercise 5.4. There is a lot more to Proposition 5.3 than we have proved. The reader is invited to decide just what has been proved and what yet needs a proof. Then, the reader is asked to develop a proof of what needs proving.

5.1 Constructing nearfields from fields

Note that an automorphism σ of a field $(F,+,\cdot)$ is a bijection $\sigma : F \to F$ such that $\sigma(x+y) = \sigma(x)+\sigma(y)$ and $\sigma(xy) = \sigma(x)\sigma(y)$ for all $x,y \in F$.

Proposition 5.5. *Let $(F,+,\cdot)$ be a finite field of characteristic $p \neq 0$, and $|F| = p^{2n}$. Then the mapping $\sigma : F \to F$ defined by $\sigma(x) = x^{p^n}$ is a field automorphism and $\sigma^2 = 1_F$.*

Proof. Easily, $\sigma(xy) = (xy)^{p^n} = x^{p^n}y^{p^n} = \sigma(x)\sigma(y)$. Also, $\sigma(x+y) = (x+y)^{p^n}$. But, by the binomial theorem,

$$(x+y)^{p^n} = x^{p^n} + \sum_{i=1}^{p^n-1} \binom{p^n}{i} x^{p^n-i}y^i + y^{p^n}.$$

But, each $\binom{p^n}{i}$ is an integer, and if $i \notin \{0,p^n\}$, then $\binom{p^n}{i}$ is a multiple of p, and so $\binom{p^n}{i}x^{p^n-i}y^i = 0$. Hence, $\sigma(x+y) = x^{p^n} + y^{p^n} = \sigma(x)+\sigma(y)$.

If $\sigma(x) = x^{p^n} = 0$, then $x = 0$, and so $\ker\sigma = \{0\}$, which makes σ injective. Since F is finite, σ is also bijective. This makes σ a field automorphism.

Certainly, $\sigma^2(0) = \sigma(\sigma(0)) = \sigma(0) = 0$. Also, each $a \in F$ is a solution to the equation

$$x^{p^{2n}} - x = 0$$

and so $a^{p^{2n}} = a$ for every $a \in F$. But

$$\sigma^2(a) = \sigma(\sigma(a)) = \sigma(a^{p^n}) = (a^{p^n})^{p^n} = a^{p^n p^n} = a^{p^{2n}} = a.$$

Hence, $\sigma^2 = 1_F$. □

Exercise 5.6. For a prime p and a positive integer n and $i \in \{1, 2, \ldots, p^n - 1\}$, prove that p is a factor of $\binom{p^n}{i}$.

Proposition 5.7. *Let $(F, +, \cdot)$ be a finite field of characteristic $p > 2$. Then*

$$U = \{a^2 \mid a \in F^*\}$$

is a subgroup of $(F^*, \cdot)$ of index 2.

Proof. Define $\mu : F \to F$ by $\mu(x) = x^2$. Then $\mu(xy) = (xy)^2 = x^2y^2 = \mu(x)\mu(y)$, and so μ is an endomorphism of the monoid $(F, \cdot, 1)$. Now $\mu(0) = 0$, and if $x \in F^*$, then $\mu(x) = x^2 \in F^*$. So, μ restricted to F^*, is an endomorphism of the group $(F^*, \cdot)$, and so $\mathrm{Im}\mu = U = \{x^2 \mid x \in F^*\}$ is a subgroup of $(F^*, \cdot)$. The kernel of μ is also a subgroup of $(F^*, \cdot)$, and $\ker \mu = \{x \in F^* \mid x^2 = 1\}$, and there are at most two elements in $\ker \mu$. Since $p > 2$, $-1 \neq 1$, and so $\ker \mu = \{\pm 1\}$. For each $a \in U$, there is a $b \in F^*$ such that $b^2 = a$. Now $-b \neq b$, and $(-b)^2 = b^2 = a$. Hence, the equation $x^2 - a = 0$ has exactly two solutions in F^*, and every $c \in F^*$ is a solution to the equation $x^2 - c^2 = 0$ with $c^2 \in U$. So, to each $a \in U$, there are exactly two elements $b, -b \in F^*$ such that $b^2 = (-b)^2 = a$. Hence, there are twice as many elements in F^* as in U. If $c \in F^* \setminus U$, then $cU \cap U = \varnothing$, and $|cU| = |U|$. Hence, U and cU are the only left cosets of U in F^*. That is, $F^* = cU \cup U$. Hence, $U = \{a^2 \mid a \in F^*\}$ is a subgroup of $(F^*, \cdot)$ of index 2. □

Proposition 5.8. *Let $(F, +, \cdot)$ be a finite field of characteristic $p > 2$, and $|F| = p^{2n}$. Then any field automorphism σ of F restricted to U of Proposition 5.7 is an automorphism of the subgroup U.*

Proof. If $a^2 \in U$, then $\sigma(a^2) = \sigma(a)\sigma(a) \in U$. □

We are ready to construct honest nearfields from various fields.

Theorem 5.9. *Let $(F,+,\cdot)$ be a field having a subgroup $(U,\cdot)$ of $(F^*,\cdot)$ of index 2. Suppose σ is a field automorphism of F of order 2 so that $\sigma|U$ is a group automorphism of U. Define $\circ$ on F by*

$$a \circ b = \begin{cases} a \cdot b, & \text{if } a \in U; \\ a \cdot \sigma(b), & \text{otherwise.} \end{cases}$$

Then $(F,+,\circ)$ is a nearfield and not a division ring.

Proof. For $a,b,c \in F$, consider $a \circ (b+c)$. If $a \in U$, then

$$a \circ (b+c) = a \cdot (b+c) = (a \cdot b) + (a \cdot c) = (a \circ b) + (a \circ c).$$

If $a \notin U$, then

$$\begin{aligned} a \circ (b+c) &= a \cdot \sigma(b+c) = a \cdot [\sigma(b) + \sigma(c)] \\ &= (a \cdot \sigma(b)) + (a \cdot \sigma(c)) = (a \circ b) + (a \circ c). \end{aligned}$$

So, $\circ$ is left distributive over $+$. Now $1 \in U$ and $\sigma(1) = 1$. So $a \circ 1 = a = 1 \circ a$ for $a \in U$, and $a \circ 1 = a \cdot \sigma(1) = a \cdot 1 = a = 1 \cdot a = 1 \circ a$ if $a \notin U$. Hence, 1 is also the identity for $\circ$. If $a \in U$, then $a^{-1} \in U$ for $(U,\cdot)$ is a subgroup of $(F^*,\cdot)$, and so $a \circ a^{-1} = a \cdot a^{-1} = 1$. Similarly, $a^{-1} \circ a = 1$. If $a \notin U$, $a \neq 0$, then $a \circ b = a \cdot \sigma(b)$. Since σ is a field automorphism, there is an $a' \in F^*$ such that $\sigma(a') = a^{-1}$. Hence, $a \circ a' = a \cdot \sigma(a') = a \cdot a^{-1} = 1$. Also, $\sigma(a'^{-1}) = a$, and $\sigma(a) = a'^{-1}$. Hence, $a' \circ a = a' \cdot \sigma(a) = a' \circ a'^{-1} = 1$. So, each $a \in F^*$ has an inverse $a' \in F^*$ with respect to $\circ$.

Let $a,b,c \in F$. There are four cases to consider to show that $\circ$ is associative. They are: (1) $a,b \in U$; (2) $a \in U$ and $b \notin U$; (3) $a \notin U$ and $b \in U$; and (4) $a,b \notin U$. First, let $a,b \in U$. Then

$$a \circ (b \circ c) = a \cdot (b \cdot c) = (a \cdot b) \cdot c$$

and

$$(a \circ b) \circ c = (a \cdot b) \circ c = (a \cdot b) \cdot c,$$

since $a \cdot b \in U$. Second, let $a \in U$ and $b \notin U$. So

$$a \circ (b \circ c) = a \cdot (b \cdot \sigma(c)) = (a \cdot b) \cdot \sigma(c)$$

and

$$(a \circ b) \circ c = (a \cdot b) \circ c = (a \cdot b) \cdot \sigma(c),$$

since $a \cdot b \notin U$. Third, with $a \notin U$ and $b \in U$, we have

$$a \circ (b \circ c) = a \cdot \sigma(b \cdot c) = a \cdot [\sigma(b) \cdot \sigma(c)] = [a \cdot \sigma(b)] \cdot \sigma(c)$$

and

$$(a \circ b) \circ c = [a \cdot \sigma(b)] \circ c = [a \cdot \sigma(b)] \cdot \sigma(c),$$

since $a \cdot \sigma(b) \notin U$. Finally, with $a, b \notin U$, we get

$$a \circ (b \circ c) = a \cdot \sigma[b \cdot \sigma(c)] = a \cdot (\sigma(b) \cdot c) = [a \cdot \sigma(b)] \cdot c,$$

since $\sigma^2 = 1_F$. And

$$(a \circ b) \circ c = (a \cdot \sigma(b)) \circ c = [a \cdot \sigma(b)] \cdot c,$$

since $a \cdot \sigma(b) \in U$. But why is $a \cdot \sigma(b) \in U$? Well, $F^*/U = \{U, cU\}$ for any $c \in F^* \setminus U$. In particular, since $b \in F^* \setminus U$, then so is $\sigma(b) \in F^* \setminus U$, hence $aU = bU = \sigma(b)U$ since $a \in F^* \setminus U$, also. Now F^*/U is a group of order 2, so we have $U = (aU)(aU) = (aU)(bU) = (aU)(\sigma(b)U) = a \cdot \sigma(b)U$. Hence, $a \cdot \sigma(b) \in U$. We can now conclude that $\circ$ is associative, and so $(F, +, \circ)$ is a nearfield.

It remains to show that $(F, +, \circ)$ is not a division ring. To do that, we will show that the right distributive law is not valid.

Let $K = U \cup \{0\}$. Let $c \in F$ such that $\sigma(c) \neq c$. Either $(K, +, \cdot)$ is a subfield of $(F, +, \cdot)$, or it is not. First, consider the case where K is a subfield. Take $a \in U$ and $b \notin K$. If $a + b \in K$, then with $a, a + b \in K$, we have that $b \in K$ since K

is a subfield. So, we may safely say that $a+b \notin K$. Hence, $a+b \in F^* \setminus U$, and therefore

$$(a+b) \circ c = (a+b) \cdot \sigma(c) = a \cdot \sigma(c) + b \cdot \sigma(c) = a \circ \sigma(c) + b \circ c.$$

So, if $(a+b) \circ c = (a \circ c) + (b \circ c)$, we have $a \circ c = a \circ \sigma(c)$, or $a \cdot c = a \cdot \sigma(c)$. With $a \in U$, we conclude that $\sigma(c) = c$, a contradiction.

Next, we consider the case that $(K, +, \cdot)$ is not a subfield. Since $K^* = K \setminus \{0\} = U$ is a multiplicative group, there are two possibilities: (1) there is an $a \in U$ with $-a \notin U$, and (2) there are $a, b \in U$ with $a+b \notin K$. If $a \in U$ and $-a \notin U$, then

$$\begin{aligned} 0 = 0 \circ c = [a + (-a)] \circ c &= [a + (-a)] \cdot \sigma(c) \\ &= a \cdot \sigma(c) + (-a) \cdot \sigma(c) = a \circ \sigma(c) + (-a) \circ c. \end{aligned}$$

If $\circ$ is right distributive over $+$, then

$$0 = [a + (-a)] \circ c = a \circ c + (-a) \circ c = a \cdot c + (-a) \cdot \sigma(c).$$

From $a \circ \sigma(c) + (-a) \circ c = a \circ c + (-a) \circ c$, we get $a \circ \sigma(c) = a \circ c$, or $a \cdot \sigma(c) = a \cdot c$. With $a \in U$, we again have $\sigma(c) = c$, a contradiction.

Finally, if $a, b \in U$ and $a+b \notin K$, then

$$(a+b) \circ c = (a+b) \cdot \sigma(c) = a \cdot \sigma(c) + b \cdot \sigma(c) = a \circ \sigma(c) + b \circ \sigma(c).$$

If $(a+b) \circ c = a \circ c + b \circ c$, then

$$a \circ c + b \circ c = a \circ \sigma(c) + b \circ \sigma(c),$$

or

$$a \cdot c + b \cdot c = a \cdot \sigma(c) + b \cdot \sigma(c),$$

which gives us

$$a \cdot (c - \sigma(c)) = -b \cdot (c - \sigma(c)).$$

$+$	0	1	2	a	$1+a$	$2+a$	$2a$	$1+2a$	$2+2a$
0	0	1	2	a	$1+a$	$2+a$	$2a$	$1+2a$	$2+2a$
1	1	2	0	$1+a$	$2+a$	a	$1+2a$	$2+2a$	$2a$
2	2	0	1	$2+a$	a	$1+a$	$2+2a$	$2a$	$1+2a$
a	a	$1+a$	$2+a$	$2a$	$1+2a$	$2+2a$	0	1	2
$1+a$	$1+a$	$2+a$	a	$1+2a$	$2+2a$	$2a$	1	2	0
$2+a$	$2+a$	a	$1+a$	$2+2a$	$2a$	$1+2a$	2	0	1
$2a$	$2a$	$1+2a$	$2+2a$	0	1	2	a	$1+a$	$2+a$
$1+2a$	$1+2a$	$2+2a$	$2a$	1	2	0	$1+a$	$2+a$	a
$2+2a$	$2+2a$	$2a$	$1+2a$	2	0	1	$2+a$	a	$1+a$

Table 1: The Addition Table for the Field $(F,+,\cdot)$ and the Nearfield $(F,+,\circ)$ of Order 9.

With $c - \sigma(c) \neq 0$, we conclude that $a = -b$, and so $a + b = 0 \in K$, a contradiction.

We can now conclude that $\circ$ is not right distributive over $+$, and that the nearfield $(F,+,\circ)$ is not a division ring. □

"Enough of this theory," you say? "It's time to see some action!" So, we shall construct the smallest honest nearfield that is universally accepted. (Recall that $(\mathbb{Z}_2,+,*)$ is not quite a nearfield.) Our example results from applying Theorem 5.9 to a field of nine elements. So, our first task will be to construct a field $(F,+,\cdot)$ with $|F| = 9$, and so $F = \mathbb{Z}_3[x]/(\alpha+\beta x+x^2)$ for some irreducible $\alpha+\beta x+x^2 \in \mathbb{Z}_3[x]$. Trial and error would quickly provide such an irreducible $\alpha+\beta x+x^2$, but the reader can be readily assured that $1+x^2 \in \mathbb{Z}_3[x]$ is irreducible.

Our field is the nine element field $\mathbb{Z}_3[x]/(1+x^2)$, and we will let

$$a = x + (1+x^2) \in \mathbb{Z}_3[x]/(1+x^2).$$

If we also identify $0 \equiv 0+(1+x^2) = (1+x^2)$, $1 \equiv 1+(1+x^2)$, and $2 \equiv 2+(1+x^2)$, then we present the addition table and the multiplication table for our field $F = \mathbb{Z}_3[x]/(1+x^2)$. See Tables 1 and 2.

$\cdot$	0	1	2	a	$1+a$	$2+a$	$2a$	$1+2a$	$2+2a$
0	0	0	0	0	0	0	0	0	0
1	0	1	2	a	$1+a$	$2+a$	$2a$	$1+2a$	$2+2a$
2	0	2	1	$2a$	$2+2a$	$1+2a$	a	$2+a$	$1+a$
a	0	a	$2a$	2	$2+a$	$2+2a$	1	$1+a$	$1+2a$
$1+a$	0	$1+a$	$2+2a$	$2+a$	$2a$	1	$1+2a$	2	a
$2+a$	0	$2+a$	$1+2a$	$2+2a$	1	a	$1+a$	$2a$	2
$2a$	0	$2a$	a	1	$1+2a$	$1+a$	2	$2+2a$	$2+a$
$1+2a$	0	$1+2a$	$2+a$	$1+a$	2	$2a$	$2+2a$	a	1
$2+2a$	0	$2+2a$	$1+a$	$1+2a$	a	2	$2+a$	1	$2a$

Table 2: The Multiplication Table for the Field $(F,+,\cdot)$.

The following table will be useful in computing the multiplication table for the nearfield $(F,+,\circ)$:

b	0	1	2	a	$1+a$	$2+a$	$2a$	$1+2a$	$2+2a$
b^2	0	1	1	2	$2a$	a	2	a	$2a$
b^3	0	1	2	$2a$	$1+2a$	$2+2a$	a	$1+a$	$2+a$

From this table we see that

$$U = \{1, 2, a, 2a\}$$

and we get the values of the automorphism $\sigma(b) = b^3$.

We can now build the multiplication table for the nearfield $(F,+,\circ)$.

Exercise 5.10. Choose another irreducible $\alpha + \beta x + x^2 \in \mathbb{Z}_3[x]$, and construct the field $\mathbb{Z}_3[x]/(\alpha + \beta x + x^2)$.

Exercise 5.11. With your field $\mathbb{Z}_3[x]/(\alpha + \beta x + x^2)$ of Exercise 5.10, construct a nearfield as we did, using Theorem 5.9 for $1 + x^2$. Compare your multiplication table to that we presented, Table 3, for the nearfield $(F,+,\circ)$.

Exercise 5.12. From the division ring $(Q,+,\cdot)$ of quaternions, construct a multiplication table for the set $S = \{\pm 1, \pm i, \pm j, \pm k\}$. Compare your multiplication table to that for $\circ$ of the nearfield $(F,+,\circ)$. Do you see any similarities?

$\circ$	0	1	2	a	$1+a$	$2+a$	$2a$	$1+2a$	$2+2a$
0	0	0	0	0	0	0	0	0	0
1	0	1	2	a	$1+a$	$2+a$	$2a$	$1+2a$	$2+2a$
2	0	2	1	$2a$	$2+2a$	$1+2a$	a	$2+a$	$1+a$
a	0	a	$2a$	2	$2+a$	$2+2a$	1	$1+a$	$1+2a$
$1+a$	0	$1+a$	$2+2a$	$1+2a$	2	a	$2+a$	$2a$	1
$2+a$	0	$2+a$	$1+2a$	$1+a$	$2a$	2	$2+2a$	1	a
$2a$	0	$2a$	a	1	$1+2a$	$1+a$	2	$2+2a$	$2+a$
$1+2a$	0	$1+2a$	$2+a$	$2+2a$	a	1	$1+a$	2	$2a$
$2+2a$	0	$2+2a$	$1+a$	$2+a$	1	$2a$	$1+2a$	a	2

Table 3: The Multiplication Table for the Nearfield $(F, +, \circ)$.

5.2 Anatomy of a nearfield

Let $(K, +, \cdot)$ be a nearfield. Then $(K^*, \cdot)$ is a group. For each $a \in K^*$, define the mapping $a^\bullet : K \to K$ by $a^\bullet(x) = ax$. If $a, b \in K^*$ and $a \neq b$, then $a^\bullet(1) \neq b^\bullet(1)$, and so $a^\bullet \neq b^\bullet$. We will show now that each $a^\bullet$ is an automorphism of the group $(K, +)$.

For $x, y \in K$,

$$a^\bullet(x+y) = a(x+y) = ax + ay = a^\bullet(x) + a^\bullet(y),$$

also,

$$a^\bullet(a^{-1}y) = a(a^{-1}y) = y.$$

Therefore, $a^\bullet$ is an epimorphism. If $a^\bullet(x) = a^\bullet(y)$, then $ax = ay$, and so $a^{-1}(ax) = a^{-1}(ay)$, which gives us $x = y$. Thus, $a^\bullet$ is a monomorphism, and so $a^\bullet$ is an automorphism of $(K, +)$.

If $a, b \in K^*$, then

$$a^\bullet \circ b^\bullet(x) = a^\bullet(b^\bullet(x)) = a(bx) = (ab)x = (ab)^\bullet(x).$$

Hence, $(ab)^\bullet = a^\bullet \circ b^\bullet$, and so the map $\phi : K^* \to \mathrm{Aut}(K, +)$ defined by $\phi(a) = a^\bullet$

is a group monomorphism, where

$$\mathrm{Aut}(K,+) = \{f : K \to K \mid f \text{ is an automorphism of } (K,+)\},$$

a subgroup of $\mathscr{U}(M(K))$. See Definition A3.16. Let $\Phi = \phi(K^*)$. So Φ is a group of automorphisms of the group $(K,+)$, i.e., a subgroup of the group $(\mathrm{Aut}(K,+),\circ)$.

There are two important properties of the group $\Phi = \{a^\bullet \mid a \in K^*\}$. The first is that the equation $a^\bullet(x) = x$ has only the solution $x = 0 \in K$, if $a \neq 1$. For if $a \neq 1$ and $a^\bullet(u) = u$, then $au = u$. If $u \neq 0$, then $a = 1$. This means that Φ is a group of *fixed point free* automorphisms of $(K,+)$. Now, let's be more careful. An automorphism f of a group $(G,\cdot)$ is *fixed point free* if (1) $f \neq 1_G$, and (2) the only $u \in G$ for which $f(u) = u$ is $u = 1$, the identity of G. (If we are considering a group $(G,+)$, then we use 0 for the identity of G.) So, a *group of fixed point free automorphisms of* $(G,\cdot)$ is a subgroup Φ of $\mathrm{Aut}(G,\cdot)$ for which each $\phi \in \Phi \setminus \{1_G\}$ is a fixed point free automorphism.

The second property of Φ is illustrated by taking any two points $u, v \in K^*$. Then $(vu^{-1})^\bullet(u) = (vu^{-1})u = v$, and $(vu^{-1})^\bullet \in \Phi$. This means that the *orbit* of Φ at any $u \in K^*$ is all of K^*. If Φ is a subgroup of $\mathrm{Aut}(G,\cdot)$, and if $u \in G$, then the *orbit of* Φ *at* u is $\Phi(u) = \{\phi(u) \mid \phi \in \Phi\}$. Certainly, the orbit of $1 \in G$ is $\Phi(1) = \{1\}$, and is called the *trivial orbit*.

Collecting our observations, we have one half of

Theorem 5.13. *For any nearfield $(K,+,\cdot)$, we have that $(K^*,\cdot)$ is isomorphic to a group Φ of fixed point free automorphisms of the group $(K,+)$, where the orbits of Φ are the trivial orbit $\Phi(0)$ and $\Phi(a) = K^*$ for any $a \in K^*$.*

Conversely, if Φ is a group of fixed point free automorphisms of a group $(K,+)$, where, for any $a \in K^* = K \setminus \{0\}$, we have the orbit $\Phi(a) = K^*$, then there is a nearfield $(K,+,\cdot)$ with K^* isomorphic to Φ.

Proof. We just need to prove the converse part, and to do that, we need a multiplication $\cdot$ on K. Of course, we will define $0 \cdot b = 0$ for all $b \in K$. Take any

$e \in K^*$, and fix it. If $a \in K^*$, then there is a unique $\phi_a \in \Phi$ such that $\phi_a(e) = a$. For if $\phi, \lambda \in \Phi$ and $\phi(e) = a = \lambda(e)$, then $e = \phi^{-1} \circ \lambda(e)$. But $\phi^{-1} \circ \lambda \in \Phi$ and $e \neq 0$, so $\phi^{-1} \circ \lambda = 1_K$, since Φ is a group of fixed point free automorphisms. From $\phi^{-1} \circ \lambda = 1_K$, we get $\phi = \lambda$. This gives the uniqueness of ϕ_a, and the existence follows from the fact that $\Phi(e) = K^*$.

With this unique $\phi_a \in \Phi$ such that $\phi_a(e) = a$, we define $\cdot$ on K by

$$a \cdot b = \begin{cases} 0, & \text{if } a = 0; \\ \phi_a(b), & \text{if } a \neq 0. \end{cases}$$

Our job now is to show that $(K, +, \cdot)$ is a nearfield.

If $a = 0$, then

$$a \cdot (b + c) = 0 = 0 + 0 = (a \cdot b) + (a \cdot c).$$

If $a \neq 0$, then

$$a \cdot (b + c) = \phi_a(b + c) = \phi_a(b) + \phi_a(c) = (a \cdot b) + (a \cdot c).$$

So, $\cdot$ is left distributive over $+$. Notice that $\phi_e = 1_K$ and that $e \cdot b = \phi_e(b) = b$. Similarly, for $b \neq 0$, $b \cdot e = \phi_b(e) = b$, and so e is the identity for $\cdot$. Since ϕ_a is an automorphism, there is an a' such that $\phi_a(a') = e$. Hence, $a \cdot a' = \phi_a(a') = e$. Also, $\phi_a^{-1}(e) = a'$, and so $\phi_{a'} = \phi_a^{-1}$. Hence, $a' \cdot a = \phi_{a'}(a) = \phi_a^{-1}(a) = e$. This means that each $a \in K^*$ has an $a' \in K^*$ such that $a \cdot a' = a' \cdot a = e$. So, a has an inverse $a^{-1} = a'$. It remains only to show associativity of $\cdot$.

Take $a, b, c \in K$. If $a = 0$, then

$$a \cdot (b \cdot c) = 0 = 0 \cdot c = (a \cdot b) \cdot c.$$

If $b = 0$, then

$$a \cdot (b \cdot c) = a \cdot 0 = \phi_a(0) = 0 = 0 \cdot c = \phi_a(0) \cdot c = (a \cdot 0) \cdot c = (a \cdot b) \cdot c.$$

It remains only to consider the case where $0 \notin \{a, b\}$. Now

$$a \cdot (b \cdot c) = \phi_a(\phi_b(c)) = (\phi_a \circ \phi_b)(c), \quad \text{and} \quad (a \cdot b) \cdot c = \phi_a(b) \cdot c.$$

From $\phi_a \circ \phi_b(e) = \phi_a(\phi_b(e)) = \phi_a(b)$, we have

$$(a \cdot b) \cdot c = \phi_a(b) \cdot c = \phi_{\phi_a(b)}(c) = \phi_a \circ \phi_b(c) = a \cdot (b \cdot c).$$

This completes the proof of the theorem. □

We are now in an excellent condition to examine equations of the form $ax = bx + c$ for a nearfield $(K, +, \cdot)$. If $b = 0$, then the equation becomes $ax = c$, or $a^\bullet(x) = c$, which, of course, has a unique solution, since $a \neq b = 0$, and so a^{-1} exists. The unique solution is $a^{-1}c$, of course. Similarly, if $a = 0$, then the equation becomes $bx = -c$, which also has the unique solution $x = b^{-1}(-c)$. Now we are ready to get serious.

Suppose $0 \notin \{a, b\}$, $a, b, c \in K$, and $a \neq b$. Solving the equation $ax = bx + c$ is equivalent to finding the values $x \in K$ for which $x = a^{-1}bx + a^{-1}c$, so we lose no generality if we assume $a = 1$. An equation $x = bx + c$ has a unique solution for $x \in K$ if and only if $-bx + x = c$ has a unique solution for $x \in K$, which is equivalent to $-b^\bullet(x) + 1_K(x) = (-b^\bullet + 1_K)(x) = c$ having a unique solution. So, when is the mapping $-b^\bullet + 1_k$ a bijection?

Definition 5.14. A nearfield $(K, +, \cdot)$ is *planar* if and only if for all $a, b, c \in K$, $a \neq b$, the equation $ax = bx + c$ has a unique solution for $x \in K$.

Continuing our discussion, we have that a nearfield $(K, +, \cdot)$ is planar if and only if $-b^\bullet + 1_K$ is a bijection for each $b \in K^* \setminus \{1\}$. We know that each $b^\bullet$, $b \neq 1$, is a fixed point free automorphism of $(K, +)$. So, if $u \neq v$, then $b^\bullet(u) \neq b^\bullet(v)$, and so $-b^\bullet(u) + u \neq -b^\bullet(v) + v$, for otherwise $v - u = b^\bullet(v) - b^\bullet(u) = b^\bullet(v - u)$. This says that $(-b^\bullet + 1_K)(u) \neq (-b^\bullet + 1_K)(v)$, and so each $-b^\bullet + 1_K$, $b \neq 1$, is injective. If K is finite, then each $-b^\bullet + 1_K$, $b \neq 1$, is surjective, and consequently bijective. This gives us

Theorem 5.15. *Every finite nearfield is planar.*

So, to find a nonplanar nearfield, we must explore among the infinite nearfields.

It wasn't until 1964 that the mathematical community became aware of the existence of a nonplanar nearfield. We shall now present J. L. Zemmer's method of constructing nonplanar nearfields [Ze].

We will let $(F, +, \cdot)$ be any field of characteristic zero, but the reader may want to let $F = \mathbb{R}$, the field of real numbers. The ring of polynomials $(F[x], +, \cdot)$ has many properties analogous to the ring of integers, and the reader may find it useful to think of the integers as we continue, and we hope that the sophisticated reader will tolerate the many liberties which we will be taking. Just like one makes the field $(\mathbb{Q}, +, \cdot)$ of rational numbers from the ring of integers $(\mathbb{Z}, +, \cdot)$, one makes the *field* $(F(x), +, \cdot)$ *of rational functions* from the ring $(F[x], +, \cdot)$. That is, with

$$F(x) = \left\{ \frac{p(x)}{q(x)} \;\middle|\; p(x), q(x) \in F[x],\ q(x) \neq 0 \right\},$$

we add and multiply elements of $F(x)$ just like we did in calculus, i.e., like we add and multiply fractions. Specifically, with

$$\frac{p(x)}{q(x)} + \frac{s(x)}{t(x)} = \frac{p(x)t(x) + q(x)s(x)}{q(x)t(x)}$$

and

$$\frac{p(x)}{q(x)} \cdot \frac{s(x)}{t(x)} = \frac{p(x)s(x)}{q(x)t(x)}.$$

then it is "easy" to see that $(F(x), +, \cdot)$ is a field, just like $(\mathbb{Q}, +, \cdot)$ is a field.

To make life easier for the reader and for the author, we will streamline the notation somewhat, as we did in §A2. For example, an element of $F(x)$ may be denoted simply by a or perhaps $a(x)$. Of course, we know that there are polynomials $p = p(x)$ and $q = q(x)$ in $F[x]$ for which $a = \frac{p}{q}$, or $a(x) = \frac{p(x)}{q(x)}$, and we will usually assume that our p/q is in reduced form, just as we do for rational numbers. That is, all common factors of the numerator and the denominator have been eliminated.

We extend the idea of the degree of a polynomial. For $a = p/q \in F(x)$, where $p, q \in F[x]$, we define

$$\delta(a) = \deg p - \deg q,$$

if $a \neq 0$, and we do not define $\delta(0)$. It is easy to see that $\delta(ab) = \delta(a) + \delta(b)$ for all $a, b \in F(x) \setminus \{0\}$.

There is an important automorphism T of the field $(F(x), +, \cdot)$. For an element $a(x) = p(x)/q(x)$ in $F(x)$, define $T(a(x)) = a(x+1) = p(x+1)/q(x+1)$. It is easy to see that $T^k(a(x)) = a(x+k)$ for each and every $k \in \mathbb{Z}$ and that $\delta(T^k(a)) = \delta(a)$ for each $k \in \mathbb{Z}$. The proof that each $T^k \in \mathrm{Aut}(F(x), +, \cdot)$ is also very easy. We are now ready to make a nonplanar nearfield from the field $(F(x), +, \cdot)$.

Define $*$ on $F(x)$ by

$$a * b = \begin{cases} 0, & \text{if } a = 0; \\ a \cdot T^{\delta(a)}(b), & \text{if } a \neq 0. \end{cases}$$

Certainly $0 * (b + c) = 0 = 0 + 0 = (0 * b) + (0 * c)$. If $a \neq 0$, then

$$\begin{aligned} a * (b + c) &= a \cdot T^{\delta(a)}(b + c) = a \cdot \{T^{\delta(a)}(b) + T^{\delta(a)}(c)\} \\ &= a \cdot T^{\delta(a)}(b) + a \cdot T^{\delta(a)}(c) = (a * b) + (a * c). \end{aligned}$$

So, $*$ is left distributive over $+$.

Turning now to associativity, let $a, b, c \in F(x)$. If $a = 0$, then

$$a * (b * c) = 0 = 0 * c = (0 * b) * c = (a * b) * c.$$

If $a \neq 0$ and $b = 0$, then

$$\begin{aligned} a * (b * c) &= a * 0 = a \cdot T^{\delta(a)}(0) = a \cdot 0 = 0 \\ &= 0 * c = [a \cdot T^{\delta(a)}(0)] * c = (a * 0) * c = (a * b) * c. \end{aligned}$$

Suppose $0 \notin \{a, b\}$. Then

$$\begin{aligned} a*(b*c) &= a \cdot T^{\delta(a)}\left[b \cdot T^{\delta(b)}(c)\right] = a \cdot \left\{T^{\delta(a)}(b) T^{\delta(a)}\left[T^{\delta(b)}(c)\right]\right\} \\ &= a \cdot T^{\delta(a)}(b) T^{\delta(a)+\delta(b)}(c) = a \cdot T^{\delta(a)}(b) T^{\delta(a)+\delta[T^{\delta(a)}(b)]}(c) \\ &= \left[a \cdot T^{\delta(a)}(b)\right] \cdot T^{\delta\left(a \cdot T^{\delta(a)}(b)\right)}(c) = \left[a \cdot T^{\delta(a)}(b)\right] * c = (a*b)*c. \end{aligned}$$

Hence, $*$ is associative.

With $1*b = 1 \cdot T^{\delta(1)}(b) = 1 \cdot T^0(b) = 1 \cdot b = b$ and $b*1 = b \cdot T^{\delta(b)}(1) = b \cdot 1 = b$, we have that 1 is the identity for $*$, as well as for $\cdot$. To see that $(F(x), +, *)$ is a nearfield, we need to see that each $a \neq 0$ has an inverse with respect to $*$. For $a \neq 0$, let $a' = T^{-\delta(a)}(a^{-1})$, where a^{-1} is the inverse for a with respect to $\cdot$. Then

$$a*a' = a \cdot T^{\delta(a)} \cdot \left(T^{-\delta(a)}(a^{-1})\right) = a \cdot a^{-1} = 1,$$

and

$$\begin{aligned} a'*a &= T^{-\delta(a)}(a^{-1}) \cdot T^{\delta(a')}(a) = T^{-\delta(a)}(a^{-1}) \cdot T^{\delta\left[T^{-\delta(a)}(a^{-1})\right]}(a) \\ &= T^{-\delta(a)}(a^{-1}) \cdot T^{\delta(a^{-1})}(a) = T^{-\delta(a)}(a^{-1}) \cdot T^{-\delta(a)}(a) \\ &= T^{-\delta(a)}(a^{-1} \cdot a) = T^{-\delta(a)}(1) = 1. \end{aligned}$$

Hence, a' is the inverse of a with respect to $*$. This makes $(F(x), +, *)$ a nearfield.

To see that this nearfield is not planar, we will see that the equation

$$X = b*X + b,$$

where $b(x) = x$, does not have a solution in $F(x)$. (This will also show that $*$ is not commutative and that $*$ is not right distributive over $+$.) We will assume that $X = p/q$, where $p, q \in F[x]$, is a solution to $X = b*X + b$, and arrive at a contradiction.

If $p/q = b*(p/q)+b$, then $p/q = x\cdot T(p/q)+x$, or

$$\frac{p(x)}{q(x)} = x\cdot\frac{p(x+1)}{q(x+1)}+x = x\cdot\left[\frac{p(x+1)+q(x+1)}{q(x+1)}\right],$$

hence,

$$p(x)q(x+1) = x[p(x+1)+q(x+1)]q(x).$$

So, $q(x)$ is a factor of $p(x)q(x+1)$, and since we may assume, without loss of generality, that $p(x)$ and $q(x)$ have no common factors, we can conclude that $q(x)$ is a factor of $q(x+1)$. Hence, $q(x+1) = k(x)q(x)$ for some $k(x) \in F[x]$. Since $\deg q(x) = \deg q(x+1) = \deg k(x) + \deg q(x)$, we may conclude that $k(x) = k \in F$. In fact, from $q(x+1) = kq(x)$, we have $k = 1$ and so $q(x+1) = q(x) = q \in F$.

So, we have $p(x) = x[p(x+1)+q]$ and $\deg p(x) = 1 + \deg\,[p(x+1)+q]$, and this cannot be, since $\deg\,[p(x+1)+q] = \deg p(x)$. With the desired contradiction, we have that the nearfield $(F(x),+,*)$ is nonplanar.

We have that each finite nearfield is planar, and we have the existence of infinite nonplanar nearfields. In each field and division ring, equations of the form $ax = bx+c$, $a \neq b$, easily have unique solutions for x. This is not so for infinite nearfields. Even though such equations have unique solutions for planar nearfields, the method for solving such equations is not the same as for fields and division rings. This is because the right distributive law is not valid for nearfields which are neither fields nor division rings.

It is really relatively difficult to construct nearfields, but nearfields are the algebraic structures in which the equations $ax = bx+c$ most naturally have relevance. They are quite close to fields, or division rings, in their definition, and most nearfields are constructed from fields. The equation $ax = bx+c$ seems most significant in algebraic structures in which the right distributive law is not valid. Later, we will examine equations of the form $ax = bx+c$ in algebraic structures called *nearrings*. But before we do this, we shall discuss an important role of nearfields in algebra and geometry.

5.3 Milestones for nearfields

Nearfields were discovered by L. E. Dickson in 1905 [Di]. Their initial role was to demonstrate the independence of various properties for a field. As it happens so often, others find good use for nice discoveries. Almost immediately, in 1907, Veblen and Wedderburn demonstrated that Dickson's nearfields were very suitable for constructing nondesarguesian planes [VW]. Nearfields have played a very significant role in describing sharply 2-transitive [C] and sharply 3-transitive groups [Wä].

In §3, we were introduced to sharply 3-transitive groups and saw their role in constructing Möbius planes. Nearfields are also useful in constructing sharply 3-transitive groups. Each nearfield yields a sharply 2-transitive group, as we shall soon see. All finite sharply 2-transitive groups come from the finite nearfields, and it is not known if there is an infinite sharply 2-transitive group that does not come from a nearfield. See [C].

In this section, we will show that every nearfield gives us a sharply 2-transitive group. We will also describe Desargues' property for affine planes and show that our example of a nearfield of order 9 gives us an affine plane which does not have Desargues' property.

For any nearfield $(F,+,\cdot)$ and for any $b,m \in F$, $m \neq 0$, define a mapping $[b,m] : F \to F$ by

$$[b,m](x) = b + mx$$

for every $x \in F$. Note that $[0,1] = 1_F$ and that $[b,m]^{-1} = [-m^{-1}b, m^{-1}]$, and so each $[b,m]$ is a unit in the monoid of mappings $(M(F), \circ, 1_F)$. Let

$$T_2(F) = \{[b,m] \mid b,m \in F,\ m \neq 0\}.$$

We are now ready to prove

Theorem 5.16. *$(T_2(F), \circ)$ is a sharply 2-transitive group on the set F.*

Proof. Certainly $[0,1] = 1_F \in T_2(F)$, and $[b,m] \circ [b',m'] = [b+mb',mm'] \in T_2(F)$ if $[b,m]$, $[b',m'] \in T_2(F)$. With $[b,m]^{-1} = [-m^{-1}b, m^{-1}] \in T_2(F)$ whenever $[b,m] \in T_2(F)$, we have that $T_2(F)$ is a subgroup of the group of units of the monoid $(M(F), \circ, 1_F)$.

Let $a, b \in F$, $a \neq b$. First, we will show that $[a, -a+b]$ is the unique element of $T_2(F)$ where

$$[a,-a+b](0) = a + (-a+b)0 = a$$

and

$$[a,-a+b](1) = a + (-a+b)1 = b.$$

For if $[a',m'](0) = a$ and $[a',b'](1) = b$, then from $a = [a',m'](0) = a' + m'0 = a'$, we get $a = a'$. Also, $b = [a',m'](1) = a' + m'1 = a + m'$, so we get $-a+b = m'$. Hence, $[a',m'] = [a,-a+b]$, as desired.

Also, let $c, d \in F$, $c \neq d$. Then let $[e,f] = [c,-c+d] \circ [a,-a+b]^{-1}$, so

$$[e,f](a) = [c,-c+d](0) = c \text{ and } [e,f](b) = [c,-c+d](1) = d.$$

This gives the existence of an $[e,f] \in T_2(F)$ for which

$$[e,f](a) = c \text{ and } [e,f](b) = d.$$

Suppose that

$$[e',f'] \in T_2(F), \; [e',f'](a) = c, \text{ and } [e',f'](b) = d.$$

Then

$$[e',f'] \circ [a,-a+b](0) = c \text{ and } [e',f'] \circ [a,-a+b](1) = d.$$

But $[c,-c+d]$ is the unique element of $T_2(F)$ with this property, so

$$[e',f'] \circ [a,-a+b] = [c,-c+d].$$

Consequently,

$$[e',f'] = [c,-c+d] \circ [a,-a+b]^{-1} = [e,f].$$

This gives the desired uniqueness and completes the proof of Theorem 5.16. □

Exercise 5.17. How effective are nearfields for the construction of sharply 3-transitive groups? Using Theorem 3.18 as a model, try to construct sharply 3-transitive groups from nearfields $(F, +, \cdot)$. If you have difficulty, what extra properties would F need?

Exercise 5.18. How effective are nearfields for the construction of affine planes? Using Theorem 3.4 as a model, try to construct an affine plane from an arbitrary nearfield $(F, +, \cdot)$. If you have any difficulties, what extra properties would F need?

Desargues' property for affine and projective planes give these planes lots of nice properties. This property can be introduced easier for affine planes than for projective planes, and so that is how we shall introduce it. Even so, it is a rather involved concept to introduce, one which the reader is asked to be actively involved. So, get several blank pieces of paper, your favorite pen or pencil, and actively follow through the procedure below, several times with variations. It may be useful to refer to the diagram of The Veblen and Wedderburn example below.

Definition 5.19. An affine plane $(P, \mathscr{L})$ is *desarguesian* provided the following steps can always be executed, theoretically at least.

Step 1. Take any point $p \in P$ and any three distinct lines $L_1, L_2, L_3 \in \mathscr{L}$ passing through p.

Step 2. Choose arbitrary points q_1, q_2, q_3 on L_1, L_2, L_3, respectively, ensuring only that $p \notin \{q_1, q_2, q_3\}$.

Step 3. Let L_4 be the line through q_1 and q_2, let L_5 be the line through q_2 and q_3, and let L_6 be the line through q_1 and q_3.

Step 4. Choose one of the lines L_1, L_2, L_3, say L_2, and then choose any point r_2 on L_2 such that $r_2 \notin \{q_2, p\}$.

Step 5. Let L_7 be the line through r_2 such that L_4 and L_7 are parallel. Similarly, let L_8 be the line through r_2 such that L_5 and L_8 are parallel.

Step 6. Let r_1 be the point where L_1 and L_7 meet, and let r_3 be the point where L_3 and L_8 meet.

Step 7. Let L_9 be the line through r_1 and r_3.

Step 8. If our affine plane $(P, \mathscr{L})$ is desarguesian, then we can be assured that L_6 and L_9 are parallel.

Exercise 5.20. Find a field F for which the affine plane $(F \times F, \mathscr{L}_F)$ of Theorem 3.4 is desarguesian.

Of course, we want to show that nearfields are useful for constructing affine planes which are not desarguesian. This is one of the reasons we have constructed in detail the nearfield $(F, +, \circ)$ in §5.1 above.

Exercise 5.21. For the nearfield $(F, +, \circ)$ in Theorem 5.9 above, consider the incidence structure $(F \times F, \mathscr{L}_F)$ modeled after those in Theorem 3.4. Show that $(F \times F, \mathscr{L}_F)$ is an affine plane.

We will now outline steps which will show that the affine plane of Exercise 5.21 is nondesarguesian. It may be helpful to be influenced by The Veblen and Wedderburn example below.

Consider the three lines $L_1 = L(2a, 1)$, defined by the equation $y = 2ax + 1$, $L_2 = L(a, 0)$, defined by the equation $y = ax$, and $L_3 = L(2 + a, 1 + 2a)$, defined by the equation $y = (2 + a)x + (1 + 2a)$. Note that the point $p = (a, 2)$ is on each of the lines L_1, L_2, and L_3. Choose $q_1 = (0, 1)$ on L_1, $q_2 = (0, 0)$ on L_2, and $q_3 = (1, 0)$ on L_3. Let L_4 be the line through q_1 and q_2, so L_4 is the "y-axis," i.e., $L_4 = L(\infty, 0)$ and has equation $x = 0$. Let L_5 be the "x-axis," i.e., $L_5 = L(0, 0)$ and has equation $y = 0$. Note that q_2 and q_3 are on L_5. If L_6 is the line through q_1 and q_3, then $L_6 = L(2, 1)$ and has equation $y = 2x + 1$.

Corresponding to Step 4, we take $r_2 = (2+2a, 1+2a)$ on L_2. If L_7 is the line through r_2 and parallel to L_4, then $L_7 = L(\infty, 2+2a)$ and has equation $x = 2+2a$. Similarly, if L_8 is the line through r_2 and parallel to L_5, then $L_8 = L(0, 1+2a)$ and has equation $y = 1+2a$.

Corresponding to Step 6, compute that L_1 and L_7 intersect at $r_1 = (2+2a, a)$, and that L_3 and L_8 intersect at $r_3 = (0, 1+2a)$. The line passing through r_1 and r_3 is $L_9 = L(1, 1+2a)$ and has equation $y = x + (1+2a)$. Note that the "slope" of L_9 is $m = 1$, and that the "slope" of L_6 is $m = 2$, so we should anticipate that L_6 and L_9 are not parallel. In fact, L_6 and L_9 intersect at the point $(2a, 1+a)$.

Had the affine plane of Exercise 5.21 been desarguesian, then we would have that L_6 and L_9 are parallel. Since they are not parallel, we must conclude that this affine plane is nondesarguesian. An illustration is presented in Figure 2.

Exercise 5.22. Hopefully, you saved your nearfield of Exercise 5.11. Use your own nearfield $(F, +, \circ)$ and construct an affine plane $(F \times F, \mathscr{L}_F)$ as in Exercise 5.18 or Theorem 3.4. Is your affine plane desarguesian?

6 Planar nearrings

If the reader sensed that working with and constructing nearfields is rather involved, then he is correct. However, nearfields have been extensively studied, and facility with these structures can now be obtained with a reasonable effort. It would be a rewarding activity to make this effort, and Wähling's book [Wä] is definitely the best source.

It is much easier, however, to construct and work with planar nearrings. One needs to develop some new techniques and points of view. These are not particularly traditional for working in algebra, combinatorics, and geometry,

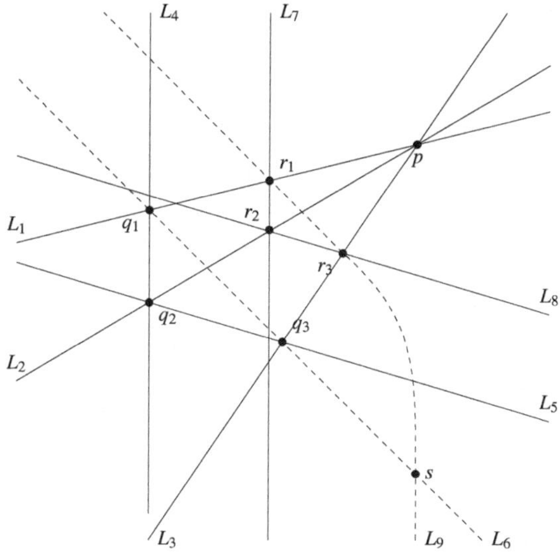

Figure 2: The Veblen and Wedderburn example.

however. But, the variety of applications seems to be quite extensive, and the horizon for future development seems to be expanding very rapidly. The techniques, points of view, and concepts have all been developed since 1967, and most of them have been developed since 1986. It is our purpose to introduce many of these techniques, points of view, and applications, and to point to horizons which promise new, interesting, and challenging developments.

There are various ways to introduce the algebraic structure of a nearring. The traditional way is to start with a group $(G,+)$, not necessarily commutative, and let $M(G)$ be the family of all mappings $f: G \to G$. One easily sees that $(M(G), \circ, 1_G)$ is a monoid. See (i) of Examples A3.2. Define an addition of functions $f, g \in M(G)$ by $f \oplus g = h$ where $h(x) = (f \oplus g)(x) = f(x) + g(x)$ for all $x \in G$. Then the algebraic structure $(M(G), \oplus, \circ)$ is a *nearring*. In general, these structures satisfy properties (A2)–(A4), (M2)–(M3), and (DR), as described in §4.

If we take $\mathscr{C}(\mathbb{R})$ to be all the continuous functions $f : \mathbb{R} \to \mathbb{R}$, as defined in (c) of Examples A2.5, and, in place of $\bullet$, we use composition $\circ$, then $(\mathscr{C}(\mathbb{R}), \oplus, \circ)$ satisfies the properties (A1)–(A4), (M2)–(M3), and (DR).

Take any group $(G, +)$. Define $\cdot$ on G by $a \cdot b = b$ for all $a, b \in G$. Then $(G, +, \cdot)$ satisfies properties (A2)–(A4), (M2), and (DL).

If R is a commutative ring with identity, and if $(R[x], +, \cdot)$ is the ring of polynomials over R (see Theorem A2.24), and if $\circ$ replaces $\cdot$, where $(p \circ q)(x) = p(q(x))$, then $(R[x], +, \circ)$ satisfies properties (A1)–(A4), (M2)–(M3), and (DR).

Structures $(M(G), \oplus, \circ)$, $(\mathscr{C}(\mathbb{R}), \oplus, \circ)$, $(G, +, \cdot)$, and $(R[x], +, \circ)$ are all *nearrings.*

Definitions 6.1. A *nearring* is an algebraic structure $(N, +, \cdot)$, where $(N, +)$ is a group, where $(N, \cdot)$ is a semigroup, and where (DL) or (DR) in §4 is valid. If (DL) is valid, then the nearring $(N, +, \cdot)$ is a *left* nearring. If (DR) is valid, then the nearring $(N, +, \cdot)$ is a *right* nearring. If (DL) and (DR) are both valid, then the nearring $(N, +, \cdot)$ is a *distributive* nearring. If (A1) is valid, and only one of (DL) or (DR) is valid, then the nearring $(N, +, \cdot)$ is an *abelian* nearring. If (A1), and both (DL) and (DR) are valid, then the nearring $(N, +, \cdot)$ is a *ring*, of course.

Although we are not interested in nearrings, in general, the subject is a very interesting one. Nearly all of the theory in this subject has been developed in the past 55 years; the theory is becoming richer continually, both by developing established areas of research and by opening new frontiers. The reader is referred to [C] for an introduction to the subject.

Our interest in nearrings will be focused on integral planar nearrings (in short, IP-nearrings), where there are numerous applications to geometry, combinatorics, coding theory, and cryptography. All of our integral planar nearrings will be left nearrings, although one can develop analogous results for right nearrings.

To better understand the definition of an IP-nearring, we introduce another technique for constructing left nearrings. Start with any group $(G,+)$ with $|G| > 1$. Let $S \subseteq G \setminus \{0\}$ and define $\cdot_S$ on G by

$$a \cdot_S b = \begin{cases} b, & \text{if } a \in S; \\ 0, & \text{otherwise.} \end{cases}$$

Then $(G,+,\cdot_S)$ is a left nearring.

For any nearring $(N,+,\cdot)$, define a relation $=_m$ on N by $a =_m b$ if and only if $ax = bx$ for all $x \in N$. It is easy to show that $=_m$ is an equivalence relation on N. If $a =_m b$, we say that a and b are *equivalent multipliers*.

Definition 6.2. A nearring $(N,+,\cdot)$ is an *integral planar nearring* (abbreviated *IP-nearring*) if (1) there are at least three equivalence classes for $=_m$, that is $|N/{=_m}| \geq 3$; (2) if $a,b,c \in N$ and $a \neq_m b$, then the equation

$$ax = bx + c$$

has a unique solution; and (3) $a =_m 0$ if and only if $a = 0$.

For each of the nearrings $(G,+,\cdot_S)$, where $|G| \geq 2$ and $S = G \setminus \{0\}$, then $|G/{=_m}| = 2$, and for $a,b,c \in N$ and $a \neq_m b$, the equation $ax = bx + c$ has a unique solution. We don't want any of these nearrings $(G,+,\cdot_S)$ to be an IP-nearring, and that is why we insist on $|N/{=_m}| \geq 3$ in Definition 6.2

The time has arrived to have some good examples of IP-nearrings. It would be difficult to overemphasize the importance of the first two examples, given next, of IP-nearrings.

Example 6.3. Let $(\mathbb{C},+,\cdot)$ be the field of complex numbers. Define $*$ and $\circ$ on $\mathbb{C}$ as follows. For $a,b \in \mathbb{C}$,

$$a * b = |a| \cdot b, \quad \text{and} \quad a \circ b = \begin{cases} 0, & \text{if } a = 0; \\ (a/|a|) \cdot b, & \text{if } a \neq 0. \end{cases}$$

It is easy to see that each of $(\mathbb{C}, +, *)$ and $(\mathbb{C}, +, \circ)$ is an IP-nearring.

The importance of $(\mathbb{C}, +, *)$ and $(\mathbb{C}, +, \circ)$ are that (1) with $(\mathbb{C}, +, *)$, we can describe all the rays in the plane in a new way, and (2) with $(\mathbb{C}, +, \circ)$, we can describe the circles in the plane in a new way. Let $\mathbb{C} * a = \{c * a \mid c \in \mathbb{C}\}$ and then $\mathbb{C} * a + b = \{c * a + b \mid c \in \mathbb{C}\}$. Then $\mathbb{C} * a$ is the ray from 0 through a, and $\mathbb{C} * a + b$ is the ray from b through $a + b$. Let $\mathbb{C}^* \circ a = \{c \circ a \mid c \in \mathbb{C}^* = \mathbb{C} \setminus \{0\}\}$ and then $\mathbb{C}^* \circ a + b = \{c \circ a + b \mid c \in \mathbb{C}^*\}$. So, $\mathbb{C}^* \circ a$ is the circle with center 0 and radius $|a|$, and $\mathbb{C}^* \circ a + b$ is the circle with center b and radius $|a|$. Therefore,

$$\mathscr{B} = \{\mathbb{C} * a + b \mid a, b \in \mathbb{C},\ a \neq 0\}$$

is the collection of all the usual rays in the complex plane, and

$$\mathscr{B}^* = \{\mathbb{C}^* \circ a + b \mid a, b \in \mathbb{C},\ a \neq 0\}$$

is the collection of all the usual circles in the complex plane.

It is helpful to understand the basic structure of IP-nearrings. Basic to this is the fact that $0 \cdot b = b \cdot 0 = 0$ for all b in an IP-nearring $(N, +, \cdot)$. That is, an IP-nearring is *zero-symmetric*.

Proposition 6.4. *For an IP-nearring $(N, +, \cdot)$, we have $0 \cdot b = b \cdot 0 = 0$ for each $b \in N$.*

Proof. For an arbitrary element a in the (left) nearring $(N, +, \cdot)$, we have

$$a \cdot 0 = a \cdot (0 + 0) = (a \cdot 0) + (a \cdot 0).$$

Hence, $a \cdot 0 = 0$. If $a \neq_m 0$, then $ax = 0x + 0$ has a unique solution if $(N, +, \cdot)$ is an IP-nearring. For any $b \in N$, we have

$$a0 = 0 = 0 \cdot 0 + 0, \text{ and } a(0b) = (a0)b = 0b = (0 \cdot 0) \cdot b = 0(0b) + 0.$$

Hence, 0 and $0b$ are solutions to the equation $ax = 0x + 0$. Uniqueness of the solution yields $0b = 0$. From the first line of the proof, we have $b0 = 0$. Hence, $0 \cdot b = b \cdot 0 = 0$ for each $b \in N$. □

With Proposition 6.4, the equation $ax = 0x + c$ becomes $ax = c$. In particular, if $a \neq_m 0$, then $ax = a$ has a unique solution, which we denote by 1_a. (This is a *right identity* for a.) It may be that $a \neq b$ but yet $1_a = 1_b$. For $a \neq_m 0$, define

$$B_a = \{b \in N^* = N \setminus \{0\} \mid b1_a = b\},$$

and let $L = \{1_a \mid a \in N^*\}$. We are now ready to prove a theorem of fundamental importance for IP-nearrings.

Theorem 6.5 (MAIN STRUCTURE THEOREM). *For each and every IP-nearring $(N, +, \cdot)$, the following are true.*

(a) $N^* = N \setminus \{0\} = \cup_{a \in N^*} B_a$.

(b) $N^* B_a = B_a$ *for any* $a \in N^*$.

(c) *For each* $a \in N^*$, *we have that* $(B_a, \cdot)$ *is a group.*

(d) $\{B_a \mid a \in N^*\}$ *is pairwise disjoint.*

(e) *For* $a, c \in N^*$, *the map* $\phi : B_a \to B_c$ *defined by* $\phi(x) = x1_c$ *is a group isomorphism.*

(f) *Each* 1_a, $a \in N^*$, *is a left identity, i.e.,* $1_a x = x$ *for all* $x \in N$.

Proof. For $a \in N^*$, $a1_a = a$, and so $a \in B_a$. This gives us (a) and the fact that $B_a \neq \varnothing$. For $x \in N^*$ and $b \in B_a$, we have $(xb)1_a = x(b1_a) = xb$, and so $xb \in B_a$. From this we conclude that $N^* B_a \subseteq B_a$ and that B_a is closed with respect to multiplication $\cdot$. Hence, each $(B_a, \cdot)$ is a semigroup.

For $a \in N^*$, $a = a1_a$, so

$$a = a1_a = (a1_a)1_a = a(1_a 1_a).$$

The equation $ax = a$ has 1_a and $1_a 1_a$ as solutions, so $1_a = 1_a 1_a$, by uniqueness. Now we have $1_a \in B_a$, and 1_a is a right identity for B_a, that is, $x1_a = x$ for every $x \in B_a$.

Suppose a' is the unique solution to $ax = 1_a$. Then

$$a(a'1_a) = (aa')1_a = 1_a 1_a = 1_a,$$

and so $a'1_a$ and a' are solutions to $ax = 1_a$. Hence, $a'1_a = a'$, and we get $a' \in B_a$. We could justifiably call a' a *right inverse* of a. But what about $b \in B_a$? Does it also have a right inverse? If so, is it in B_a, also? The equation $bx = 1_b$ has a unique solution b', and $b' \in B_b$, but could we have $b \in B_a \cap B_b$ with $B_a \neq B_b$? If $c \in B_a \cap B_b$, then $c1_a = c = c1_b$. So, 1_a and 1_b are solutions to $cx = c$, giving us $1_a = 1_b$ and $B_a = B_b$. Thus, $bb' = 1_b = 1_a$, which tells us that each $b \in B_a$ has a right inverse $b' \in B_a$. It also tells us that $B_a \cap B_b = \varnothing$ or $B_a = B_b$. This gives us (d).

Each $(B_a, \cdot)$ is a semigroup with a right identity 1_a, and each $b \in B_a$ has a right inverse in B_a, that is, there is a $b' \in B_a$ such that $bb' = 1_a$. Such semigroups are groups, which the following lemma will show.

Lemma. *Let $(S, \cdot)$ be a semigroup with a right identity e. Suppose each $s \in S$ has a right inverse $s' \in S$. Then $(S, \cdot)$ is a group.*

Proof. If $b \in S$ and $bb = b$, then $b = be = b(bb') = (bb)b' = bb' = e$. But $(s's)(s's) = s'(ss')s = s'es = s's$, so $s's = e$. Consequently, $s's = ss' = e$ for each $s \in S$. Also, $es = (ss')s = s(s's) = se = s$, and so $es = se = s$ for each $s \in S$. Hence, $(S, \cdot)$ is a group. □

Applying this lemma to each $(B_a, \cdot)$, we have (c) and then (b).

As a consequence of (b), the map $\phi : B_a \to B_c$ defined by $\phi(x) = x1_c$ is well defined. For $x, y \in B_a$,

$$\phi(xy) = (xy)1_c = x(y1_c) = x[1_c(y1_c)] = (x1_c)(y1_c) = \phi(x)\phi(y),$$

and so ϕ is a group homomorphism.

For $b \in B_a$, $b1_a = b$, and so $b(1_a 1_c) = (b1_a)1_c = b1_c$. The equation $bx = b1_c$ has $1_a 1_c$ and 1_c as solutions, and so $1_a 1_c = 1_c$.

If $x1_c = y1_c$, $x, y \in B_a$, then

$$y = y1_a = y(1_c 1_a) = (y1_c)1_a = (x1_c)1_a = x(1_c 1_a) = x1_a = x.$$

This makes ϕ a monomorphism.

For $y \in B_c$, we get $y1_a \in B_a$ and

$$\phi(y1_a) = (y1_a)1_c = y(1_a 1_c) = y1_c = y.$$

Hence, ϕ is an epimorphism. Altogether, ϕ is an isomorphism. We have (e).

Consider the equation $1_a x = 1_a d$. It certainly has d as a solution. But $1_a(1_a d) = (1_a 1_a)d = 1_a d$, and so $1_a d$ is also a solution. Hence, $1_a d = d$. Since $d \in N^*$ is arbitrary, and since $1_a 0 = 0$, we have that 1_a is a left identity for all of N. This gives us (f) and completes the proof of Theorem 6.5. □

It would be natural, and simpler, to define an IP-nearring as a nearring $(N,+,\cdot)$ for which $a \neq b$ and $a,b,c \in N$ imply that the equation $ax = bx + c$ has a unique solution. Doing so would just make the nearring $(N,+,\cdot)$ a planar nearfield.

Theorem 6.6. *Let $(N,+,\cdot)$ be a nearring. Suppose $a,b,c \in N$ and $a \neq b$ imply that the equation*

$$ax = bx + c$$

has a unique solution. Also, suppose $|N| \geq 2$. Then $(N,+,\cdot)$ is a planar nearfield.

Proof. Suppose $a \neq b$, but $a =_m b$. Then $ax = bx$ for all $x \in N$, and so for $c \in N^* = N \setminus \{0\}$, the equation $ax = bx + c$ has no solution. So, if $a \neq b$, then $a \neq_m b$. If $|N| \geq 3$, then $|N/{=_m}| \geq 3$, and so $(N,+,\cdot)$ is an IP-nearring. Hence, $1_a = 1_b$ for arbitrary $a,b \in N^*$, and so $N^* = B_a$. This makes $(N^*,\cdot)$ a group, and so $(N,+,\cdot)$ is a nearfield. The nearfield $(N,+,\cdot)$ is planar by hypothesis. If $|N| = 2$, then $(N,+)$ is isomorphic to $(\mathbb{Z}_2,+)$, and we take $N = \{0,1\}$. Since $0 \neq 1$, we must have that $(N,+,\cdot)$ is the field $(\mathbb{Z}_2,+,\cdot)$, which is a planar nearfield. □

Let's apply Theorem 6.5 to the two classical examples of IP-nearrings found in Example 6.3. First, with $(\mathbb{C}, +, *)$, we have $a =_m b$ if and only if $|a| = |b|$. So the equivalence classes of $=_m$ in $\mathbb{C}^*$ are exactly the circles with center 0. The set $L = \{1_a \mid a \in N^*\}$ is the unit circle $|z| = 1$. For $a \in \mathbb{C}^*$, we have $1_a = a/|a|$, and so B_a is the ray from 0 through a and 1_a, with 0 removed. Notice how $\mathbb{C}^*$ is the disjoint union of all these groups $(B_a, *)$, that is, the B_a's partition $\mathbb{C}^*$.

Next, consider $(\mathbb{C}, +, \circ)$. We have $a =_m b$ if and only if $a/|a| = b/|b|$. Notice now that the equivalence classes of $=_m$ are exactly the B_a for $(\mathbb{C}, +, *)$. The set L is the positive real axis. For $a \in \mathbb{C}^*$, $1_a = |a|$, and so B_a is the circle with center 0 and radius $|a|$. Again, notice how $\mathbb{C}^*$ is the disjoint union of all the groups $(B_a, \circ)$.

It is good if you sense that $(\mathbb{C}, +, *)$ and $(\mathbb{C}, +, \circ)$ are closely related. Their relationship is the model, the prototype, of a *double planar nearring* $(\mathbb{C}, +, *, \circ)$ to be discussed in §20.

7 How to make an integral planar nearring

Making fields is not too difficult; making division rings is not so easy; making nearfields is rather complicated; and making integral planar nearrings is really quite easy. The basic ideas have already been presented in §5.2.

Let's start with an IP-nearring $(N, +, \cdot)$. For each $a \in N^* = N \setminus \{0\}$, we define $a^\bullet : N \to N$ by $a^\bullet(x) = ax$ for every $x \in N$. Just as was done in §5.2, each $a^\bullet$ is an automorphism of the group $(N, +)$. However, if $(N, +, \cdot)$ is not a nearfield, then we can have $a \neq b$, but yet $a =_m b$. In fact, $a^\bullet = b^\bullet$ if and only if $a =_m b$. If $b \in B_a$, then $b1_c \in B_c$ and $b^\bullet = (b1_c)^\bullet$, since $(b1_c)(x) = b(1_c x) = bx$ for every $x \in N$. If $b, c \in B_a$ and $b^\bullet = c^\bullet$, then $b^\bullet(1_a) = c^\bullet(1_a)$, so $b = b1_a = c1_a = c$. The equivalence relation $=_m$ partitions N^* into equivalence classes, and we have just seen that an equivalence class $[b]_{=_m}$ has a representative in each and every group B_a, $a \in N^*$.

As with nearfields, $\Phi = \{a^\bullet \mid a \in N^*\}$ is a subgroup of $\text{Aut}(N,+)$. In fact, each B_a is isomorphic to Φ with isomorphism defined by $\phi_a : B_a \to \Phi$ where $\phi_a(b) = b^\bullet$ for each $b \in B_a$.

To make a nearfield planar, we needed that each $-b^\bullet + 1_N$, $b^\bullet \neq 1_N$, be a bijection of N. This was to get a solution to each equation

$$x = bx + c.$$

To get uniqueness, we needed that each $b^\bullet$, $b \neq 1_N$, be a fixed point free automorphism. The same is true for IP-nearrings, and the proof is the same.

We have just outlined the proof of

Theorem 7.1. *To each integral planar nearring $(N,+,\cdot)$, there is the nontrivial group*

$$\Phi = \{a^\bullet \mid a \in N^*\}$$

of fixed point free automorphisms of the group $(N,+)$. If $b^\bullet \in \Phi$ and $b^\bullet \neq 1_N$, then $-b^\bullet + 1_N$ is a bijection. Moreover, if $c \in N^*$, then Φ and B_c are isomorphic groups.

Let's start now with a group $(N,+)$ having a nontrivial group of automorphisms $(\Phi,\circ)$. That is, Φ is a nontrivial subgroup of $\text{Aut}(N,+)$. Suppose that Φ is a nontrivial group of fixed point free automorphisms, and that $-\phi + 1_N$ is bijective whenever $\phi \in \Phi$ and $\phi \neq 1_N$. We call the pair (N,Φ) a *Ferrero pair*. For each $a \in N$, we have the *orbit* $\Phi(a) = \{\phi(a) \mid \phi \in \Phi\}$ *of* Φ *at* a. The orbit $\Phi(0) = \{0\}$ is the *trivial orbit* of Φ. There are a few important basic facts about a Ferrero pair that we should formalize.

Theorem 7.2. *Let (N,Φ) be a Ferrero pair. Then the following are true.*

(a) *For each $a \in N^* = N \setminus \{0\}$, we have $|\Phi(a)| = |\Phi|$.*

(b) *If $b \in \Phi(a)$, then $\Phi(a) = \Phi(b)$, and conversely.*

(c) *The orbits of Φ form a partition of N.*

Proof. Define $f : \Phi \to \Phi(a)$ by $f(\phi) = \phi(a)$. Easily, f is surjective. If $f(\phi) = f(\mu)$, then $\phi(a) = \mu(a)$, and so $\mu^{-1} \circ \phi(a) = a$. But $\mu^{-1} \circ \phi \in \Phi$, a group of fixed point free automorphisms of $(N, +)$, and $a \in N^*$, so $\mu^{-1} \circ \phi = 1_N$. Hence, $\mu = \phi$, and now we have that f is injective. This gives us (a).

If $b \in \Phi(a)$, then $b = \phi(a)$ for some $\phi \in \Phi$. Hence, $\Phi(b) = \Phi(\phi(a)) = \Phi(a)$, since Φ is a group. If $\Phi(a) = \Phi(b)$, then $b = 1_N(b) \in \Phi(b) = \Phi(a)$. Now we have (b).

An $a \in N^*$ belongs to $\Phi(a)$, and $0 \notin \Phi(b)$ for each $b \in N^*$. Hence, $N^* = \cup_{a \in N^*}$. If $c \in \Phi(a) \cap \Phi(b)$, then $\Phi(a) = \Phi(c) = \Phi(b)$, from (b). Hence, $\Phi(a) = \Phi(b)$, or else $\Phi(a) \cap \Phi(b) = \varnothing$. Now we have (c) and a completion to the proof of Theorem 7.2. □

7.3. The Ferrero Planar Nearring Factory. We are now ready to construct an IP-nearring from a Ferrero pair. The process is called *The Ferrero Planar Nearring Factory*. Mastery of the process gives you full authority to construct IP-nearrings of the highest quality.

With a Ferrero pair (N, Φ), let Φ act on N^* by computing all the orbits $\Phi(a)$, $a \in N^*$. For each orbit $\Phi(a)$, choose a representative $e_i \in \Phi(a)$, so $\Phi(e_i) = \Phi(a)$, and $\mathscr{C} = \{\Phi(e_i) \mid i \in I\}$ is the collection of all orbits of Φ at points of N^*, and where I is an index set. That is, for each $i \in I$, there is a unique orbit $\Phi(e_i) \in \mathscr{C}$, and if $a \in N^*$, then $\Phi(a) = \Phi(e_j)$ for some $j \in I$. From (c) of Theorem 7.2, we know that $\mathscr{C}$ is a partition of N^*.

If $a \in N^*$, there is a unique $\phi_a \in \Phi$ such that $\phi_a(e_i) = a$. Define $\cdot$ on N by

$$a \cdot b = \begin{cases} 0, & \text{if } a = 0; \\ \phi_a(b), & \text{if } a \neq 0. \end{cases}$$

We shall see that $(N, +, \cdot)$ is an integral planar nearring.

First, we will demonstrate that $\cdot$ is left distributive over $+$. Let $a, b, c \in N$. If $a = 0$, then

$$a \cdot (b + c) = 0 \cdot (b + c) = 0 = 0 + 0 = (0 \cdot b) + (0 \cdot c) = (a \cdot b) + (a \cdot c).$$

If $a \neq 0$, then

$$a \cdot (b+c) = \phi_a(b+c) = \phi_a(b) + \phi_a(c) = (a \cdot b) + (a \cdot c).$$

So, $\cdot$ is left distributive over $+$.

For associativity of $\cdot$, again take $a, b, c \in N$. There are three cases to consider. First, suppose $a = 0$. Then

$$a \cdot (b \cdot c) = 0 = 0 \cdot c = (a \cdot b) \cdot c.$$

Second, suppose $b = 0$. Then

$$a \cdot (b \cdot c) = a \cdot 0 = 0 = 0 \cdot c = (a \cdot 0) \cdot c = (a \cdot b) \cdot c.$$

Finally, suppose $0 \notin \{a, b\}$. Then

$$a \cdot (b \cdot c) = \phi_a[\phi_b(c)] = \phi_a \circ \phi_b(c),$$

and

$$(a \cdot b) \cdot c = [\phi_a(b)] \cdot c = \phi_{\phi_a(b)}(c).$$

Now $\phi_a(b) \in N^*$, and so $\phi_a(b) \in \Phi(e_j)$ for a unique $j \in I$, and there is a unique $\phi_{\phi_a(b)} \in \Phi$ such that $\phi_{\phi_a(b)}(e_j) = \phi_a(b)$. But if $\phi_a(b) \in \Phi(e_j)$, then $b \in \Phi(e_j)$, since

$$\phi_a^{-1} \in \Phi \text{ and } \phi_a^{-1}(\phi_a(b)) = b \in \Phi(\phi_a(b)) = \Phi(e_j).$$

Also,

$$\phi_a \circ \phi_b(e_j) = \phi_a(\phi_b(e_j)) = \phi_a(b).$$

Hence, $\phi_{\phi_a(b)} = \phi_a \circ \phi_b$. Consequently,

$$\phi_a \circ \phi_b(c) = \phi_{\phi_a(b)}(c),$$

so $a \cdot (b \cdot c) = (a \cdot b) \cdot c$. With the associativity of $\cdot$, we have that $(N, +, \cdot)$ is a nearring.

If $a \in N^*$, then $a \neq_m 0$. If $a, b \in N^*$, then $a =_m b$ if and only if $\phi_a = \phi_b$. Since $|\Phi| \geq 2$, we have $|N/{=_m}| \geq 3$.

We now focus on the planarity equation

$$ax = bx + c, \tag{7:1}$$

with $a \neq_m b$. Again, there are various cases to consider, namely (i) $a = 0$, (ii) $b = 0$, and (iii) $0 \notin \{a, b\}$.

If $a = 0$, then (7:1) becomes $0 = bx + c$, or $\phi_b(x) = -c$. Since ϕ_b is an automorphism, $x = \phi_b^{-1}(-c)$ is the only solution. If $b = 0$, then (7:1) becomes $ax = c$, or $\phi_a(x) = c$. Similarly, $\phi_a^{-1}(c)$ is the only solution.

If $0 \notin \{a, b\}$, then (7:1) becomes $\phi_a(x) = \phi_b(x) + c$. Applying ϕ_a^{-1} to the equation, we get $x = \phi_a^{-1} \circ \phi_b(x) + \phi_a^{-1}(c)$, which is equivalent to

$$-\phi_a^{-1} \circ \phi_b(x) + x = \phi_a^{-1}(c)$$

or

$$(-\phi_a^{-1} \circ \phi_b + 1_N)(x) = \phi_a^{-1}(c).$$

Since $\phi_a \neq \phi_b$, we have $\phi_a^{-1} \circ \phi_b \neq 1_N$, and so $-\phi_a^{-1} \circ \phi_b + 1_N$ is a bijection, because (N, Φ) is a Ferrero pair. This means that there is a unique value for x such that

$$(-\phi_a^{-1} \circ \phi_b + 1_N)(x) = \phi_a^{-1}(c),$$

which in turn means that $ax = bx + c$ has a unique solution. Hence, $(N, +, \cdot)$ is an integral planar nearring. □

Let's examine the two classical examples of Example 6.3 in light of what we have done here. For $(\mathbb{C}, +, *)$, we have $\Phi = \{a^\bullet \mid a \in \mathbb{C}^*\}$, and since $a^\bullet(x) = a * x = |a| \cdot x$, we see that Φ is isomorphic to the multiplicative group of positive real numbers. Had we started with the Ferrero pair $(\mathbb{C}, \Phi)$, then the orbit $\Phi(a) = \{ra \mid r \in \mathbb{R}_+\}$, which is all positive scalar multiples of a. We could take $a/|a|$ as a representative, and so L is the unit circle $T = \{e^{i\theta} \mid \theta \in \mathbb{R}\}$.

For $(\mathbb{C},+,\circ)$, we have $\Gamma = \{a^\bullet \mid a \in \mathbb{C}^*\}$, with $a^\bullet(x) = a \circ x = (a/|a|) \cdot x$, and so Γ is isomorphic to the unit circle group $T = \{e^{i\theta} \mid \theta \in \mathbb{R}\}$. Had we started with the Ferrero pair $(\mathbb{C},\Gamma)$, then the orbits $\Gamma(a) = \{e^{i\theta}a \mid \theta \in \mathbb{R}\}$, which is the circle with center 0 and radius $|a|$. We could take $|a|$ as the representative of $\Gamma(a)$.

At the beginning of this section, we promised the reader that constructing IP-nearrings was really quite easy. All one needs is a Ferrero pair. And where does one get a Ferrero pair? The easiest way to get a Ferrero pair is from another Ferrero pair, and there are lots of these hanging around. For example, if (N,Φ) is a Ferrero pair, and if Γ is a nontrivial subgroup of Φ, then (N,Γ) is also a Ferrero pair. This method has been very useful in constructing examples.

Remember, each field, each division ring, and each planar nearfield is an IP-nearring. The multiplicative group $(K^*,\cdot)$ of any of these structures $(K,+,\cdot)$ has $\Phi = \{a^\bullet \mid a \in K^*\}$, which gives a Ferrero pair (K,Φ). For $(\mathbb{C},+,*)$ of Example 6.3, we essentially take $\Phi = \mathbb{R}_+$, the positive real numbers. For $(\mathbb{C},+,\circ)$ of Example 6.3, we essentially take $\Phi = T$, the unit circle.

We close this section with three examples of how to use The Ferrero Planar Nearring Factory.

Example 7.4. Our first example comes from a Ferrero pair (F,Φ), where $(F,+,\cdot)$ is the field of order 9 illustrated in §5. Looking at the multiplication table for $(F,+,\cdot)$, one sees that $\{a,\ 2,\ 2a,\ 1\}$ is a subgroup of F^*, and so we take $\Phi = \{a^\bullet,\ 2^\bullet,\ (2a)^\bullet,\ 1^\bullet\}$. The two nontrivial orbits of Φ are

$$\Phi(1) = \{a,2,2a,1\} \quad \text{and} \quad \Phi(1+a) = \{2+a,2+2a,1+2a,1+a\}.$$

Let 1 and $1+a$ be the representatives of these two orbits. With the multiplication $*$ on F defined as in **7.3**, we get an IP-nearring $(F,+,*)$, where $*$ has the multiplication table as shown in Table 4.

$*$	0	1	2	a	$1+a$	$2+a$	$2a$	$1+2a$	$2+2a$
0	0	0	0	0	0	0	0	0	0
1	0	1	2	a	$1+a$	$2+a$	$2a$	$1+2a$	$2+2a$
2	0	2	1	$2a$	$2+2a$	$1+2a$	a	$2+a$	$1+a$
a	0	a	$2a$	2	$2+a$	$2+2a$	1	$1+a$	$1+2a$
$1+a$	0	1	2	a	$1+a$	$2+a$	$2a$	$1+2a$	$2+2a$
$2+a$	0	a	$2a$	2	$2+a$	$2+2a$	1	$1+a$	$1+2a$
$2a$	0	$2a$	a	1	$1+2a$	$1+a$	2	$2+2a$	$2+a$
$1+2a$	0	$2a$	a	1	$1+2a$	$1+a$	2	$2+2a$	$2+a$
$2+2a$	0	2	1	$2a$	$2+2a$	$1+2a$	a	$2+a$	$1+a$

Table 4: The Multiplication Table for the IP-nearring $(F,+,*)$.

Exercise 7.5. For the IP-nearring $(F,+,*)$ above, construct the multiplication tables for the groups B_1 and B_{1+a}.

Example 7.6. Our second example comes from a Ferrero pair $(\mathbb{Z}_{13},\Phi)$, where $\mathbb{Z}_{13}$ is the field of 13 elements. One can easily verify that $\{5,\ 12,\ 8,\ 1\}$ is a subgroup of $\mathbb{Z}_{13}^*$, and this gives us $\Phi = \{5^\bullet,\ 12^\bullet,\ 8^\bullet,\ 1^\bullet\}$. The three nontrivial orbits of Φ are

$$\Phi(1) = \{5,12,8,1\},\quad \Phi(2) = \{10,11,3,2\},\quad \text{and}\quad \Phi(4) = \{7,9,6,4\}.$$

Let 1, 2, and 4 be representatives of these orbits. With The Ferrero Planar Nearring Factory we get an IP-nearring $(\mathbb{Z}_{13},+,*)$ where $*$ has the multiplication table as shown in Table 5.

Exercise 7.7. For the IP-nearring $(\mathbb{Z}_{13},+,*)$ above, construct the multiplication tables for the groups B_1, B_2, and B_4.

This next theorem shows that The Ferrero Planar Nearring Factory can be used for Ferrero pairs which do not come from fields.

Theorem 7.8. *Let $(A,+)$ be any abelian group without an element of order 2. Also, suppose each equation $2x = b$, $b \in A$, has a unique solution for x in A. Then $\phi : A \to A$,*

*	0	1	2	3	4	5	6	7	8	9	10	11	12
0	0	0	0	0	0	0	0	0	0	0	0	0	0
1	0	1	2	3	4	5	6	7	8	9	10	11	12
2	0	1	2	3	4	5	6	7	8	9	10	11	12
3	0	8	3	11	6	1	9	4	12	7	2	10	5
4	0	1	2	3	4	5	6	7	8	9	10	11	12
5	0	5	10	2	7	12	4	9	1	6	11	3	8
6	0	8	3	11	6	1	9	4	12	7	2	10	5
7	0	5	10	2	7	12	4	9	1	6	11	3	8
8	0	8	3	11	6	1	9	4	12	7	2	10	5
9	0	12	11	10	9	8	7	6	5	4	3	2	1
10	0	5	10	2	7	12	4	9	1	6	11	3	8
11	0	12	11	10	9	8	7	6	5	4	3	2	1
12	0	12	11	10	9	8	7	6	5	4	3	2	1

Table 5: The Multiplication Table for the IP-nearring $(\mathbb{Z}_{13},+,*)$.

defined by $\phi(a) = -a$, *is a fixed point free automorphism and* $\phi^2 = 1_A$. *Hence,* (A,Φ) *is a Ferrero pair where* $\Phi = \{\phi, 1_A\}$.

Proof. Certainly,

$$\phi(a+b) = -(a+b) = (-b)+(-a) = (-a)+(-b) = \phi(a)+\phi(b).$$

If $\phi(a) = \phi(b)$, then $-a = -b$, and so $a = b$. Finally, $\phi(-a) = -(-a) = a$. Hence, ϕ is an automorphism, and $\phi^2(a) = \phi(\phi(a)) = \phi(-a) = -(-a) - a$. So $\phi^2 = 1_A$.

If $\phi(a) = a$, then $-a = a$, and so $a+a = 0$. Since A has no element of order 2, we have $a = 0$, and so ϕ is fixed point free.

Finally, consider the mapping $-\phi + 1_A$. If $(-\phi + 1_A)(x) = (-\phi + 1_A)(y)$, then $x+x = y+y$. From this follows $(-y+x)+(-y+x) = 0$, and since A has no element of order 2, we get $-y+x = 0$ and $x = y$. This makes $-\phi + 1_A$ an injection. For $b \in A$, consider the equation $b = (-\phi + 1_A)(x) = x+x = 2x$. By hypothesis, this equation has a solution for x in A, so $-\phi + 1_A$ is a surjection,

$*$	0	1	2	3	4	5	6	7	8
0	0	0	0	0	0	0	0	0	0
1	0	1	2	3	4	5	6	7	8
2	0	1	2	3	4	5	6	7	8
3	0	1	2	3	4	5	6	7	8
4	0	1	2	3	4	5	6	7	8
5	0	8	7	6	5	4	3	2	1
6	0	8	7	6	5	4	3	2	1
7	0	8	7	6	5	4	3	2	1
8	0	8	7	6	5	4	3	2	1

Table 6: The Multiplication Table for the IP-nearring $(\mathbb{Z}_9, +, *)$.

and therefore a bijection. Now we are assured that (A, Φ) is a Ferrero pair. □

Example 7.9. Consider the group $(\mathbb{Z}_9, +)$ from the ring $(\mathbb{Z}_9, +, \cdot)$. $\mathbb{Z}_9$ has no element of order 2, and since $\phi(a) = -a$ is injective, we have that $-\phi + 1_{\mathbb{Z}_9}$ is a bijection. So, for $\Phi = \{\phi, 1_{\mathbb{Z}_9}\}$, we have that $(\mathbb{Z}_9, \Phi)$ is a Ferrero pair. The nontrivial orbits of Φ are

$$\Phi(1) = \{1,8\},\ \Phi(2) = \{2,7\},\ \Phi(3) = \{3,6\},\ \text{and}\ \Phi(4) = \{4,5\}.$$

Let 1, 2, 3, and 4 be the representatives of these orbits. Applying The Factory, one gets an IP-nearring $(\mathbb{Z}_9, +, *)$ where $*$ has the multiplication table shown in Table 6.

Exercise 7.10. Apply Theorem 7.8 to a suitable abelian group of your choosing to construct an IP-nearring.

Exercise 7.11. Choose a finite field $(K, +, \cdot)$, and then choose a nontrivial subgroup Φ' of K^*. With the resulting Ferrero pair (F, Φ), construct a multiplication table for an IP-nearring $(F, +, *)$.

Chapter III

Planar nearrings and combinatorics

There are numerous applications of finite IP-nearrings to combinatorics. Most of these have been discovered in recent years. There are probably many more applications awaiting discovery, and so the authors invite the reader to make his/her own discoveries and contributions. In this chapter, we introduce the reader to many of the applications. Some might argue that a lot of the next chapter, Chapter IV, also falls under the heading of combinatorics. This chapter, together with Chapter IV, will introduce the reader to most of the combinatorial applications of IP-nearrings known at this time.

8 Balanced incomplete block designs, part 1

Throughout §1, we introduced various balanced incomplete block designs (BIBD). We noted that finite affine planes, finite projective planes, and finite Möbius planes are BIBDs, and we have provided ways to construct affine planes, projective planes, and Möbius planes, and so we have techniques for constructing special types of BIBDs. Actually, nearly everything we do within the bounds of these notes is related to BIBDs, and so we present here another method for their construction.

The method of constructing BIBDs we present here is modeled and motivated by $(\mathbb{C},+,\circ)$ from Example 6.3. For an arbitrary finite IP-nearring $(N,+,\cdot)$, we keep an eye on the geometry of $(\mathbb{C},+,\circ)$ and hope to get some geometry from $(N,+,\cdot)$. So, we construct the blocks

$$N^*a+b=\{na+b \mid n\in N^*=N\setminus\{0\}\},$$

where $a,b\in N$, and $a\neq 0$. We consider all the blocks

$$\mathscr{B}^*=\{N^*a+b \mid a,b\in N,\ a\neq 0\}.$$

It is our goal in this section to prove that $(N,\mathscr{B}^*)$ is a BIBD.

Also in §1, we strongly suggested that there was a relationship among the parameters v, b, r, k, and λ of a BIBD. This relationship is of fundamental importance. It was stated and proved in Theorem 2.15, and is restated here for emphasis and reference.

Theorem 8.1. *Let $(V,\mathscr{B})$ be a BIBD with parameters v, b, r, k, and λ. Then*

$$vr=bk \text{ and } \lambda(v-1)=r(k-1).$$

We will need to count the number of blocks $N^*a+b\in\mathscr{B}^*$. To know when we have duplication, we use

Theorem 8.2. *Let $(N,+,\cdot)$ be a finite IP-nearring. For $0\notin\{a,c\}$, $N^*a+b=N^*c+d$ if and only if $N^*a=N^*c$ and $b=d$.*

Proof. If $N^*a=N^*c$, and $b=d$, then $N^*a+b=N^*a+d=N^*c+d$. Now, that was easy.

If $N^*a+b=N^*c+d$, then $N^*a=N^*c+(d-b)$. If we can prove that this forces $d-b=0$, then we will have $N^*a=N^*c$ and our desired conclusion.

Suppose $N^*a = N^*b + c$ with $0 \notin \{a,b\}$. If we can conclude that $c = 0$, then our proof will be finished. Suppose $c \neq 0$. For $u \in N^*$, we have $u(N^*a) = N^*a$, and $u(N^*b + c) = N^*b + uc$. Hence, $N^*a = N^*b + uc$. Let $1_c = u_1, u_2, \ldots, u_k$ be distinct representatives of $N^*/=_m$. Since $k \geq 2$, we obtain $N^*b + c = N^*b + u_2c$, because each is equal to N^*a. With $u_2 \neq 1_c$, we have $N^*b = N^*b + (u_2c - c)$ and $u_2c - c \neq 0$. (Why?) Let $t = u_2c - c$, and so we have $N^*b = N^*b + t$ with $t \neq 0$.

Multiplying $N^*b = N^*b + t$ on the left by u_i, $1 \leq i \leq k$, we get $N^*b = N^*b + u_it$ for each i, $1 \leq k$. Hence, $N^*b = N^*b + N^*t$. Since N is finite, if follows that $|N^*b| = |N^*t| = |b + N^*t|$. But $b + N^*t \subseteq N^*b$, so we conclude that $N^*b = b + N^*t$. Therefore, $b = b + u_it$ for some i, and so $u_it = 0$, which cannot be, since $t \neq 0$. So, assuming that $t \neq 0$ leads to a contradiction. This means that we cannot have $N^*a = N^*b + c$ with $0 \notin \{a,b,c\}$. If we insist that $N^*a = N^*b + c$ with $0 \notin \{a,b\}$, then we must accept the fact that $c = 0$. Having met our goal, we are finished with our proof of the theorem. □

With a fresh dose of courage, you should now be ready for the main result of this section.

Theorem 8.3. *Let $(N, +, \cdot)$ be a finite IP-nearring. Let*

$$\mathscr{B}^* = \{N^*a + b \mid a, b \in N,\ a \neq 0\}.$$

Then $(N, \mathscr{B}^*)$ is a BIBD with parameters

$$v = |N|,\ k = |N^*/=_m|,\ b = v(v-1)/k,\ r = v - 1,\ \text{and}\ \lambda = k - 1.$$

Proof. From $(N, +, \cdot)$, we have $\Phi = \{n^\bullet \mid n \in N^*\}$. See Theorem 7.1. Since

$$N^*a = \{na \mid n \neq 0\} = \{n^\bullet(a) \mid n \in N^*\} = \Phi(a),$$

we have

$$k = |N^*a + b| = |N^*a| = |\Phi(a)| = |\Phi| = |N^*/=_m|.$$

(Recall that $a =_m b$ if and only if $a^\bullet = b^\bullet$.) Now $N^*a = N^*a' = \Phi(a')$ for each of the k values $a' \in N^*a$. Hence, there are $|N^*|/k = (v-1)/k$ distinct blocks of the form N^*a, $a \neq 0$, and there are v distinct blocks N^*a+b, one for each $b \in N$. This follows from Theorem 8.2. Hence, there are $v(v-1)/k = b$ distinct blocks $N^*a+b \in \mathscr{B}^*$.

Suppose $N^*a = \{a_1, a_2, \ldots, a_k\}$. Let $x \in N$. For each i, $1 \leq i \leq k$, there is a unique $y_i \in N$ such that $a_i + y_i = x$. Hence, x belongs to exactly the k blocks $N^*a + y_i$, $1 \leq i \leq k$. There are $(v-1)/k$ distinct blocks N^*a $a \neq 0$. Hence, x belongs to $k[(v-1)/k] = v-1 = r$ distinct blocks in $\mathscr{B}^*$. Certainly, $v = |N|$. We now have everything but λ.

Take $x, y \in N$, $x \neq y$. Now $x, y \in N^*a+t$ if and only if $0, y-x \in N^*a+(t-x)$. So, we need only count the number of blocks containing 0 and z in N, $z \neq 0$. We will show, for arbitrary $z \in N^*$, that 0 and z belong to exactly $k-1$ distinct blocks $N^*a+b \in \mathscr{B}^*$. Let $1_e = u_1, u_2, \ldots, u_k$ be the distinct representatives of the equivalence classes of $N^*/{=_m}$, where $e \in N$. Certainly, $0 \in N^*a - u_ia$ for each $a \in N^*$ and each u_i, $1 \leq i \leq k$. If $0 \in N^*a+t$, then $0 = u_ia+t$ for some u_i, and so $t = -u_ia$. Hence, $N^*a+t = N^*a-u_ia$. So, the blocks containing 0 are exactly the blocks $N^*a - u_ia$, $a \in N^*$, $1 \leq i \leq k$.

For $1 < i \leq k$, the equation $1_ex = u_ix + z$ has a unique solution a_i for x. If $a_i = 0$, then $0 = z$, so we are assured that each $a_i \neq 0$. From $1_ea_i = u_ia_i + z$, we get $z = -(u_ia_i) + 1_ea_i = u_i(-a_i) + a_i$. Hence, $z \in N^*(-a_i) + a_i$, which also contains 0. Hence, 0 and z belong to each $N^*(-a_i) + a_i$, $2 \leq i \leq k$. Do we really have $k-1$ distinct blocks here? If $i \neq j$ and $N^*(-a_i) + a_i = N^*(-a_j) + a_j$, then by Theorem 8.2, we get $a_i = a_j$, and $u_i(-a_i) + a_i = z = u_j(-a_j) + a_j$. Hence, $u_i(-a_i) = u_j(-a_j)$. Consequently, $u_ia_i = u_ja_j$. Now $u_i \neq_m u_j$, so the equation $u_ix = u_jx + 0$ has 0 and $a_j = a_i$ as solutions, so $a_i = a_j = 0$. But $a_i = a_j \neq 0$, and so $a_i \neq a_j$. This gives us $k-1$ distinct blocks containing both 0 and z. But, are there more? Our next step is to show that these are all of them.

If $0, z \in N^*a+b$, then $N^*a+b = N^*a - u_ia$ for some i, $1 \leq i \leq k$, since these are exactly the blocks containing 0. This means that $z = u_j a - u_i a$ for some u_j, where $i \neq j$, since $z \neq 0$. From this, we have

$$-u_j a + z = -u_i a, \; u_j(-a) + z = u_i(-a), \text{ and } u_i(-a) = u_j(-a) + z.$$

Since $(B_{u_i}, \cdot)$ is a group, there is a $u_i' \in N^*$ such that $(u_i' u_i)y = y$ for all $y \in N$, and so $1_e[u_i(-a)] = (u_j u_i')[u_i(-a)] + z$. Hence, $u_i(-a)$ is a solution to the equation $1_e x = (u_j u_i')x + z$, and since $u_j u_i' =_m u_l$ for some l, $1 < l \leq k$, we have that $u_i(-a)$ is a solution to the equation $1_e x = u_l x + z$. But a_l is the unique solution to this equation. Hence, $a_l = u_i(-a) = -(u_i a)$. So,

$$0, z \in N^* a + b = N^* a - u_i a = N^*(u_i a) + u_i(-a) = N^*(-a_l) + a_l.$$

This means that $N^* a + b$ is one of the $N^*(-a_s) + a_s$, that is, one of the $k-1$ blocks already accounted for. So $\lambda = k-1$ is the number of distinct blocks containing 0 and z, and therefore, there are exactly $\lambda = k-1$ distinct blocks containing an arbitrary pair $x, y \in N$, $x \neq y$. Now we have an end to the proof of this theorem. □

We now have a very powerful way to make BIBDs. Just take any finite IP-nearring $(N, +, \cdot)$, construct all $N^* a + b$, and be assured that $(N, \mathscr{B}^*)$ is a BIBD, where $\mathscr{B}^* = \{N^* a + b \mid a, b \in N, \; a \neq 0\}$. The Ferrero Planar Nearring Factory of **7.3** tells us how to make an IP-nearring, and immediately after **7.3**, we were told were to find the raw materials for The Factory. It would be nice to know when a BIBD of the form $(N, \mathscr{B}^*)$ where N is a finite IP-nearring. We can sharpen Theorem 2.16 to sometimes tell when $(N, \mathscr{B}^*)$ is not circular. From , we will get an upper bound for the number of points k is a circle when $(N, \mathscr{B}^*)$ is circular an $|N| < \infty$.

Theorem 8.4. *Let $(N, +, \cdot)$ be a finite IP-nearring. Suppose $(N, \mathscr{B}^*)$ is circular. Then $k \leq (3 + \sqrt{4v - 7})/2$.*

Proof. Since $(N, \mathscr{B}^*)$ is a BIBD with $\lambda = k-1$, we $k \leq (3+\sqrt{4v-7})/2$ from Corollary 2.18. □

Now, let's make a BIBD with our new tool.

Example 8.5. Let $(N,+) = (\mathbb{Z}_7,+)$, from the field $(\mathbb{Z}_7,+,\cdot)$. Let $\Phi = \{2^\bullet, 4^\bullet, 1^\bullet\}$ which is isomorphic to the subgroup $\{2,4,1\}$ of Z_7^*. Then, with $N = \mathbb{Z}_7$, $N^*1 = \{2,4,1\}$ and $N^*3 = \{6,5,3\}$. The blocks of $\mathscr{B}^* = \{N^*a+b \mid a,b \in N,\ a \neq 0\}$ are

$$\begin{array}{llll}
N^*1+0=\{2,4,1\}, & N^*1+4=\{6,1,5\}, & N^*3+0=\{6,5,3\}, & N^*3+4=\{3,2,0\}, \\
N^*1+1=\{3,5,2\}, & N^*1+5=\{0,2,6\}, & N^*3+1=\{0,6,4\}, & N^*3+5=\{4,3,1\}, \\
N^*1+2=\{4,6,3\}, & N^*1+6=\{1,3,0\}, & N^*3+2=\{1,0,5\}, & N^*3+6=\{5,4,2\}. \\
N^*1+3=\{5,0,4\}, & & N^*3+3=\{2,1,6\}, &
\end{array}$$

Notice that we have just constructed the example of Example 1.9. You should now be able to solve

Exercise 8.6. You should now be able to solve Problem 1.7. Give it a try.

Exercise 8.7. Construct a few more BIBDs using Theorem 8.3.

Statisticians use BIBDs in the design of complex experiments. They measure a BIBD by the formula

$$E = \frac{\lambda v}{rk}$$

and call the number E the *efficiency* of a BIBD $(V, \mathscr{B})$ with parameters v, b, r, k, and λ. Some statisticians have declared a BIBD to be "good" if $0.75 \leq E \leq 1.00$, while others require $0.90 \leq E \leq 1.00$ for this attribute.

Exercise 8.8. Compute the efficiency of your BIBDs of Exercise 8.6 and Exercise 8.7. Are they "good?"

Exercise 8.9. Construct a BIBD which is "good" by the stricter criterion.

Exercise 8.10. Are the BIBDs of an affine plane, a projective plane, and/or a Möbius plane "good?"

9 Nets

Let's take another look at the affine planes we constructed from a field $(F,+,\cdot)$. Let $(F\times F,\mathscr{L}_F)$ be the affine plane of Theorem 3.4. $\mathscr{L}_F$ consists of $\mathscr{L}(\infty)$ together with all the $\mathscr{L}(m)$, $m\in F$. In fact, $\mathscr{L}_F=\mathscr{L}(\infty)\cup[\cup_{m\in F}\mathscr{L}(m)]$. Moreover, for $s,t\in F'=F\cup\{\infty\}$ and $s\neq t$, we get $\mathscr{L}(s)\cap\mathscr{L}(t)=\varnothing$. So, $\{\mathscr{L}(s)\mid s\in F'\}$ is a partition of $\mathscr{L}_F$.

If $s,t\in F'$ and $s\neq t$, and if $L\in\mathscr{L}(s)$ and $M\in\mathscr{L}(t)$, then $|L\cap M|=1$. That is, two nonparallel lines intersect at exactly one point. For any $s\in F'$, $\mathscr{L}(s)$ is a partition of $F\times F$. That is, $\mathscr{L}(s)$ is a family of parallel lines, and each point $x\in F\times F$ belongs to exactly one line $L\in\mathscr{L}(s)$. Also, if $L\in\mathscr{L}_F$, then $|L|=|F|$, which is to say that every line has the same number of points. Finally, if $x,y\in F$ and $x\neq y$, then there is a unique $L\in\mathscr{L}_F$ with $x,y\in L$.

Suppose we want to relax the above properties slightly. Specifically, suppose we have $(P,\mathscr{L})$, where P is a nonempty finite set and $\mathscr{L}$ is a nonempty family of subsets of P. Perhaps the following are true.

It may be that $\mathscr{L}$ has a partition $\{\mathscr{L}(m_1),\dots,\mathscr{L}(m_r)\}$. That is, $\mathscr{L}=\cup_{i=1}^{r}\mathscr{L}(m_i)$, and if $i\neq j$, then $\mathscr{L}(m_i)\cap\mathscr{L}(m_j)=\varnothing$. Perhaps $|L\cap M|=1$ whenever $L\in\mathscr{L}(m_i)$ and $M\in\mathscr{L}(m_j)$ and $i\neq j$. Suppose each $\mathscr{L}(m_i)$ is a partition of P, so, if $L,M\in\mathscr{L}(m_i)$ and $L\neq M$, then $L\cap M=\varnothing$, and if $x\in P$, then there is a unique $L\in\mathscr{L}(m_i)$ such that $x\in L$. Perhaps each $L\in\mathscr{L}$ has the same number of elements, that is, if $L,M\in\mathscr{L}$, then $|L|=|M|$. Finally, it may be that for every $x,y\in P$, $x\neq y$, there is at most one $L\in\mathscr{L}$ with $x,y\in L$. Whenever we come this close to getting an affine plane, we have a *net*.

Definition 9.1. Suppose $\mathscr{L}$ is a nonempty family of subsets of a finite set P with the following properties.

(a) If $L,M\in\mathscr{L}$, then $|L|=|M|$.

(b) If $x,y\in P$ and $x\neq y$, then there is at most one $L\in\mathscr{L}$ with $x,y\in L$.

(c) $\mathscr{L}$ has a partition $\{\mathscr{L}(m_1),\ldots,\mathscr{L}(m_r)\}$ where each $\mathscr{L}(m_i)$, $1\le i\le r$, is itself a partition of P.
(d) If $L\in\mathscr{L}(m_i)$ and $M\in\mathscr{L}(m_j)$ and $i\neq j$, then $|L\cap M|=1$.

Then $(P,\mathscr{L})$ is a *net*. If each $L\in\mathscr{L}$ has $|L|=s$, then s and r are the parameters of the net $(P,\mathscr{L})$, and one sometimes refers to $(P,\mathscr{L})$ as an (s,r)-*net*.

Of course, every finite affine plane $(P,\mathscr{L})$ is a net, but there is not much sense in having nets if all of them are affine planes. So, we need a net that is not an affine plane. We will soon see that every finite IP-nearring N will yield a net $(N\times N,\mathscr{L}_N)$, and if N is not a nearfield, then the net will not be an affine plane. When we are assured that this is true, then the reader will be invited to explicitly construct an example of a net that is not an affine plane.

Our development should be familiar by now. Start with a finite IP-nearring $(N,+,\cdot)$. Let $P=N\times N$. Also, let $\mathbf{M}^*=\{1_e=m_1,m_2,\ldots,m_k\}$ be a set of representatives of $N^*/{=_m}$, and so $\mathbf{M}=\{0,m_1,m_2,\ldots,m_k\}$ is a set of representatives of $N/{=_m}$, with $|\mathbf{M}|=k+1$.

If $m\in\mathbf{M}$, define, for each $b\in N$,

$$L(m,b)=\{(x,mx+b)\mid x\in N\}$$

and

$$\mathscr{L}(m)=\{L(m,b)\mid b\in N\}.$$

Also, for each $a\in N$, define

$$L(\infty,a)=\{(a,y)\mid y\in N\}$$

and

$$\mathscr{L}(\infty)=\{L(\infty,a)\mid a\in N\}.$$

Let

$$\mathscr{L}_N=\mathscr{L}(0)\cup\mathscr{L}(m_1)\cup\cdots\cup\mathscr{L}(m_k)\cup\mathscr{L}(\infty).$$

We will see that $(N\times N,\mathscr{L}_N)$ is a net.

Theorem 9.2. *If $(N,+,\cdot)$ is a finite IP-nearring, then $(N\times N,\mathscr{L}_N)$ is a net.*

Proof. If $L(u,c)\in\mathscr{L}$, then $|L(u,c)|=|N|$. Hence (a) of Definition 9.1 is valid. Suppose $x=(a,b)$, $y=(c,d)\in N\times N$. If $a=c$, then $x,y\in L(\infty,a)$, and conversely. If $b=d$, then $x,y\in L(0,b)$, and conversely. Suppose $a\neq c$ and $b\neq d$. If $x,y\in L(m,B)$ for some $m\in\mathbf{M}^*$ and some $B\in N$, then $x=(a,b)=(a,ma+B)$ and $y=(c,d)=(c,mc+B)$. Hence, $b=ma+B$ and $d=mc+B$. Consequently, $-ma+b=B=-mc+d$, and so $d-b=mc-ma=m(c-a)$. There may or may not be an m such that $m(c-a)=d-b$, but if there is one, there is only one, since $m\in\{m_1,\dots,m_k\}$ and $\Phi=\{m_1^\bullet,\dots,m_k^\bullet\}$ has each $m_i^\bullet$ fixed point free if $1<i\leq k$. In summary, $x,y\in N$ with $x\neq y$ imply that there is at most one $L\in\mathscr{L}$ with $x,y\in L$. This gives us (b) of Definition 9.1. (How do we know that if there is an $L(m,B)\in\mathscr{L}(m)$ with $x,y\in L(m,B)$, then there is no other $L(m,B')\in\mathscr{L}(m)$ with $B\neq B'$ and $x,y\in L(m,B')$?)

Take $\mathscr{L}(\alpha),\mathscr{L}(\beta)\in\{\mathscr{L}(0),\mathscr{L}(m_1),\dots,\mathscr{L}(m_k),\mathscr{L}(\infty)\}$ with $\mathscr{L}(\alpha)\neq\mathscr{L}(\beta)$. We want to show that $\mathscr{L}(\alpha)\cap\mathscr{L}(\beta)=\varnothing$. There are four cases to consider: (1) $\alpha=0$ and $\beta=\infty$; (2) $\alpha=0$ and $\beta\in\mathbf{M}^*$; (3) $\alpha=\infty$ and $\beta\in\mathbf{M}^*$; and (4) $\alpha,\beta\in\mathbf{M}^*$ with $\alpha\neq\beta$.

If $\alpha=0$ and $\beta=\infty$, and if $L\in\mathscr{L}(0)\cap\mathscr{L}(\infty)$, then $L=L(0,b)=L(\infty,a)$ for some $a,b\in N$. But $(u,b)\in L(0,b)$ for all $u\in N$, and for any $u\neq a$, $(u,b)\notin L(\infty,a)$. Hence, $\mathscr{L}(0)\cap\mathscr{L}(\infty)=\varnothing$.

If $L\in\mathscr{L}(0)\cap\mathscr{L}(m)$ for $m\in\mathbf{M}^*$, then there are $b,m,B\in N$ such that $L=L(0,b)=L(m,B)$. But $L(0,b)$ consists of all (u,b) for $u\in N$, and if $x\in L(m,B)$, then $x=(u,mu+B)$. As u varies through the values of N, $mu+B$ does also. That is, $mu+B=b$ for only one value $u\in N$. Hence, $\mathscr{L}(0)\cap\mathscr{L}(m)=\varnothing$.

For the third case, suppose $L\in\mathscr{L}(\infty)\cap\mathscr{L}(m)$ for some $m\in\mathbf{M}^*$. Since $L\in\mathscr{L}(\infty)$, $L=L(\infty,a)$ for some $a\in N$. So $(c,d)\in L(\infty,a)$ if and only if $c=a$. But $L\in\mathscr{L}(m)$ implies $L=L(m,b)$ for some $b\in N$, and there are points in $L(m,b)$ of the form (c,d) with $c\neq a$. Hence, $\mathscr{L}(\infty)\cap\mathscr{L}(m)=\varnothing$.

Finally, take $m,m'\in\mathbf{M}^*$ with $m\neq m'$. Suppose $L\in\mathscr{L}(m)\cap\mathscr{L}(m')$. Then there are $b, b' \in N$ such that $L = L(m,b) = L(m',b')$. For any $x \in N$, only the point $(x, mx+b) \in L(m,b)$ has first coordinate x, and only the point $(x, m'x+b')$ in $L(m',b')$ has first coordinate x. Hence, if $L(m,b) = L(m',b')$, then $(x, mx+b)$ and $(x, m'x+b')$ are the same for each $x \in N$. But there is a unique $u \in N$ such that $mu+b = m'u+b'$, since $m, m' \in \mathbf{M}^*$ and $m \neq m'$. So, $L(m,b)$ cannot equal $L(m',b')$ if $m \neq m'$. From this we conclude that $\mathscr{L}(m) \cap \mathscr{L}(m') = \varnothing$.

We can now conclude that $\{\mathscr{L}(0), \mathscr{L}(m_1), \ldots, \mathscr{L}(m_k), \mathscr{L}(\infty)\}$ is a partition of $\mathscr{L}_N$, and we have (c) of Definition 9.1.

There are four cases to consider to get (d) of Definition 9.1. They are: (1) $L \in \mathscr{L}(0)$ and $M \in \mathscr{L}(m)$, $m \in \mathbf{M}^*$; (2) $L \in \mathscr{L}(0)$ and $M \in \mathscr{L}(\infty)$; (3) $L \in \mathscr{L}(\infty)$ and $M \in \mathscr{L}(m)$, $m \in \mathbf{M}^*$, and (4) $L \in \mathscr{L}(m)$, $M \in \mathscr{L}(m')$, $m, m' \in \mathbf{M}^*$, and $m \neq m'$.

For $L \in \mathscr{L}(0)$ and $M \in \mathscr{L}(m)$, $m \in \mathbf{M}^*$, we have $L = L(0,a)$ for some $a \in N$, and $M = L(m,b)$ for some $b \in N$. If $(s,t) \in L(0,a) \cap L(m,b)$, i.e.

$$(s,t) \in \{(x,a) \mid x \in N\} \cap \{(x, mx+b) \mid x \in N\},$$

then $(s,t) = (s,a) = (s, ms+b)$. There is only one $s \in N$ such that $ms+b = a$, and so $(s,a) \in L(0,a) \cap L(m,b)$, and we have $|L \cap M| = 1$.

With $L \in \mathscr{L}(0)$ and $M \in \mathscr{L}(\infty)$, we have $L = L(0,a)$ for some $a \in N$ and $M = L(\infty, b)$ for some $b \in N$. If $(s,t) \in L(0,a) \cap L(\infty,b)$, i.e.

$$(s,t) \in \{(x,a) \mid x \in N\} \cap \{(b,y) \mid y \in N\},$$

then $(s,t) = (b,a)$, and certainly $(b,a) \in L(0,a) \cap L(\infty,b)$, so we have $|L \cap M| = 1$.

Take $L \in \mathscr{L}(\infty)$ and $M \in \mathscr{L}(m)$ with $m \in \mathbf{M}^*$. Then $L = L(\infty,a)$ and $M = L(m,b)$ for some $a, b \in N$. If $(s,t) \in L(\infty,a) \cap L(m,b)$, i.e.

$$(s,t) \in \{(a,y) \mid y \in N\} \cap \{(x, mx+b) \mid x \in N\},$$

then $s = a$ and $t = ma+b$. Certainly, $(a, ma+b) \in L(\infty,a) \cap L(m,b)$, and so $|L \cap M| = 1$.

Finally, with $L \in \mathscr{L}(m)$, $M \in \mathscr{L}(m')$, $m, m' \in \mathbf{M}^*$, and $m \neq m'$, we have $L = L(m,b)$ and $M = L(m',b')$ for some $b, b' \in N$. If $(s,t) \in L(m,b) \cap L(m',b')$, i.e.

$$(s,t) \in \{(x, mx+b) \mid x \in N\} \cap \{(x, m'x+b' \mid x \in N\},$$

then $(s,t) = (s, ms+b) = (s, m's+b')$. Since $m \neq m'$ and $m, m' \in \mathbf{M}^*$, there is a unique $s \in N$ such that $ms+b = m's+b'$, and so $|L \cap M| = 1$. This assures us that (d) of Definition 9.1 is valid. Altogether, we have an end to the proof of Theorem 9.2. □

A problem on constructing a net from an IP-nearring $(N,+,\cdot)$ which has Ferrero pair (N, Φ), where $(N,+)$ is the additive group of a field $(N,+,*)$ and Φ is isomorphic to a subgroup of N^*, is that the net is just a part of the affine plane constructed from the field $(N,+,*)$. This becomes more of a problem when we construct mutually orthogonal Latin squares in §10. So, in applying Theorem 9.2, it is of interest to have a nearring $(N,+,\cdot)$ where $(N,+)$ is not the additive group of a field.

Exercise 9.3. In Example 7.9, we gave an example of an IP-nearring $(\mathbb{Z}_9,+,*)$, and $(\mathbb{Z}_9,+)$ is not the additive group of a field. Applying Theorem 9.2, one could construct the net $(\mathbb{Z}_9 \times \mathbb{Z}_9, \mathscr{L})$, where $\mathscr{L} = \mathscr{L}(0) \cup \mathscr{L}(1) \cup \mathscr{L}(8) \cup \mathscr{L}(\infty)$. Explicitly construct $\mathscr{L}(1)$ and $\mathscr{L}(8)$ and save your results for §10.

Exercise 9.4. In Exercise 9.3, each $|L(m,b)| = 9$, and there is a field $(F,+,\cdot) = (\mathbb{Z}_3 \times \mathbb{Z}_3,+,\cdot)$ with $|F| = 9$. It may be that $(\mathbb{Z}_9 \times \mathbb{Z}_9, \mathscr{L})$ of Exercise 9.3 is, in some way, the same as a net constructed from $(\mathbb{Z}_3 \times \mathbb{Z}_3,+)$. Find the net $(N \times N, \mathscr{L})$, using Theorem 9.2 with $N = \mathbb{Z}_3 \times \mathbb{Z}_3$, and with $|\mathscr{L}|$ as small as you can, which could not be the same as a net constructed from a field. (Hint: Study Example 7.9.)

10 Mutually orthogonal Latin squares

With a field F, we constructed an affine plane $(F \times F, \mathscr{L}_F)$. If F is finite with $|F| = q$, we then constructed a family of $q-1$ mutually orthogonal Latin squares. We want to do something similar for a finite IP-nearring.

Let $N = \{0 = n_1, n_2, \ldots, n_q\}$ be the elements of a finite IP-nearring $(N, +, \cdot)$ with $|N| = q$. Let $\mathbf{M} = \{0 = m_0, m_1, \ldots, m_k\}$ be a set of representatives of $N/{=_m}$ with $|\mathbf{M}| = k+1$. Now $|N \times N| = q^2$ and $N \times N = \{(n_i, n_j) \mid 1 \leq i \leq q, 1 \leq j \leq q\}$. Next, we construct a $k+2$ by q^2 matrix M. Label the $k+2$ rows with $\mathscr{L}(0)$, $\mathscr{L}(m_1), \ldots, \mathscr{L}(m_k)$, $\mathscr{L}(\infty)$, and label the q^2 columns of M with

$$\begin{array}{cc} (n_1,n_1),(n_1,n_2),\ldots,(n_1,n_q), & (n_2,n_1),(n_2,n_2),\ldots,(n_2,n_q), \\ \cdots \quad \cdots \quad \cdots & (n_q,n_1),(n_q,n_2),\ldots,(n_q,n_q). \end{array}$$

We speak of the $(\mathscr{L}(u), (n_s, n_t))$-entry of M as the element of N which we put in the row labeled with $\mathscr{L}(u)$ and the column labeled with (n_s, n_t).

Now $\mathscr{L}(u) = \{L(u,b) \mid b \in N\}$ is a partition of $N \times N$, and so there is a unique $b \in N$ for which $(n_s, n_t) \in L(u,b)$. This $b \in N$ is the element we use for the $(\mathscr{L}(u), (n_s, n_t))$-entry of M. See §3.4 and the example of Example 3.29.

Take any row labeled $\mathscr{L}(m_i)$, $1 \leq i \leq k$, of the matrix M. Suppose the $\mathscr{L}(m_s)$-row is

$$\begin{pmatrix} b^s_{11} & b^s_{12} & \cdots & b^s_{1q} & b^s_{21} & b^s_{22} & \cdots & b^s_{2q} & \cdots & b^s_{q1} & b^s_{q2} & \cdots & b^s_{qq} \end{pmatrix}.$$

As in §3.4, the superscript s is not an exponent, but a bookkeeping device. The $\mathscr{L}(n_s)$-row has q^2 entries which gives us the matrix matrix

$$A_s = \begin{pmatrix} b^s_{11} & b^s_{12} & \cdots & b^s_{1q} \\ b^s_{21} & b^s_{22} & \cdots & b^s_{2q} \\ \vdots & \vdots & \ddots & \vdots \\ b^s_{q1} & b^s_{q2} & \cdots & b^s_{qq} \end{pmatrix},$$

and so we have a family $\mathscr{A} = \{A_1, A_2, \ldots, A_k\}$ of k square q by q matrices. As in §3.4, we are led to

Theorem 10.1. *For a finite IP-nearring $(N, +, \cdot)$ with $|N| = q$ and $|N^*/{=_m}| = k$, the k matrices in the family $\mathscr{A} = \{A_1, \ldots, A_k\}$ are mutually orthogonal Latin squares.*

Proof. Our proof is nearly identical to the proof of Theorem 3.30. It is common to leave such proofs to the reader, but, in a break with tradition, we'll go through it again here, since Theorem 3.30 really is a special case of Theorem 10.1.

So, our first step is to show that each A_s is a Latin square. A_s is constructed from the $\mathscr{L}(m_s)$-row of M, $1 \le s \le k$, so $m_x \notin \{0, \infty\}$. We used the "lines" $L(m_s, b)$, $b \in N$, to construct the $\mathscr{L}(m_s)$-row of M. Since $m_s \notin \{0, \infty\}$, we are using "nonvertical" and "nonhorizontal" lines.

The columns of M are labeled with (n_i, n_j) with $i, j \in \{1, 2, \ldots, k\}$, but for $1 \le i \le q$, we have $(n_i, n_1), (n_i, n_2), \ldots, (n_i, n_q)$ as consecutive labels of columns of M. These are exactly the "points" of the "vertical" line $x = n_i$, i.e., $L(\infty, n_i)$. As such, each line $x = n_i$ will intersect each $L(u, b)$ exactly once, if $u \neq \infty$. To see this, recall that for $u \neq \infty$, $L(u, b) = \{(x, ux + b) \mid x \in N\}$, and so the only candidate for an element of $L(u, b) \cap L(\infty, n_i)$ is $(n_i, un_i + b)$, and certainly $(n_i, un_i + b) \in L(u, b) \cap L(\infty, n_i)$.

The matrix A_s is constructed from the $\mathscr{L}(m_s)$-row. The entries in the ith row

$$(b^s_{i1}, b^s_{i2}, \ldots, b^s_{iq})$$

are in the columns labeled (n_i, n_1), ..., (n_i, n_q). As such, $(n_i, n_j) = (n_i, un_i + b^s_{ij})$. So, if $b^s_{ij} = b^s_{ik}$, then $un_i + b^s_{ij} = un_i + b^s_{ik}$, and so $n_j = n_k$. This assures us that $|\{b^s_{i1}, \ldots, b^s_{iq}\}| = q$. Hence, the ith row of A_s consists of q distinct elements of N, since each $b^s_{ij} \in N$. This is one of our requirements for A_s to be a Latin square.

Consider now the jth column of A_s. It consists of $\{b^s_{1j}, b^s_{2j}, \ldots, b^s_{qj}\}$ with each $b^s_{ij} \in N$, and b^s_{ij} is in the column labeled (n_i, n_j). Hence, we have $(n_i, n_j) =$

$(n_i, un_i + b^s_{ij})$. If $b^s_{ij} = b^s_{kj}$, then $(n_i, n_j) = (n_i, un_i + b^s_{ij})$ and $(n_k, n_j) = (n_k, un_k + b^s_{kj})$, and since $n_j = un_i + b^s_{ij} = un_k + b^s_{kj}$, we get $n_i = n_k$, since $u \notin \{0, \infty\}$. This means that $(n_i, n_j) = (n_k, n_j)$, and so $|\{b^s_{1j}, b^s_{2j}, \ldots, b^s_{qj}\}| = q$. This is the other requirement for A_s to be a Latin square. Hence, for $2 \leq s \leq q$, A_s is a Latin square.

Now consider two distinct Latin squares $A_s, A_t \in \mathscr{A}$. Let $A_s = (b^s_{ij})$ and $A_t = (b^t_{ij})$. The matrix A_s was constructed from the $\mathscr{L}(m_s)$-row of M, namely

$$\begin{pmatrix} b^s_{11} & b^s_{12} & \cdots & b^s_{1q} & b^s_{21} & b^s_{22} & \cdots & b^s_{2q} & \cdots & b^s_{q1} & b^s_{q2} & \cdots & b^s_{qq} \end{pmatrix}$$

and A_t was constructed from the $\mathscr{L}(m_t)$-row of M, namely

$$\begin{pmatrix} b^t_{11} & b^t_{12} & \cdots & b^t_{1q} & b^t_{21} & b^t_{22} & \cdots & b^t_{2q} & \cdots & b^t_{q1} & b^t_{q2} & \cdots & b^t_{qq} \end{pmatrix}.$$

To see that A_s and A_t are orthogonal, we must construct the matrix $A_s * A_t$ of Problem 1.30, and here

$$A_s * A_t = \begin{pmatrix} (b^s_{11}, b^t_{11}) & (b^s_{12}, b^t_{12}) & \cdots & (b^s_{1q}, b^t_{1q}) \\ (b^s_{21}, b^t_{21}) & (b^s_{22}, b^t_{22}) & \cdots & (b^s_{2q}, b^t_{2q}) \\ \vdots & \vdots & \ddots & \vdots \\ (b^s_{q1}, b^t_{q1}) & (b^s_{q2}, b^t_{q2}) & \cdots & (b^s_{qq}, b^t_{qq}) \end{pmatrix}.$$

Then we must be assured that

$$\{(b^s_{ij}, b^t_{ij}) \mid i, j \in \{1, 2, \ldots, q\}\} = \{(x, y) \mid x, y \in N\} = N \times N.$$

Suppose $(b^s_{ij}, b^t_{ij}) = (b^s_{kl}, b^t_{kl})$. What does it mean for $b^s_{ij} = b^s_{kl}$?

Now b^s_{ij} is in the $\mathscr{L}(m_s)$-row and in the (n_i, n_j)-column of M, and (n_i, n_j) is in $L(m_s, b^s_{ij})$. Similarly, b^s_{kl} is in the $\mathscr{L}(m_s)$-row and in the (n_k, n_l)-column of M, and (n_k, n_l) is in $L(m_s, b^s_{kl})$. With $b^s_{ij} = b^s_{kl}$ we have $L(m_s, b^s_{ij}) = L(m_s, b^s_{kl})$, and so $(n_i, n_j), (n_k, n_l) \in L(n_s, b^s_{ij})$.

Next, what does it mean for $b^t_{ij} = b^t_{kl}$? As in the preceding paragraph, it means that $(n_i, n_j), (n_k, n_l) \in L(m_t, b^t_{ij})$. Since $m_s \neq m_t$ follows from $A_s \neq A_t$, we have that the two distinct lines $L(m_s, b^s_{ij})$ and $L(m_t, b^t_{ij})$ have in common the points (n_i, n_j) and (n_k, n_l). Since $(N \times N, \mathscr{L}_N)$ is a net, we have no choice but to conclude that $(n_i, n_j) = (n_k, n_l)$ and that $(i, j) = (k, l)$. This means that the mapping $\phi : N \times N \to \{(b^s_{ij}, b^t_{ij}) \mid i, j \in \{1, 2, \ldots, q\}\}$, defined by sending (n_i, n_j) to (b^s_{ij}, b^t_{ij}) is a bijection, which in turn completes the proof that A_i and A_j are orthogonal, and also completes the proof of the theorem. □

Exercise 10.2. In Exercise 9.3, you were asked to construct $\mathscr{L}(1)$ and $\mathscr{L}(8)$ of a net. With these, you are now requested to construct a pair of orthogonal Latin squares of order 9.

11 Duality

Before we talk about duality, we need to broaden our viewpoint of geometry. We usually think of a geometric structure as a pair $(P, \mathscr{B})$, where P is a nonempty set of "points," and $\mathscr{B}$ is a nonempty set of subsets of P. Members of $\mathscr{B}$ are often called "lines," "circles," "rays," or "blocks," etc. For a point x in P and a block $B \in \mathscr{B}$, we are often concerned about the question, "Is x a member of B?" That is, we want to know if $x \in B$ is true or false. In geometric language, we want to know if x is *on* B, or, equivalently, if B is *on* x. We can extend these ideas further.

Let P and $\mathscr{B}$ be nonempty sets, and let $I \subseteq P \times \mathscr{B}$. So, I is a relation from P to $\mathscr{B}$. For $x \in P$ and $B \in \mathscr{B}$, we wonder if $(x, B) \in I$, that is, if xIB.

Definitions 11.1. Let P and $\mathscr{B}$ denote nonempty sets with $I \subseteq P \times \mathscr{B}$, so I is a relation from P to $\mathscr{B}$. Then $(P, \mathscr{B}, I)$ is an *incidence structure*, and I is an *incidence relation*. If $(x, B) \in I$, we often write xIB and say that x is *on* B, but we also might say that B is *on* x.

Definition 11.2. Let $(P,\mathscr{B},I)$ be an incidence structure. Let $J = I^{-1}$, i.e., $J = \{(B,x) \mid (x,B) \in I\}$. Then the incidence structure $(\mathscr{B},P,J) = (\mathscr{B},P,I^{-1})$ is the *dual* of $(P,\mathscr{B},I)$.

With the ideas of an incidence structure and the dual of an incidence structure, it will be of interest to broaden the definition of a projective plane.

Definition 11.3. A *projective plane* is an incidence structure $(P,\mathscr{L},I)$, where P is a nonempty set whose elements are called "points," where $\mathscr{L}$ is a nonempty set whose elements are called "lines," and where I has the following properties.

(a) If $x,y \in P$ and $x \neq y$, then there is a unique $L \in \mathscr{L}$ with xIL and yIL.

(b) If $L,M \in \mathscr{L}$ and $L \neq M$, then there is a unique $x \in P$ with xIL and xIM.

(c) If $L \in \mathscr{L}$, then there are at least three distinct $x,y,z \in P$ such that xIL, yIL, and zIL.

(d) If $L \in \mathscr{L}$, then there is an $x \in P$ such that xIL is false, that is, $(x,L) \notin I$.

In Definition 3.9, if we let $I = \in$, then Definition 3.9 is certainly a special case of Definition 11.3. But, have we gained anything by introducing Definition 11.3?

Theorem 11.4. *Let $(P,\mathscr{L},I)$ be a projective plane. Then its dual $(\mathscr{L},P,J)$ is also a projective plane.*

Proof. We first want to show that requirement (a) is satisfied. Take $L,M \in \mathscr{L}$ with $L \neq M$. Since $(P,\mathscr{L},I)$ is a projective plane, there is a unique $x \in P$ with xIL and xIM. So, there is a unique $x \in P$ with LJx and MJx. Hence, (a) is satisfied.

Next, we want to show that (b) is satisfied. Take $x,y \in P$ with $x \neq y$. Since $(P,\mathscr{L},I)$ is a projective plane, then there is a unique $L \in \mathscr{L}$ with xIL and yIL. So, there is a unique $L \in \mathscr{L}$ with LJx and LJy. Hence, (b) is satisfied.

Next, we want to show that (d) is satisfied. Take $x \in P$. Is there an $L \in \mathscr{L}$ such that $(L,x) \notin J$? If not, then $(L,x) \in J$ for every $L \in \mathscr{L}$, which is to say, $(x,L) \in I$ for every $L \in \mathscr{L}$. But $(P,\mathscr{L},I)$ is a projective plane. Could we have a point x on every line L of a projective plane? We could use a few elementary facts about projective planes. □

Proposition 11.5. *Let $(P,\mathscr{L},I)$ be a projective plane. If $x \in P$, then there is a $T \in \mathscr{L}$ such that $(x,T) \notin I$. Also, there are at least three distinct lines $L,M,N \in \mathscr{L}$ such that $(x,L),(x,M),(x,N) \in I$.*

Proof. The proof will be given in the language of geometry. Each student is invited to translate this language to the language used in Definition 11.3.

If $x \in P$ and $M \in \mathscr{L}$, then by (c) of Definition 11.3, there is a $y \in P$ such that $y \neq x$. Let $L = L(x,y) \in \mathscr{L}$ be the unique line on x and y. Next, let $z \in P$ be any point not on L. There is a unique line $T = T(y,z) \in \mathscr{L}$ on both y and z. By uniqueness of the L on x and y, we have $T \neq L$, and x is not on T, so $(x,T) \notin I$.

There are at least three distinct points x_1, x_2, and x_3 on T. Let $L_i = L(x,x_i)$ be the unique line in $\mathscr{L}$ on both x and x_i, $i = 1,2,3$. When we show that $i \neq j$ implies $L_i \neq L_j$, then we will let $L = L_1$, $M = L_2$, and $N = L_3$, and our proof will be complete.

Without loss of generality, we may take $L_1 = L_2$. Hence, x, x_1, and x_2 are three distinct points on L_1, with x_1 and x_2 on T. So, $L_1 = T$ and so x is on T, which we cannot tolerate. Hence, $i \neq j$ implies $L_i \neq LPj$.

Returning to the proof of Theorem 11.4, we now know that a point x of a projective plane cannot be on every line, and so, given a point x, there is a line T not on x. That is, $(x,T) \notin I$, and so $(T,x) \notin J$. Hence, (d) of Definition 11.3 is satisfied.

By Proposition 11.5, we know that each point x of a projective plane has at least three distinct lines L, M, and N so that $(x,L),(x,M),(x,N) \in I$, which means that $(L,x),(M,x),(N,x) \in J$. Hence, (c) of Definition 11.3 is satisfied, and now we know that $(\mathscr{L}, P, J)$ is a projective plane, and we have an end to the proof of Theorem 11.4. □

We are now ready to present an interesting family of projective planes. For any field F, let

$$P = \{(1;a;b),\ (0;1;b),\ (0;0;1) \mid a,b \in F\}$$

and

$$\mathscr{L} = \{[1,a,b],\ [0,1,b],\ [0,0,1] \mid a,b \in F\}.$$

Notice how a "point" and a "line" appear to be about the same thing. At the present time, we do not give any particular meaning or interpretation to points $\mathbf{a} = (a;b;c)$ or lines $A = [a,b,c]$. The reader is asked to guess what the motivation for P and $\mathscr{L}$ is, as we develop the proof of the next theorem. Perhaps we should state that

$$\left.\begin{array}{c} (x_1;x_2;x_3) = (y_1;y_2;y_3) \\ \text{and} \\ [x_1,x_2,x_3] = [y_1,y_2,y_3] \end{array}\right\} \iff \text{each } x_i = y_i,\ 1 \le i \le 3.$$

For $a,b,c \in F$, with $\{a,b,c\} \neq \{0\}$, we can make a point $p(a,b,c) \in P$ and a line $L(a,b,c) \in \mathscr{L}$ by setting

$$p(a,b,c) = (a/d;b/d;c/d) \ \text{ and } \ L(a,b,c) = [a/d,b/d,c/d],$$

where

$$d = \begin{cases} a, & \text{if } a \neq 0; \\ b, & \text{if } a = 0 \text{ and } b \neq 0; \\ c, & \text{if } a = b = 0. \end{cases}$$

Notice that $p(a,b,c) = p(\lambda a, \lambda b, \lambda c)$ and $L(a,b,c) = L(\lambda a, \lambda b, \lambda c)$ for any nonzero $\lambda \in F^*$.

We have our nonempty set of points P and our nonempty set of lines $\mathscr{L}$. What we need now is an incidence relation $I \subseteq P \times \mathscr{L}$. For $\mathbf{a} = (a_1;a_2;a_3) \in P$ and $L = [x_1, x_2, x_3] \in \mathscr{L}$, we say that $\mathbf{a}IL$ if and only if

$$\mathbf{a} \bullet L = L \bullet \mathbf{a} = a_1x_1 + a_2x_2 + a_3x_3 = 0.$$

The reader can think of $\mathbf{a} \bullet L = L \bullet \mathbf{a}$ as the "dot product" of $\mathbf{a}$ and L.

With considerable patience, we are prepared to develop a proof to the following theorem.

Theorem 11.6. *For a field F, and for P, $\mathscr{L}$, and I as defined above, we get that $(P, \mathscr{L}, I)$ is a projective plane.*

Proof. Since the proof is rather long and tedious, we break it down rather formally in accordance with Definition 11.3. Our first step will be to show that two distinct points are on a unique line. In doing so, we will almost show that two distinct lines are on a unique point.

(1) *Let $\mathbf{a} = (a_1;a_2;a_3)$, $\mathbf{b} = (b_1;b_2;b_3) \in P$ be two distinct points. Then there is a unique line $L = \mathbf{L}(\mathbf{a}, \mathbf{b}) \in \mathscr{L}$ with $\mathbf{a}IL$ and $\mathbf{b}IL$.*

Proof of **(1)**. Let $\mathbf{a} = (a_1;a_2;a_3)$, $\mathbf{b} = (b_1;b_2;b_3) \in P$ be two distinct points. Define

$$\mathbf{L}(\mathbf{a}, \mathbf{b}) = L(a_2b_3 - b_2a_3, -(a_1b_3 - b_1a_3), a_1b_2 - b_1a_2).$$

First, we will show that $\mathbf{a} \bullet \mathbf{L}(\mathbf{a}, \mathbf{b}) = \mathbf{b} \bullet \mathbf{L}(\mathbf{a}, \mathbf{b}) = 0$, that is, $\mathbf{a}I\mathbf{L}(\mathbf{a}, \mathbf{b})$ and $\mathbf{b}I\mathbf{L}(\mathbf{a}, \mathbf{b})$ if $\mathbf{L}(\mathbf{a}, \mathbf{b}) \in \mathscr{L}$. (We have yet to show that

$$a_2b_3 - b_2a_3 = a_1b_3 - b_1a_3 = a_1b_2 - b_1a_3 = 0$$

is impossible.) Easy computation gives us

$$\mathbf{a} \bullet \mathbf{L}(\mathbf{a}, \mathbf{b}) = a_1(a_2b_3 - b_2a_3) - a_2(a_1b_3 - b_1a_3) + a_3(a_1b_2 - b_1a_2) = 0$$

and

$$\mathbf{b} \bullet \mathbf{L}(\mathbf{a},\mathbf{b}) = b_1(a_2b_3 - b_2a_3) - b_2(a_1b_3 - b_1a_3) + b_3(a_1b_2 - b_1a_2) = 0.$$

The messy part comes in showing that $\mathbf{L}(\mathbf{a},\mathbf{b})$ is unique. While we are doing this, we will also show that $\mathbf{L}(\mathbf{a},\mathbf{b}) \in \mathscr{L}$.

Suppose $\mathbf{a} \bullet L = \mathbf{b} \bullet L = 0$ for some $L \in \mathscr{L}$. Now $L = [x_1,x_2,x_3] = L(x_1,x_2,x_3)$ for some $x_1,x_2,x_3 \in F$. From $\mathbf{a} \bullet L = \mathbf{b} \bullet L = 0$, we get

$$\begin{cases} a_1x_1 + a_2x_2 + a_3x_3 = 0, \\ b_1x_1 + b_2x_2 + b_3x_3 = 0. \end{cases} \tag{11:1}$$

Since $\mathbf{a} \neq \mathbf{b}$, one of the following five cases must be true: (1) $a_1 \neq b_1$; (2) $a_1 = b_1 = 1$ and $a_2 \neq b_2$; (3) $a_1 = b_1 = 1$, $a_2 = b_2$, and $a_3 \neq b_3$; (4) $a_1 = b_1 = 0$ and $a_2 \neq b_2$; and (5) $a_1 = b_1 = 0$, $a_2 = b_2 = 1$, and $a_3 \neq b_3$. We will examine each of these cases individually.

For (1), we may assume $a_1 = 0$ and $b_1 = 1$. Then

$$\mathbf{L}(\mathbf{a},\mathbf{b}) = L(a_2b_3 - b_2a_3, a_3, -a_2).$$

If $a_2 = 0$, then

$$\mathbf{L}(\mathbf{a},\mathbf{b}) = L(-b_2a_3, a_3, 0) = L(-b_2, 1, 0).$$

Also, (11:1) gives us $a_3x_3 = 0$ and $x_1 = -b_2x_2 - b_3x_3$. Since $a_1 = a_2 = 0$, we cannot have $a_3 = 0$. Hence, $x_3 = 0$, and $x_1 = -b_2x_2$. Since $x_2 \neq 0$,

$$L = L(x_1,x_2,x_3) = L(-b_2x_2, x_2, 0) = L(-b_2, 1, 0) = \mathbf{L}(\mathbf{a},\mathbf{b}).$$

This shows that $\mathbf{L}(\mathbf{a},\mathbf{b}) \in \mathscr{L}$ since $1 \neq 0$.

If $a_1 = 0$, $b_1 = 1$, and $a_2 \neq 0$, then $x_2 = -a_3x_3/a_2$ and $x_1 = (b_2a_3/a_2 - b_3)x_3$. If $x_3 = 0$, then $x_1 = x_2 = 0$, so $x_3 \neq 0$. Hence,

$$\begin{aligned} L(x_1,x_2,x_3) &= L(b_2a_3/a_2 - b_3, -a_3/a_2, 1) \\ &= L(b_2a_3 - a_2b_3, -a_3, a_2) \\ &= L(a_2b_3 - b_2a_3, a_3, -a_2) = \mathbf{L}(\mathbf{a},\mathbf{b}). \end{aligned}$$

This shows that $\mathbf{L}(\mathbf{a},\mathbf{b}) \in \mathscr{L}$ since $a_2 \neq 0$.

So, for (1), we get $\mathbf{L}(\mathbf{a},\mathbf{b}) = L(x_1,x_2,x_3) = L$.

Next, consider (2), with $a_1 = b_1 = 1$ and $a_2 \neq b_2$. Now

$$\mathbf{L}(\mathbf{a},\mathbf{b}) = L(a_2b_3 - b_2a_3, -(b_3 - a_3), b_2 - a_2).$$

From (11:1), we obtain

$$(b_2 - a_2)x_2 + (b_3 - a_3)x_3 = 0.$$

Since $b_2 - a_2 \neq 0$, we have $x_2 = -(b_3 - a_3)x_3/(b_2 - a_2)$, and $x_1 = -b_2x_2 - b_3x_3$. If $x_3 = 0$, then $x_2 = 0$ and so $x_1 = 0$. So, $x_3 \neq 0$. Hence,

$$\begin{aligned} L &= L(x_1,x_2,x_3) \\ &= L([b_2(b_3 - a_3)/(b_2 - a_2) - b_3]x_3,\ -(b_3 - a_3)x_3/(b_2 - a_2),\ x_3) \\ &= L(b_2(b_3 - a_3) - b_3(b_2 - a_2),\ -(b_3 - a_3),\ b_2 - a_2) \\ &= L(a_2b_3 - b_2a_3,\ -(b_3 - a_3),\ b_2 - a_2) = \mathbf{L}(\mathbf{a},\mathbf{b}). \end{aligned}$$

For (2), we again get $\mathbf{L}(\mathbf{a},\mathbf{b}) = L(x_1,x_2,x_3) = L$, and that $\mathbf{L}(\mathbf{a},\mathbf{b}) \in \mathscr{L}$ since $b_2 - a_2 \neq 0$.

Continuing, for (3), with $a_1 = b_1 = 1$, $a_2 = b_2$, and $a_3 \neq b_3$, we obtain $(b_3 - a_3)x_3 = 0$. If $a_3 = b_3$, then $\mathbf{a} = \mathbf{b}$. So, $b_3 - a_3 \neq 0$, and $x_3 = 0$.

Now

$$\mathbf{L}(\mathbf{a},\mathbf{b}) = L(a_2(b_3 - a_3), -(b_3 - a_3), 0) = L(-a_2, 1, 0),$$

and $L(x_1,x_2,x_3) = L(x_1,x_2,0)$. With $x_3 = 0$ and $a_2 = b_2$, (11:1) yields $x_1 = -a_2x_2$, and so $L(x_1,x_2,x_3) = L(-a_2x_2,x_2,0)$. If $x_2 = 0$, then $x_1 = x_2 = x_3 = 0$. So, $x_2 \neq 0$, and $L = L(x_1,x_2,x_3) = L(-a_2,1,0) = \mathbf{L}(\mathbf{a},\mathbf{b})$, and $\mathbf{L}(\mathbf{a},\mathbf{b}) \in \mathscr{L}$ since $1 \neq 0$.

For (4), $a_1 = b_1$, and $a_2 \neq b_2$. There will be no loss in assuming that $a_2 = 0$ and $b_2 = 1$. From (11:1), we have $a_3x_3 = 0$ and $x_2 = -b_3x_3$. With $a_1 = a_2 = 0$, we have $a_3 \neq 0$, so $x_3 = 0$ and $x_2 = 0$. Consequently, $x_1 \neq 0$, and $L(x_1,x_2,x_3) = L(x_1,0,0) = L(1,0,0)$. Also, $\mathbf{L}(\mathbf{a},\mathbf{b}) = L(-b_2a_3,0,0) = L(-a_3,0,0) = L(1,0,0) = L(x_1,x_2,x_3) = L$. With $1 \neq 0$, we are assured that $\mathbf{L}(\mathbf{a},\mathbf{b}) \in \mathscr{L}$.

For (5), $a_1 = b_1 = 0$, $a_2 = b_2 = 1$, and $a_3 \neq b_3$, we obtain $(b_3 - a_3)x_3 = 0$ from (11:1). Since $a_3 \neq b_3$, we have $x_3 = 0$. But $x_2 = -a_3x_3$, so $x_2 = 0$, also. Consequently, $L(x_1,x_2,x_3) = L(x_1,0,0) = L(1,0,0)$, since $x_1 \neq 0$. With $b_3 - a_3 \neq 0$, we have $\mathbf{L}(\mathbf{a},\mathbf{b}) = L(b_3 - a_3, 0, 0) = L(1,0,0)$. Again, $\mathbf{L}(\mathbf{a},\mathbf{b}) = L(x_1,x_2,x_3) = L$ and $\mathbf{L}(\mathbf{a},\mathbf{b}) \in \mathscr{L}$, since $1 \neq 0$.

In all five cases, we obtain $\mathbf{L}(\mathbf{a},\mathbf{b}) = L$, and so $\mathbf{L}(\mathbf{a},\mathbf{b})$ is the unique line in $\mathscr{L}$ on both **a** and **b**. □

The next step is to show that two distinct lines are on a unique point.

(2) *Let $A = [a_1,a_2,a_3]$, $B = [b_1,b_2,b_3] \in \mathscr{L}$ be two distinct lines. Then there is a unique point $\mathbf{x} = (x_1;x_2;x_3) \in P$ such that $\mathbf{x}IA$ and $\mathbf{x}IB$.*

Proof of **(2)**. Let $A = [a_1,a_2,a_3]$ and $B = [b_1,b_2,b_3]$ be two distinct lines in $\mathscr{L}$. Define $\mathbf{p}(A,B) = p(a_2b_3 - b_2a_3,\ -(a_1b_3 - b_1a_3),\ a_1b_2 - b_1a_2)$. As in the proof of **(1)**, we have that

$$\mathbf{p}(A,B) \bullet A = \mathbf{p}(A,B) \bullet B = 0.$$

We must be assured that $\mathbf{p}(A,B) \in P$ and that $\mathbf{p}(A,B)$ is the unique point on both A and B.

Suppose $\mathbf{x}IA$ and $\mathbf{x}IB$ for some $\mathbf{x} = (x_1;x_2;x_3) \in P$. Then $\mathbf{x} \bullet A = \mathbf{x} \bullet B = 0$, which means that

$$\begin{cases} a_1x_1 + a_2x_2 + a_3x_3 = 0, \\ b_1x_1 + b_2x_2 + b_3x_3 = 0. \end{cases} \tag{11:2}$$

Notice that (11:2) is the same as (11:1).

Since $A \neq B$, there are the same five cases as in the proof of **(1)**. The proof parallels that of **(1)** almost exactly to get (a) that $\mathbf{x} = \mathbf{p}(A,B)$, and (b) that $\mathbf{p}(A,B) \in P$. So, we will conclude that two distinct lines $A, B \in \mathscr{L}$ are on a unique point $\mathbf{p}(A,B) \in P$. □

We proceed now to show that each line is on at least three distinct points.

(3) *Let $A = [a_1, a_2, a_3] \in \mathscr{L}$. Then there are at least three distinct points $x, y, z \in P$ such that xIA, yIA, and zIA.*

Proof of **(3)**. We must find points $\mathbf{s} = (s_1; s_2; s_3) \in P$ so that $\mathbf{s}IA$, that is, $\mathbf{s} \bullet A = 0$. In more detail,

$$a_1 s_1 + a_2 s_2 + a_3 s_3 = 0 \tag{11:3}$$

must be true.

There are three cases to be considered: (1) $a_1 = 1$; (2) $a_1 = 0$ and $a_2 = 1$; and (3) $a_1 = a_2 = 0$ and $a_3 = 1$.

Case (3) is the easiest, so we will do that one first. Simply let $\mathbf{x} = (1;0;0)$, $\mathbf{y} = (0;1;0)$ and $\mathbf{z} = (1;1;0)$. These are three distinct points, and clearly, $\mathbf{s} \bullet A = 0$ for any $\mathbf{s} \in \{\mathbf{x}, \mathbf{y}, \mathbf{z}\}$.

Case (2) is almost as easy. Let

$$\mathbf{x} = (1;0;0), \quad \mathbf{z} = (1;-a_3;1), \quad \text{and} \quad \mathbf{y} = \begin{cases} (0;0;1), & \text{if } a_3 = 0; \\ (0;1;-a_3^{-1}), & \text{if } a_3 \neq 0. \end{cases}$$

With case (2), we have $a_1 = 0$ and $a_2 = 1$. So, to satisfy (11:3) is to satisfy $s_2 + a_3 s_3 = 0$. Easily, $\mathbf{x} = (1;0;0)$ satisfies $s_2 + a_3 s_3 = 0$, as does $\mathbf{z} = (1;-a_3;1)$. If $a_3 = 0$, then $\mathbf{y} = (0;0;1)$ satisfies $s_2 + a_3 s_3 = 0$. If $a_3 \neq 0$, then $\mathbf{y} = (0;1;-a_3^{-1})$ satisfies $s_2 + a_3 s_3 = 0$. In either case, $\mathbf{x}$, $\mathbf{y}$, and $\mathbf{z}$ are three distinct points.

We must turn our attention now to case (1). Here, (11:1) becomes

$$s_1 + a_2 s_2 + a_3 s_3 = 0. \tag{11:4}$$

Let

$$\mathbf{x} = \begin{cases} (0;1;0), & \text{if } a_2 = 0; \\ (1;-a_2^{-1};0), & \text{if } a_2 \neq 0; \end{cases}$$

$$\mathbf{y} = \begin{cases} (0;0;1), & \text{if } a_3 = 0; \\ (1;0;-a_3^{-1}), & \text{if } a_3 \neq 0; \end{cases}$$

and

$$\mathbf{z} = \begin{cases} (0;1;1), & \text{if } a_2 = -a_3; \\ (1;-(a_2+a_3)^{-1};-(a_2+a_3)^{-1}), & \text{if } a_2 \neq -a_3. \end{cases}$$

We must check all eight subcases.

If $a_2 = 0$ and $a_3 = 0$, then $a_2 = -a_3$, and so $a_2 = 0$, $a_3 = 0$ and $a_2 \neq a_3$ is impossible. So, for $a_2 = 0$ and $a_3 = 0$, then

$$\mathbf{x} = (0;1;0), \mathbf{y} = (0;0;1) \text{ and } \mathbf{z} = (0;1;1).$$

Also, (11:4) becomes $s_1 = 0$. Hence, **x**, **y**, and **z** are three distinct points on A.

If $a_2 = 0$ and $a_3 \neq 0$, then $a_2 = -a_3$ is impossible. So, we have $a_2 \neq a_3$. Then

$$\mathbf{x} = (0;1;0), \mathbf{y} = (1;0;-a_3^{-1}), \text{ and } \mathbf{z} = (1;-a_3^{-1};-a_3^{-1}),$$

while (11:4) becomes $s_1 + a_3 s_3 = 0$. Clearly, **x**, **y**, and **z** are then three distinct points on A.

This takes care of the four cases when $a_2 = 0$. There are also the four cases when $a_2 \neq 0$.

Suppose $a_2 \neq 0$ and $a_3 = 0$. Then $a_2 = -a_3$ is impossible, so we take $a_2 = -a_3$. Then

$$\mathbf{x} = (1;-a_2^{-1};0), \mathbf{y} = (0;0;1), \text{ and } \mathbf{z} = (1;-a_2^{-1};-a_2^{-1}),$$

while (11:4) becomes $s_1 + a_2 s_2 = 0$. We can easily see that **x**, **y**, and **z** are three distinct points on A.

Suppose $a_2 \neq 0$ and $a_3 \neq 0$. If $a_2 = -a_3$, then (11:4) becomes

$$s_1 - a_3 s_2 + a_3 s_3 = 0,$$

and we have

$$\mathbf{x} = (1;-a_2^{-1};0), \mathbf{y} = (1;0;-a_3^{-1}), \text{ and } \mathbf{z} = (0;1;1).$$

Since $1-a_3(-a_2^{-1})+a_30=1+a_3a_2^{-1}=1-a_2a_2^{-1}=0$, $1-a_30+a_3(-a_3^{-1})=0$, and $0-a_3+a_3=0$, we have that $\mathbf{x},\mathbf{y},\mathbf{z}\in P$ are three distinct points.

Finally, suppose $a_2\neq 0$, $a_3\neq 0$, and $a_2\neq -a_3$. Then

$$\mathbf{x}=(1;-a_2^{-1};0),\ \mathbf{y}=(1;0;-a_3^{-1}),\ \text{and}\ \mathbf{z}=(1;-(a_2+a_3)^{-1};-(a_2+a_3)^{-1}).$$

Now $1+a_2(-a_2^{-1})+a_30=0$, $1+a_20+a_3(-a_3^{-1})=0$, and

$$\begin{aligned}&1+a_2[-(a_2+a_3)^{-1}]+a_3[-(a_2+a_3)^{-1}]\\&=1-a_2(a_2+a_3)^{-1}-a_3(a_2+a_3)^{-1}\\&=1-(a_2+a_3)(a_2+a_3)^{-1}=0,\end{aligned}$$

so $\mathbf{x}$, $\mathbf{y}$, and $\mathbf{z}$ are three distinct points on A. This completes **(3)**. □

To complete the proof of Theorem 11.6, there remains to show that (d) of Definition 11.3 is satisfied.

(4) *If $A=[a_1,a_2,a_3]\in\mathscr{L}$, then there is an $\mathbf{x}=(x_1;x_2;x_3)\in P$ such that $(\mathbf{x},A)\notin I$.*

Proof of **(4)**. If **(4)** is not true, then $(\mathbf{x},A)\in I$ for each $\mathbf{x}\in P$, that is, $\mathbf{x}\bullet A=0$ for each $\mathbf{x}\in P$. This means

$$a_1x_1+a_2x_2+a_3x_3=0 \tag{11:5}$$

for every $\mathbf{x}=(x_1;x_2;x_3)\in P$.

Since $A\in\mathscr{L}$, one of the values a_1, a_2, a_3 is not 0. If $a_1\neq 0$, then we have $a_1\cdot 1+a_2\cdot 0+a_3\cdot 0=a_1\neq 0$. So, let $\mathbf{x}=(1;0;0)$ and conclude that $\mathbf{x}\bullet A\neq 0$, and so $(\mathbf{x},A)\notin I$. Similarly, if $a_2\neq 0$, let $\mathbf{x}=(0;1;0)$, and if $a_3\neq 0$, let $\mathbf{x}=(0;0;1)$. In all cases, $\mathbf{x}\bullet A\neq 0$, $\mathbf{x}\in P$, and so $(\mathbf{x},A)\notin I$. □

With **(1)**, **(2)**, **(3)**, and **(4)**, we can conclude that $(P,\mathscr{L},I)$ is indeed a projective plane. □

What does Theorem 11.6 mean?

Example 11.7. Let $F = \mathbb{R}$, the field of real numbers. Consider $E = \mathbb{R}^3 = \{(a,b,c) \mid a,b,c \in \mathbb{R}\}$, which is the traditional three-dimensional euclidean space. For $\mathbf{x} \in P$, $\mathbf{x} = (x_1;x_2;x_3)$, we can think of $(x_1;x_2;x_3)$ representing all scalar multiples $\lambda(x_1,x_2,x_3) = (\lambda x_1, \lambda x_2, \lambda x_3)$, $\lambda \in \mathbb{R}$. So, $\mathbf{x}$ represents the line in E passing through the points $(0,0,0)$ and (x_1,x_2,x_3). Consequently, P would be the family of all lines in E passing through $(0,0,0)$. These are our "points" in the projective plane $(P,\mathscr{L},I)$ of Theorem 11.6.

For $L = [x_1,x_2,x_3] \in \mathscr{L}$, the "points" on the "line" L are those elements $\mathbf{a} = (a_1;a_2;a_3) \in P$ such that

$$\mathbf{a} \bullet L = a_1x_1 + a_2x_2 + a_3x_3 = 0.$$

That is, the "points" on the "line" L are those lines in E which are perpendicular to the line through $(0,0,0)$ and (a_1,a_2,a_3). This means, L represents the plane in E passing through $(0,0,0)$ and perpendicular to the line through $(0,0,0)$ and (a_1,a_2,a_3). Hence, $\mathscr{L}$ would be the family of all planes in E passing through $(0,0,0)$.

With this in mind, it is easy to visualize, or believe, that $(P,\mathscr{L},I)$ is a projective plane. Two distinct planes in E passing through $(0,0,0)$ are on a unique line through $(0,0,0)$; two distinct lines in E which pass through $(0,0,0)$ are on a unique plane through $(0,0,0)$. Any plane through $(0,0,0)$ has on it three distinct lines through $(0,0,0)$, and for any plane through $(0,0,0)$, there is a line through $(0,0,0)$ which is not in that plane. This is the essence of what we have proven in Theorem 11.6.

Exercise 11.8. Let $F = \mathbb{Z}_2$. Compute the projective plane $(P,\mathscr{L},I)$ of Theorem 11.6. Where have you essentially seen this projective plane before?

Exercise 11.9. Let $F = \mathbb{Z}_3$. Compute the projective plane $(P,\mathscr{L},I)$ of Theorem 11.6.

Exercise 11.10. Let F be a finite field with $|F| = q$. Let $(P,\mathscr{L},I)$ be the projective plane of Theorem 11.6. Compute $|P|$ and $|\mathscr{L}|$.

12 Transversal designs

There is a curious geometric structure that can be made from any of our affine planes of Theorem 3.4. For any field F, start with the affine plane $(F \times F, \mathscr{L}_F)$ of Theorem 3.4. From $\mathscr{L}_F$, we want to take away a family of parallel lines $\mathscr{L}(m)$, $m \in F \cup \{\infty\}$. To be specific, and to help in visualization, we remove $\mathscr{L}(\infty)$. So, let $\mathscr{B}_F = \mathscr{L}_F \setminus \mathscr{L}(\infty)$. The lines in $\mathscr{L}(\infty)$ form a partition of $F \times F$, so $\mathscr{L}(\infty) = \{L(\infty, a) \mid a \in F\}$, $\cup_{a \in F} L(\infty, a) = F \times F$, and if $a, b \in F$ with $a \neq b$, then $L(\infty, a) \cap L(\infty, b) = \varnothing$.

Consider the structure $(F \times F, \mathscr{B}_F, \in, \mathscr{L}(\infty))$. We have (a) $|L(\infty, a)| = |L(\infty, b)|$ for any $a, b \in F$, and (b) $\mathscr{L}(\infty)$ is a partition of $F \times F$. Next, take any $L \in \mathscr{B}_F$, so $L = L(m, b)$ for some $m, b \in F$. Since $(F \times F, \mathscr{L}_F)$ is an affine plane, we have that (c) $|L \cap L(\infty, a)| = 1$ for each $L(\infty, a) \in \mathscr{L}(\infty)$. Finally, if $x, y \in F \times F$ and $x \neq y$, then x and y belong to a unique line $M \in \mathscr{L}_F$. Either $M = L(\infty, a) \in \mathscr{L}(\infty)$ for some $a \in F$, or $M = L(m, b) \in \mathscr{B}_F$ for some $m, b \in F$. So, we may say that (d) the number of blocks from $\mathscr{B}_F$ on both x and y is $1 - \delta(x, y)$ where

$$\delta(x, y) = \begin{cases} 1, & \text{if } M \in \mathscr{L}(\infty); \\ 0, & \text{if } M \notin \mathscr{L}(\infty). \end{cases}$$

This motivates the idea of a *transversal design*.

Definitions 12.1. Suppose $\mathscr{G}$ is a family of subsets of X, where $(X, \mathscr{B}, I)$ is an incidence structure. Then $(X, \mathscr{B}, I, \mathscr{G})$ is a *transversal design* if the following are true.

(a) If $G_1, G_2 \in \mathscr{G}$, then $|G_1| = |G_2|$.

(b) $\mathscr{G}$ is a partition of X.

(c) If $L \in \mathscr{B}$ and $G \in \mathscr{G}$, then there is exactly one point $x \in G$ such that $(x, L) \in I$.

(d) If $x \in G_1, y \in G_2, x \neq y$, and $G_1, G_2 \in \mathcal{G}$, then the number of blocks $B \in \mathcal{B}$ for which xIB and yIB is $1 - \delta(x,y)$, where

$$\delta(x,y) = \begin{cases} 1, & \text{if } G_1 = G_2; \\ 0, & \text{if } G_1 \neq G_2. \end{cases}$$

If $0 < |X| < \infty$, then $|X| = |\mathcal{G}||G|$ for any $G \in \mathcal{G}$. If $s = |G|$ and $r = |\mathcal{G}|$, then we refer to $(X, \mathcal{B}, I, \mathcal{G})$ as a *transversal design with r plots of size s*, or simply a $TD[r;s]$. In this case, we refer to the $G \in \mathcal{G}$ as *plots* , and $|G| = s$ is the *size* of a plot.

We have already proven

Theorem 12.2. *For any field F, $(F \times F, \mathcal{B}_F, \in, \mathcal{L}(\infty))$ is a transversal design. If $|F| = q < \infty$, then $(F \times F, \mathcal{B}_F, \in, \mathcal{L}(\infty)$ is a $TD[q;q]$.*

Nets also give us transversal designs, but in a different way. We first extend the idea of a net.

Definitions 12.3. An incidence structure $(P, \mathcal{B}, I)$ is a *net* if the following conditions are met.

(a) If $L, M \in \mathcal{B}$, then $|L| = |M|$.
(b) If $x, y \in P$ and $x \neq y$, then there is at most one $L \in \mathcal{B}$ with xIL and yIL.
(c) $\mathcal{B}$ has a partition $\mathbf{B} = \{\mathcal{B}_i \mid i \in T\}$, for some index set T, where each $\mathcal{B}_i$ is a partition of P, that is, each $x \in P$ is incident to exactly one $L \in \mathcal{B}_i$.
(d) If $L \in \mathcal{B}_i$, $M \in \mathcal{B}_j$, $\mathcal{B}_i, \mathcal{B}_j \in \mathbf{B}$, and $\mathcal{B}_i \neq \mathcal{B}_j$, then there is exactly one $x \in P$ such that $(x, L), (x, M) \in I$.

If $0 < |P| < \infty$ and each $L \in \mathcal{B}$ has $|L| = s$, and if $|\mathbf{B}| = r$, then $(P, \mathcal{B}, I)$ is an *(s,r)-net*.

An application of duality is now given.

Theorem 12.4. *Let $(P,\mathscr{B},I)$ be a net with $|P|<\infty$ and $|\mathscr{B}|<\infty$. Then the dual $(\mathscr{B},P,J)$ yields a transversal design $(\mathscr{B},P,J,\mathbf{B})$, where $\mathbf{B}$ is the partition of $\mathscr{B}$ provided by* (c) *of Definitions* 12.3. *If $(P,\mathscr{B},I)$ is an (s,r)-net, then $(\mathscr{B},P,J,\mathbf{B})$ is a $TD[r;s]$.*

Proof. Since $(P,\mathscr{B},I)$ is a net, $\mathscr{B}$ has a partition $\mathbf{B}=\{\mathscr{B}_i \mid i\in T\}$ for some index set T, and where each $\mathscr{B}_i$ is a partition of P. Hence, (b) of Definitions 12.1 is satisfied. Suppose $\mathscr{B}_i,\mathscr{B}_j\in\mathbf{B}$. So, $\mathscr{B}_i$ and $\mathscr{B}_j$ are each a partition of P. Since $(P,\mathscr{B},I)$ is a net, we have, for $L\in\mathscr{B}_i\subseteq\mathscr{B}$ and $M\in\mathscr{B}_j\subseteq\mathscr{B}$, that $|L|=|M|$, and since each $\mathscr{B}_i$ is a partition of P, it follows that $|\mathscr{B}_i||L|=|P|=|\mathscr{B}_j||M|$. Hence, $|\mathscr{B}_i|=|\mathscr{B}_j|$, and this gives us (a) of Definitions 12.1.

Suppose $x\in P$ and $\mathscr{B}_i\in\mathbf{B}$. Since $\mathscr{B}_i$ is a partition of P, there is exactly one $L\in\mathscr{B}_i$ such that $(x,L)\in I$, and so $(L,x)\in J=I^{-1}$ for exactly one $L\in\mathscr{B}_i$. This gives us (c) of Definitions 12.1.

Finally, take $\mathscr{B}_i,\mathscr{B}_j\in\mathbf{B}$. Either $\mathscr{B}_i=\mathscr{B}_j$ or $\mathscr{B}_i\neq\mathscr{B}_j$. If $\mathscr{B}_i=\mathscr{B}_j$, and if $L,M\in\mathscr{B}_i=\mathscr{B}_j$, with $L\neq M$, then, since $\mathscr{B}_i=\mathscr{B}_j$ is a partition of P, it is impossible that $(x,L),(x,M)\in I$. Hence, $(L,x),(M,x)\in J=I^{-1}$ is impossible for each $x\in P$. So, the number of $x\in P$ on both L and M is 0. If $\mathscr{B}_i\neq\mathscr{B}_j$, $L\in\mathscr{B}_i$, $M\in\mathscr{B}_j$, then $L\neq M$. By (d) of Definitions 12.3, there is exactly one $x\in P$ such that $(x,L),(x,M)\in I$, and so there is exactly one $x\in P$ such that $(L,x),(M,x)\in J=I^{-1}$. Hence, (d) of Definitions 12.1 is satisfied.

Having satisfied (a), (b), (c), and (d) of Definitions 12.1, we can proudly conclude that $(\mathscr{B},P,J,\mathbf{B})$ is a transversal design. □

13 Tactical configurations and association schemes

This is your lucky section. Here, we introduce a very complicated structure, partially in preparation to introduce partially balanced incomplete block designs, and partially in preparation for an application to coding theory. This structure is an *association scheme*. We also introduce the idea of a *tactical configuration*, which is a generalization of a balanced incomplete block design.

Definition 13.1. Let $(N,\mathscr{T},\in)$ be an incidence structure. Suppose each $B\in\mathscr{T}$ is incident to the same number k of points $x\in N$, and suppose each $x\in N$ is incident to the same number r of blocks $B\in\mathscr{T}$. Then $(N,\mathscr{T},\in)$ is a *tactical configuration*.

Note that every BIBD is a tactical configuration.

Definition 13.2. We will now introduce the idea of an *association scheme*. Start with a tactical configuration $(N,\mathscr{T},\in)$ where N and therefore $\mathscr{T}$ are finite. We must have a partition of the family of all two-element subsets. So, if $|N|=v$, we select from $\mathscr{P}(N)=\{A\mid A\subseteq N\}$ the $\binom{v}{2}$ subsets of N each with exactly two elements. Let $\mathscr{P}_2(N)=\{A\subseteq N\mid |A|=2\}$, and suppose $\mathscr{A}=\{A_1,A_2,\dots,A_m\}$ is a partition of $\mathscr{P}_2(N)$. So, each A_j is a family of two-element subsets of N, and for each $\{x,y\}\subseteq N$, $x\neq y$, there is exactly one $A_i\in\mathscr{A}$ with $\{x,y\}\in A_i$.

Let $I_m=\{1,2,\dots,m\}$ and $(h,i,j)\in I_m^3$. For any $\{x,y\}\in A_h$, define two sets

$$A_{hi}(x,y)=\{z\in N\mid \{x,z\}\in A_i\}$$

and

$$B_{hj}(x,y)=\{z\in N\mid \{y,z\}\in A_j\}.$$

Then define

$$P^h_{ij}(x,y)=|A_{hi}(x,y)\cap B_{hj}(x,y)|.$$

It may be that there is a number $p^h_{ij}\in\mathbb{Z}$ so that if $\{x,y\}\in A_h$, then

$$P^h_{ij}(x,y)=p^h_{ij}=P^h_{ij}(y,x).$$

If the numbers p^h_{ij} exist for every $(h,i,j)\in I_m^3$, then $(N,\mathscr{T},\in,\mathscr{A})$ is an *association scheme* on N.

The numbers p^h_{ij} express the following idea. If $\{x,y\}\in A_h$, then there are exactly p^h_{ij} elements $z\in N$ such that both $\{x,z\}\in A_i$ and $\{y,z\}\in A_j$.

Example 13.3. Let $N = \{0,1,\ldots,8\}$, and let $\mathscr{B} = \{B_1, B_2, \ldots, B_9\}$ where

$$\begin{aligned} B_1 &= \{0,1,8,2,7\}, & B_4 &= \{3,4,2,5,1\}, & B_7 &= \{6,7,5,8,4\}, \\ B_2 &= \{1,2,0,3,8\}, & B_5 &= \{4,5,3,6,2\}, & B_8 &= \{7,8,6,0,5\}, \\ B_3 &= \{2,3,1,4,0\}, & B_6 &= \{5,6,4,7,3\}, & B_9 &= \{8,0,7,1,6\}. \end{aligned}$$

Then $(N, \mathscr{B}, \in)$ is a tactical configuration with $k = 5$ and $r = 5$.

We need a partition $\mathscr{A} = \{A_1, A_2, \ldots, A_m\}$ of $\mathscr{P}_2(N)$. Such a partition is given for $m = 4$ and

$$\begin{aligned} A_1 &= \{\{0,1\},\{0,8\},\{1,2\},\{2,3\},\{3,4\},\{4,5\},\{5,6\},\{6,7\},\{7,8\}\}, \\ A_2 &= \{\{0,2\},\{0,7\},\{1,3\},\{2,4\},\{3,5\},\{4,6\},\{5,7\},\{6,8\},\{8,1\}\}, \\ A_3 &= \{\{0,3\},\{0,6\},\{1,4\},\{2,5\},\{3,6\},\{4,7\},\{5,8\},\{7,1\},\{8,2\}\}, \\ A_4 &= \{\{0,4\},\{0,5\},\{1,5\},\{2,6\},\{3,7\},\{4,8\},\{6,1\},\{7,2\},\{8,3\}\}. \end{aligned}$$

One can, with patience, compute that $p_{ij}^h = 1$ if (h,i,j) is any permutation of $(1,1,2)$, $(1,2,3)$, $(1,3,4)$, or $(1,4,4)$. Otherwise, $p_{ij}^h = 0$.

In Example 6.3, we defined two IP-nearrings $(\mathbb{C},+,*)$ and $(\mathbb{C},+,\circ)$. In §8, we used $(\mathbb{C},+,\circ)$ to motivate constructing $(N, \mathscr{B}^*, \in)$ for any IP-nearring $(N,+,\cdot)$. If $|N| < \infty$, then Theorem 8.3 assures us that $(N, \mathscr{B}^*, \in)$ is a BIBD.

Next, we take motivation from $(\mathbb{C},+,*)$ of Example 6.3. Let $(N,+,\cdot)$ be an IP-nearring, and for $a, b \in N$, $a \neq 0$, construct $Na + b = \{na + b \mid n \in N\}$. Let $\mathscr{B} = \{Na + b \mid a, b \in N,\ a \neq 0\}$. We will soon see that $(N, \mathscr{B}, \in)$ is a tactical configuration. Before doing so, we need a way to count the number of blocks $Na + b \in \mathscr{B}$.

Theorem 13.4. *Let $(N,+,\cdot)$ be an IP-nearring with blocks*

$$\mathscr{B} = \{Na + b \mid a, b \in N, a \neq 0\}.$$

Let $Na + b$, $Na' + b' \in \mathscr{B}$. Then $Na + b = Na' + b'$ if and only if either

(a) *$Na = Na'$, $(Na,+)$ is a subgroup of $(N,+)$, and $b' - b \in Na$;*

or

(b) $Na = Na'$, $(Na,+)$ *is not a subgroup of* $(N,+)$, *and* $b = b'$.

Proof. If (a) is true, start with $Na = Na'$ and with $(Na,+)$ being a subgroup of $(N,+)$. Then $Na + x = Na + y$ if and only if $x - y \in Na$. But $b' - b \in Na$, so $Na + b' = Na + b$. Since $Na = Na'$, we have $Na' + b' = Na + b$.

If (b) is true, then with $Na = Na'$ and $b = b'$, we easily have $Na + b = Na' + b = Na' + b'$.

We must now focus on the converse. Suppose $Na + b = Na' + b'$. Then $Na = Na' + (b' - b) = Na' + c$, where $c = b' - b$. Either $(Na,+)$ is a subgroup of $(N,+)$ or it is not. Suppose Na is a subgroup. Then from $Na = Na' + c$, we get $Na' = Na - c$, a coset of Na containing 0, since $0 = 0a' \in Na'$. Therefore, $Na - c = Na$, since $0 \in Na - c$ and Na is the only coset of Na containing 0. Hence, $Na = Na - c = Na'$, and $c = b' - b \in Na$. So, if $(Na,+)$ is a subgroup of $(N,+)$, then (a) is valid.

The other alternative is that $(Na,+)$ is not a group. If $c = 0$, then $b = b'$, and we have $Na = Na' + c = Na'$ and $b = b'$. This is exactly (b). To finish the proof of the theorem, we assume that Na is not a subgroup and $c \neq 0$. This will lead to a contradiction. We will show that when we insist that $Na = Na' + c$ and $c \neq 0$, then $(Na,+)$ must be a group, a subgroup of $(N,+)$.

From $Na = Na' + c$, we obtain $0 + c = c \in Na' + c = Na$. Hence, $Nc = Na$. Similarly, from $Na - c = Na + (-c) = Na'$, we have $-c \in Na'$, so $N(-c) = Na'$. Now we have $Nc = N(-c) + c$. Take $m, n \in N$ so that $mc - nc \neq 0$ and $0 \notin \{mc, nc\}$. Then, multiplying $Nc = N(-c) + c$ on the left by m and n, we get $Nc = N(-c) + mc$ and $Nc = N(-c) + nc$. From $N(-c) + mc = N(-c) + nc$, we have $N(-c) = N(-c) + (nc - mc)$, and so $nc - mc \in N(-c)$. This means that there is an $s \in N$ such that $s(-c) = -sc = nc - mc$, hence $sc = mc - nc \in Nc$. Also, with $mc \in Nc = N(-c) + nc$, we obtain $mc = t(-c) + nc$ for some $t \in N$, and so $mc - nc = t(-c) \in N(-c)$. We have $0 \neq mc - nc \in Nc \cap N(-c)$, and since

$Nc \cap N(-c) = \{0\}$ if $Nc \neq N(-c)$, we must conclude that $Nc = N(-c)$.

If $uc \in Nc$, then there is a $w \in N$ so that $uc = w(-c) = -(wc)$. Hence, $-(uc) = wc \in Nc$.

The equation $Nc = N(-c) + c$ becomes $Nc = Nc + c$, since $Nc = N(-c)$. For any $u, v \in N$, we have $Nc = Nc + uc$ and $Nc = Nc + vc$. From $Nc + uc = Nc + vc$, we obtain $Nc = Nc + (vc - uc)$, which in turn gives us $vc - uc \in Nc$.

So, for $x, y \in Nc$, we have $-y \in Nc$ and $x - y \in Nc$. From $x, -y \in Nc$, we get $x - (-y) = x + y \in Nc$. So, $(Nc, +)$ is a subgroup of $(N, +)$, as promised. This completes the proof of Theorem 13.4. □

A particular "basic block" $Na \in \mathscr{B}$ is either a subgroup of N or it is not. Those which are will often be denoted by Ng_i, where $i \in S$ and S is an index set with $|S| = s$. Similarly, those $Na \in \mathscr{B}$ which are not subgroups will often be denoted by Ne_j, where $j \in T$ and T is an index set with $|T| = t$.

Theorem 13.5. *For any IP-nearring $(N, +, \cdot)$, the incidence structure $(N, \mathscr{B}, \in)$ is a tactical configuration with $k = |N/{=_m}| = |Na|$ for any nonzero $a \in N$, and $r = s + tk$ where s and t are defined in the preceding paragraph.*

Proof. Since each $|Na| = |N/{=_m}|$ for any $a \neq 0$, $a \in N$, we have each $|Na + b| = |N/{=_m}|$ for every $Na + b \in \mathscr{B}$. Thus, $|Na + b| = k = |N/{=_m}| = |Na|$ for any nonzero $a \in N$.

Since each Ng_i is a subgroup, an arbitrary $x \in N$ belongs to exactly the coset $Ng_i + x$ of Ng_i. So, x belongs to the s blocks $Ng_i + x$, $i \in S$. There are $s = |S|$ such $Ng_i + x$. For each $j \in T$, $|Ne_j| = k$, and for each of the k distinct elements $u_j \in Ne_j$, there is a unique $y_j \in N$ such that $x = u_j + y_j \in Ne_j + y_j$, and x cannot belong to any other $Ne_j + u$. Hence, x belongs to k distinct blocks $Ne_j + y_j$ for each of the t basic blocks Ne_j. Altogether, x belongs to $s + tk$ distinct blocks in $\mathscr{B}$. We know these blocks are distinct by the force of Theorem 13.4. □

Corollary 13.6. *If N is a finite IP-nearring with $|N| = v$, then the tactical configuration $(N, \mathscr{B}, \in)$ has $|\mathscr{B}| = tv + (vs/k)$.*

Proof. Apply Theorem 13.4. Each of the s subgroups Ng_i gives v/k blocks $Ng_i + u$. So, the Ng_i collectively give $(vs)/k$ blocks altogether. Each of the t basic blocks Ne_i give v distinct blocks $Ne_j + u$, one for each $u \in N$. So, the Ne_j collectively give tv blocks. Putting everything together, we get $|\mathscr{B}| = tv + (vs/k)$. □

There are finite IP-nearrings $(N, +, \cdot)$ where $s = |S| = 1$. For an example, see Example 7.9. With such IP-nearrings, one can easily construct an association scheme.

Start with a finite IP-nearring $(N, +, \cdot)$ with $|N| = v$ and $s = 1$. That is, Ng denotes the only basic block in $\mathscr{B}$ which is a subgroup of N. We want to construct an association scheme $(N, \mathscr{B}, \in, \mathscr{A})$. We already have $(N, \mathscr{B}, \in)$, we need $\mathscr{A}$. Let $\mathscr{A} = \{A_1, A_2\}$ where

$$A_1 = \{\{x,y\} \in \mathscr{P}_2(N) \mid x - y \in Ng\}$$

and

$$A_2 = \mathscr{P}_2(N) \setminus A_1 = \{\{x,y\} \in \mathscr{P}_2(N) \mid x - y \notin Ng\}.$$

For each of the eight $(h,i,j) \in I_2^3$, we need to show the existence of the numbers p_{ij}^h.

Exactly what does p_{ij}^h represent? It says, for any pair $\{x,y\} \in A_h$, there are exactly p_{ij}^h distinct elements $z \in N$ for which $\{x,z\} \in A_i$ and $\{y,z\} \in A_j$.

So, suppose $\{x,y\} \in A_1$, that is, suppose that $h = 1$. Then $x - y \in Ng$. This is true if and only if $Ng + x = Ng + y$. If $\{x,t\} \in A_1$, then $t - x$, $x - y \in Ng$, so $(t - x) + (x - y) = t - y \in Ng$. Hence, $\{y,t\} \in A_1$, and conversely. So, $\{x,t\} \in A_1$ if and only if $\{y,t\} \in A_1$. Also, $\{x,t\} \notin A_1$ if and only if $\{y,t\} \notin A_1$. This says that $\{x,t\} \in A_2$ if and only if $\{y,t\} \in A_2$.

Since $x, y \in Ng + x = Ng + y$, $x \neq y$, and $|Ng| = k + 1$, we see that there are exactly $k - 1$ elements $z \in N$ such that $\{x,z\} \in A_1$ and $\{y,z\} \in A_1$. Hence, $p_{11}^1 = k - 1$. These are the $k - 1$ elements of $(Ng + x) \setminus \{x,y\}$. All $t \in N \setminus (Ng + x)$

have the property that $\{x,t\} \notin A_1$ and $\{y,t\} \notin A_1$. Hence, for such a t, we have $\{x,t\} \in A_2$ and $\{y,t\} \in A_2$, and these are the only such elements. Therefore, $p^1_{22} = v-(k+1) = v-k-1$.

If $\{x,t\} \in A_1$, then $\{y,t\} \in A_1$, and so $\{y,t\} \in A_2$ is impossible. Hence, $p^1_{12} = 0$. Similarly, if $\{x,t\} \in A_2$, then $\{y,t\} \in A_2$, and so $\{y,t\} \in A_1$ is impossible. Therefore, $p^1_{21} = 0$.

We now consider the case where $h = 2$. So, take $\{x,y\} \in A_2$. Therefore, $x-y \notin Ng$. If $\{x,z\} \in A_1$ and $\{y,z\} \in A_1$, then $x-z$ and $z-y$ are contained in Ng, and so $(x-z)+(z-y) = x-y \in Ng$. This puts $\{x,y\} \in A_1$, which cannot be. Hence, $p^2_{11} = 0$. So, if $\{x,z\} \in A_1$, then $\{y,z\} \in A_2$. There are exactly the k elements $z \in (Ng+x) \setminus \{x\}$ for which $\{x,z\} \in A_1$ is true, and $\{y,z\} \in A_2$ is true for each of these. Hence, $p^2_{12} = k$. Similarly, if $\{y,z\} \in A_1$, then $\{x,z\} \in A_2$, and these are exactly the k elements $z \in (Ng+y) \setminus \{y\}$ for which $\{y,z\} \in A_1$ is true, and $\{x,z\} \in A_2$ is true for each of these. Therefore, $p^2_{21} = k$.

Of the $v = |N|$ elements of N, there are $v-2$ elements in $N \setminus \{x,y\}$. Of these, there are exactly the k elements in

$$A = \{u \in N \setminus \{x,y\} \mid \{x,u\} \in A_1\},$$

and there are exactly k elements in

$$B = \{v \in N \setminus \{x,y\} \mid \{y,v\} \in A_1\}.$$

If $z \in A \cap B$, then $\{x,z\} \in A_1$ and $\{y,z\} \in A_1$, and we have seen that this cannot be. Hence, $A \cap B = \varnothing$. There are $v-2-2k$ elements in $(N \setminus \{x,y\}) \setminus (A \cup B)$, and for any one of them, call it t, it is true that $\{x,t\} \in A_2$ and $\{y,t\} \in A_2$. Hence, these $v-2-2k$ elements in $(N \setminus \{x,y\}) \setminus (A \cup B) = N \setminus [(Ng+x) \cup (Ng+y)]$ are exactly the elements for which $\{x,t\}$, $\{y,t\} \in A_2$ is true. Hence, $p^2_{22} = v-2-2k$.

We summarize with

Theorem 13.7. *For a finite IP-nearring $(N,+,\cdot)$ with exactly one basic block Ng being a subgroup of N, and for $\mathscr{A} = \{A_1, A_2\}$ as defined above, we get that $(N, \mathscr{B}, \in, \mathscr{A})$*

is an association scheme. If $|N| = v$ *and* $|Ng| = k+1$*, we have*

$$\begin{array}{lll} p^1_{11} = k-1; & p^1_{12} = p^1_{21} = 0; & p^1_{22} = v-1-k; \\ p^2_{11} = 0; & p^2_{12} = p^2_{21} = k; & p^2_{22} = v-2-2k. \end{array}$$

Exercise 13.8. Take $(N,+,\cdot)$ to be the IP-nearring $(\mathbb{Z}_9,+,*)$ of Example 7.9. Construct A_1 explicitly, and then compute all the p^h_{ij}.

Exercise 13.9. Perhaps you had success with Exercise 7.10. If so, see if Theorem 13.7 is applicable. If not, then go back to Exercise 7.10 and find an example for which Theorem 13.7 is applicable. Then compute A_1 and all the p^h_{ij}.

14 Partially balanced incomplete block designs

We introduced tactical configurations in §13, and then we built upon this idea to get association schemes. Certain association schemes yield *partially balanced incomplete block designs* (PBIBD). Partially balanced incomplete block designs were also developed by statisticians in their efforts to design elaborate agricultural experiments.

Association schemes are certainly complicated enough, but PBIBDs are even more complex.

Definitions 14.1. Suppose $(N, \mathscr{T}, \in, \mathscr{A})$ is an association scheme. This structure is also a *partially balanced incomplete block design* if the following two conditions are satisfied.

(a) To each $A_i \in \mathscr{A}$, there is a number $n_i \in \mathbb{N}$ so that for each $x \in N$, there are exactly n_i distinct $y \in N$ such that $\{x,y\} \in A_i$.

(b) To each $A_i \in \mathscr{A}$, there is a number $\lambda_i \in \mathbb{Z}$ such that $\{x,y\} \in A_i$ implies x and y belong to exactly λ_i distinct blocks $B \in \mathscr{T}$.

A PBIBD has the parameters v, b, r, and k of the tactical configuration, the parameters p_{ij}^{h} of the association scheme, and the additional parameters n_i and λ_i.

Example 14.2. Consider the association scheme of Example 13.3. It is also a PBIBD with the new parameters $n_i = 2$ for each i, and $(\lambda_1, \lambda_2, \lambda_3, \lambda_4) = (4,3,2,1)$.

Exercise 14.3. In Exercise Exercise 13.8, the reader was asked to partially construct an association scheme. This association scheme is also a PBIBD. Determine the parameters n_1, n_2, λ_1, and λ_2.

From requirement (b) of Definitions 14.1, we see that we will need to count the number of blocks of the tactical configuration $(N, \mathscr{T}, \in)$ which contain a pair $\{x,y\}$. The next few results will be useful toward that end.

Proposition 14.4. *Let $(N,+,\cdot)$ be a finite IP-nearring, with basic blocks $Ng_1, \ldots, Ng_s$, $Ne_1, \ldots, Ne_t$. (See §13.) Suppose $z \in Ne_j$ and $z \neq 0$. Then $\{0,z\}$ is a subset of exactly $k = |Ne_j|$ distinct blocks of $\mathscr{B} = \{Na+b \mid a,b \in N,\ a \neq 0\}$, and none of these are of the form $Ng_i + b$.*

Proof. With $z \in Ne_j$ and $z \neq 0$, we have $Ne_j = Nz$. Of the blocks $Ng_i + b$, the only ones containing 0 are the Ng_i, and z cannot belong to any of these since $Ng_i \cap Ne_j = \{0\}$ for any $Ng_i, Ne_j \in \mathscr{B}$. Easily, $0, z \in Nz \cap (N(-z)+z)$, and z belongs to at least these two blocks Nz and $N(-z)+z$.

The IP-nearring $(N,+,\cdot)$ has a Ferrero pair (N,Φ). See Theorem 7.1 and Theorem 7.2. We will obtain a bijection between $\Phi \setminus \{1_N\}$ and the remaining blocks containing 0 and z. For $\phi \in \Phi \setminus \{1_N\}$, $-\phi + 1_N$ is a bijection. So, there is a unique $a_\phi \in N$ such that

$$z = (-\phi + 1_N)(a_\phi) = -\phi(a_\phi) + a_\phi = \phi(-a_\phi) + a_\phi.$$

Hence, $z \in N(-a_\phi) + a_\phi$, and also $0 \in N(-a_\phi) + a_\phi$. So, to each $\phi \in \Phi \setminus \{1_N\}$, we get a block $N(-a_\phi) + a_\phi \in \mathscr{B}$ with $0, z \in N(-a_\phi) + a_\phi$.

If $N(-a_\phi)+a_\phi = N(-a_\lambda)+a_\lambda$, then $a_\phi = a_\lambda$ since $0,z \notin Ng_i+b$ for any elements $g_i, b \in N$. From $\phi(-a_\phi)+a_\phi = z = \lambda(-a_\lambda)+a_\lambda$ we obtain $\phi(a_\phi) = \lambda(a_\lambda) = \lambda(a_\phi)$. This gives us $\phi = \lambda$ since Φ is fixed point free. Hence, the map $T(\phi) = N(-a_\phi)+a_\phi$ is injective.

Suppose now that $0,z \in Na+b \in \mathscr{B}$, where $b \notin \{0,z\}$. Then $0 = \phi a+b$ for some $\phi \in N$, and so $0 \neq \phi a = -b \in Na$. We then have $Na = N(-b)$, and $0,z \in N(-b)+b$. This means that $z = m(-b)+b = -\mu(b)+b = (-\mu+1_N)(b)$ for some $\mu \in \Phi\setminus\{1_N\}$, since $z \neq 0$. Consequently, $b = a_\mu$ and $Na+b = N(-b)+b = N(-a_\mu)+a_\mu$. This means that $T(\phi) = N(-a_\phi)+a_\phi$ is the desired bijection promised in the second paragraph of our proof. These $|\Phi|-1$ blocks $N(-a_\phi)+a_\phi$, $\phi \in \Phi\setminus\{1_N\}$, together with the two blocks Nz and $N(-z)+z$, give us exactly $|\Phi|+1 = |Ne_j|$ blocks containing both 0 and z. (Note that $Ne_j = \Phi(e_j)\cup\{0\}$.) □

Lemma 14.5. *If $(N,+,\cdot)$ is an IP-nearring, $g \in N^* = N\setminus\{0\}$, and $(Ng,+)$ is a subgroup of $(N,+)$, then $(Ng,+,\cdot)$ is a nearfield.*

Proof. For $g \in N^*$, let B_g be as in Theorem 6.5. Then $g \in B_g$, and so

$$N^*g \subseteq N^*B_g = B_g$$

by Theorem 6.5. But $N^*g\cup\{0\} = Ng$. Also, $B_g = B_g g \subseteq N^*g = Ng\setminus\{0\}$, and since $(B_g,\cdot)$ is a group, we have that $(N^*g,\cdot)$ is a group. Hence, $(Ng,+,\cdot)$ is a nearfield. The nearfield $(Ng,+,\cdot)$ is constructed from the Ferrero pair (Ng,Φ) where $\Phi = \{u^\bullet \mid u \in Ng\setminus\{0\}\}$. See Theorem 7.1. Hence, $(Ng,+,\cdot)$ is a planar nearfield. □

Corollary 14.6. *Let $(N,+,\cdot)$ be an IP-nearring. Suppose $z \in Ng$, $z \neq 0$, and that $(Ng,+)$ is a subgroup of $(N,+)$. Then, the only block $Na+b \in \mathscr{B}$ containing both 0 and z is Ng.*

Proof. As in the proof of Proposition 14.4, $0,z \in Nz\cap(N(-z)+z)$, but since $z \in Ng$, we have $Ng = Nz$, and so $Nz = N(-z)$ and $N(-z)+z = Nz$.

If $0, z \in Na+b$, where $b \notin \{0,z\}$, then the last paragraph of the proof of Proposition 14.4 shows that $Na+b = N(-a_\mu)+a_\mu$, where $z = \mu(-a_\mu)+a_\mu$ for some $\mu \in N^*$. That is, $\mu a_\mu + z = a_\mu$, so a_μ is the unique solution to the equation

$$x = \mu x + z. \tag{14:6}$$

Each $\mu \in N$ has a $\mu' \in Ng$ so that $\mu =_m \mu'$, so (14:6) could be replaced by

$$x = \mu' x + z, \tag{14:7}$$

where $\mu' \in Ng$. Since $(Ng, +, \cdot)$ is a planar nearfield, the equation (14:7) has a unique solution in Ng, and this solution is also a solution to (14:6), and so each $a_\mu \in Ng$. Since $a_\mu \in Ng$, we have each $N(-a_\mu)+a_\mu = Ng$. So, all the blocks containing both 0 and z are equal to Ng, and this is what we wanted to prove. □

For $0 \neq z \in N$, we have $z \in Na$ for exactly one basic block $Na = Nz \in \mathscr{B}$. If $Nz = Ng$, a subgroup of N, then there is only the one block $Nz = Na = Ng$ containing both 0 and z. If $Na = Ne$, and if Ne is not a subgroup of N, then there are exactly $k = |Ne|$ distinct blocks in $\mathscr{B}$ containing both 0 and z. Our next result makes these facts more significant.

Proposition 14.7. *Let N be an IP-nearring. Suppose $x, y \in N$, and $x \neq y$. Then the number of distinct blocks in $\mathscr{B}$ containing both x and y is the number of distinct blocks in $\mathscr{B}$ containing both 0 and $y-x$.*

Proof. Let

$$\mathscr{B}\{x,y\} = \{Na+b \in \mathscr{B} \mid x, y \in Na+b\}$$

and

$$\mathscr{B}\{0,y-x\} = \{Na+b \in \mathscr{B} \mid 0, y-x \in Na+b\}.$$

Define $\Gamma : \mathscr{B}\{x,y\} \to \mathscr{B}\{0,y-x\}$ by $\Gamma(Na+b) = Na+(b-x)$. Since $x, y \in Na+b$, we get $0, y-x \in Na+(b-x) \in \mathscr{B}\{0,y-x\}$. Now define $\Lambda : \mathscr{B}\{0,y-x\} \to \mathscr{B}\{x,y\}$

by $\Lambda(Na+b) = Na+(b+x)$. If $0, y-x \in Na+b$, then $x, y \in Na+(b+x)$. Notice that $\Lambda \circ \Gamma(Na+b) = Na+b$ and $\Gamma \circ \Lambda(Na+b) = Na+b$. Hence, $\Gamma = \Lambda^{-1}$, and so $\mathscr{B}\{x,y\}$ and $\mathscr{B}\{0,y-x\}$ have the same number of elements. □

The association schemes of Theorem 13.7 are PBIBDs.

Theorem 14.8. *Let $(N, \mathscr{B}, \in, \mathscr{A})$ be an association scheme of Theorem 13.7. So, $(N,+,\cdot)$ is a finite IP-nearring with exactly one basic block Ng that is a subgroup of N. With $\mathscr{A} = \{A_1, A_2\}$, where*

$$A_1 = \{\{x,y\} \subseteq N \mid x-y \in Ng\}, \quad \textit{and} \quad A_2 = \{\{x,y\} \subseteq N \mid x-y \notin Ng\},$$

the association scheme $(N, \mathscr{B}, \in, \mathscr{A})$ is a PBIBD with parameters as defined in Theorem 13.7 and with $n_1 = k$, $n_2 = v-1-k$, $\lambda_1 = 1$, and $\lambda_2 = k+1$, where $|N| = v$ and $|Ng| = k+1$.

Proof. Take any $x \in N$. Then for any other $y \in N$, $\{x,y\} \in A_1$ if and only if $x-y \in Ng$, which is true if and only if $Ng+x = Ng+y$, that is, x and y belong to the same coset of Ng. The elements of N may be displayed as

$$x; \quad z_1, z_2, \ldots, z_k; \quad u_1, u_2, \ldots, u_{v-1-k},$$

where $Ng+x = \{x, z_1, z_2, \ldots, z_k\}$ and $N \setminus (Ng+x) = \{u_1, u_2, \ldots, u_{v-1-k}\}$. So, there are exactly $n_1 = k$ elements $y \in N$ such that $\{x,y\} \in A_1$. They are $z_1, z_2, \ldots, z_k$. Also, there are exactly $n_2 = v-1-k$ elements $y \in N$ such that $\{x,y\} \in A_2$. They are $u_1, u_2, \ldots, u_{v-1-k}$. This gives us (a) of Definitions 14.1.

Take any pair $\{x,y\} \subseteq N$, $x \neq y$. Either $\{x,y\} \in A_1$ or $\{x,y\} \in A_2$. If $\{x,y\} \in A_1$, then $x-y \in Ng$ and $x-y \neq 0$. The number of blocks in $\mathscr{B}$ containing both x and y is the same as the number of blocks in $\mathscr{B}$ containing both 0 and $x-y$. By Corollary 14.6, this number is $\lambda_1 = 1$. If $\{x,y\} \in A_2$, then $x-y \notin Ng$ and $x-y \neq 0$. Hence, $0, x-y \in Ne$ where Ne is not a subgroup of N. Again, the number of blocks in $\mathscr{B}$ containing both x and y is the number of blocks in $\mathscr{B}$

containing both 0 and $x-y$. By Proposition 14.4, this number is $\lambda_2 = k+1$. This gives us (b) of Definitions 14.1 and finishes our proof of Theorem 14.8. (Note: In Theorem 14.8, $k = |N^*g|$, but in Proposition 14.4, $k = |Ng| = |N^*g|+1$.) □

Exercise 14.9. In Exercise 13.8 and Exercise 13.9, you were asked to construct association schemes. These association schemes are also PBIBDs. Compute the parameters n_i and λ_i of the PBIBDs.

Exercise 14.10. In Example 13.3, an association scheme was displayed. It is also a PBIBD. Determine the parameters n_i and λ_i.

15 Balanced incomplete block designs, part 2

In §8, we took any finite IP-nearring $(N,+,\cdot)$ and constructed a BIBD $(N,\mathscr{B}^*)$. The blocks N^*a+b were motivated by $(\mathbb{C},+,\circ)$ from Example 6.3. In §13, we took motivation from $(\mathbb{C},+,*)$, also of Example 6.3, to construct $(N,\mathscr{B},\in)$. With these $(N,\mathscr{B},\in)$, we always get a tactical configuration, and sometimes a PBIBD. Sometimes we also get a BIBD.

Theorem 15.1. *Let $(N,+,\cdot)$ be a finite IP-nearring. Suppose each basic block $Ne \in \mathscr{B}$ is not a subgroup of N. Then $(N,\mathscr{B},\in)$ is a BIBD with parameters $v = |N|$, $k = |Ne|$, $b = v(v-1)/(k-1)$, $r = k(v-1)/(k-1)$ and $\lambda = k$.*

Proof. Since none of the basic blocks Ne are subgroups of N, the s of Theorem 13.5 is 0. Also, $t = (v-1)/(k-1)$. Hence, $b = v(v-1)/(k-1)$ and $r = k(v-1)/(k-1)$. The number of blocks in $\mathscr{B}$ containing both x and y is the same as the number of blocks in $\mathscr{B}$ containing both 0 and $y-x$, by Proposition 14.7. Since $y-x \in Ne$ for some $Ne \in \mathscr{B}$, and since Ne is not a subgroup, then the number of blocks in $\mathscr{B}$ containing both 0 and $y-x$ is $|Ne| = k$, by Proposition 14.4. □

Theorem 15.2. *Let* $(N,+,\cdot)$ *be a finite IP-nearring. Suppose that each basic block* $Ng \in \mathscr{B}$ *is a subgroup of* N. *Then* $(N,\mathscr{B},\in)$ *is a BIBD with parameters* $v = |N|$, $k = |Ng|$, $b = v(v-1)/k(k-1)$, $r = (v-1)/(k-1)$, *and* $\lambda = 1$.

Proof. Since each of the basic blocks Ng is a subgroup of N, the t of Theorem 13.5 is 0. Also, $s = (v-1)/(k-1)$. Hence, $b = v(v-1)/k(k-1)$ and $r = (v-1)/(k-1)$. The number of blocks in $\mathscr{B}$ containing both x and y is the same as the number of blocks in $\mathscr{B}$ containing both 0 and $y-x$, by Proposition 14.7. Since $0 \neq y-x \in Ng$ for some $Ng \in \mathscr{B}$ and since Ng is a subgroup of N, we know that the number of blocks in $\mathscr{B}$ containing both 0 and $y-x$ is 1. Hence, $\lambda = 1$, as claimed. □

Exercise 15.3. You should now be able to solve Problem 1.1.

Exercise 15.4. How was the example in Example 1.3 obtained?

Exercise 15.5. How was the example in Example 1.9 obtained?

Exercise 15.6. From the finite field $(\mathbb{Z}_{13},+,\cdot)$, how many BIBDs can you make?

16 Elements of graph theory

Planar nearrings have some very nice graphs. We won't need much graph theory for our purposes.

Definitions 16.1. Consider a finite incidence structure $\Gamma = (V,\mathscr{E},I)$. We say that Γ is a *graph* with *vertices* the elements of V, and with *edges* the elements of $\mathscr{E}$, provided

(1) each edge $E \in \mathscr{E}$ is incident with exactly two distinct vertices of V, and

(2) for each pair of distinct vertices from V, there is at most one edge of $\mathscr{E}$ incident with each element of this pair of vertices.

If $E \in \mathscr{E}$ and $(x,E),(y,E) \in I$, with $x \neq y$, we write $v(E) = \{x,y\}$. If $x,y \in V$ and $x \neq y$, there may or may not be an edge $E \in \mathscr{E}$ with $v(E) = \{x,y\}$. If there is, we write $e\{x,y\} = E$ and also $e\{x,y\} \in \mathscr{E}$. If there is no such edge, we write $e\{x,y\} \notin \mathscr{E}$. Given a vertex x of a graph $\Gamma = (V,\mathscr{E},I)$, we say that a vertex $y \in V \setminus \{x\}$ is *adjacent* to x if $e\{x,y\} \in \mathscr{E}$. The graph Γ is *complete* if $e\{x,y\} \in \mathscr{E}$ for every pair $x,y \in V$, $x \neq y$. The graph is *null* if $e\{x,y\} \notin \mathscr{E}$ for every pair $x,y \in V$, $x \neq y$.

For $x \in V$, let $V_\Gamma(x) = \{y \in V \setminus \{x\} \mid e\{x,y\} \in \mathscr{E}\}$. The *valency of* x is $v_\Gamma(x) = |V_\Gamma(x)|$, so $v_\Gamma(x)$ is the number of vertices which are adjacent to x. If there is a constant v_Γ such that $v_\Gamma(x) = v_\Gamma$ for each $x \in V$, then v_Γ is the *valency of* Γ, and Γ is said to be *regular*.

If $\Gamma = (V,\mathscr{E},I)$ is a regular graph which is neither complete nor null, it may be that there are constants α and β such that, for $x,y \in V$, $x \neq y$,

$$|V_\Gamma(x) \cap V_\Gamma(y)| = \begin{cases} \alpha, & \text{if } e\{x,y\} \in \mathscr{E}; \\ \beta, & \text{if } e\{x,y\} \notin \mathscr{E}. \end{cases} \tag{16:8}$$

That is, the number of vertices adjacent to both x and y depends only upon whether x and y are adjacent or not. If this is the case, we say that Γ is *strongly regular*.

We need some examples of these concepts.

Examples 16.2. (a) Let $(G,+)$ be a finite group with proper subgroup $(H,+)$. For V we take G; for $\mathscr{E}$ we take $\mathscr{E} = \{\{x,y\} \mid x \neq y,\ H+x = H+y\}$; and for I we take $\in$. For any $x \in G$, the vertices adjacent to x are exactly the other elements in the coset $H+x$, that is, those elements in $(H+x) \setminus \{x\}$. Since H is a proper subgroup, $G \setminus (H+x) \neq \varnothing$, and any $y \in G \setminus (H+x)$ is not adjacent to x. so $\Gamma = (G,\mathscr{E},\in)$ is not complete, and it is not a null graph. Of course, $v_\Gamma(x) = |(H+x) \setminus \{x\}| = |H| - 1$ for each $x \in G$, and so $|H| - 1$ is the valency of Γ and Γ is regular. If $x,y \in V = G$, $x \neq y$, then either $H+x = H+y$ or $H+x \neq H+y$.

If $H+x=H+y$, then $|V_\Gamma(x)\cap V_\Gamma(y)|=|H|-2$, and if $H+x\neq H+y$, then $|V_\Gamma(x)\cap V_\Gamma(y)|=0$. So Γ is strongly regular.

(b) Let $(G,+)$ be a finite group, and let $H=\{0\}$. Define $\mathscr{E}$ as in (a). Then $\Gamma=(G,\mathscr{E},\in)$ is a null graph.

(c) Let $(G,+)$ be a finite group, and let $H=G$. Define $\mathscr{E}$ as in (a). Then $\Gamma=(G,\mathscr{E},\in)$ is a complete graph.

(d) Take any finite IP-nearring $(N,+,\cdot)$ and fix $N^*c\in\mathscr{B}^*$. Choose any $N^*r\in\mathscr{B}^*$, and define

$$E_c^r=\{N^*r+c' \mid c'\in N^*c\}.$$

We construct $\Gamma_c^r=(N^*c,\mathscr{E},I)$ by saying

$$e\{c_1,c_2\}\in\mathscr{E}\quad\text{if and only if}\quad (N^*r+c_1)\cap(N^*r+c_2)\neq\varnothing.$$

Then we have $(c_i,e\{c_1,c_2\})\in I$ for $i=1,2$, and $(c',e\{c_1,c_2\})\notin I$ if $c'\notin\{c_1,c_2\}$.

Exercise 16.3. Start with the field $(\mathbb{Z}_{13},+,\cdot)$. Take $\Phi=\{5^\bullet,12^\bullet,8^\bullet,1^\bullet\}$ and construct an IP-nearring $(N,+,\cdot)$ from the Ferrero pair $(\mathbb{Z}_{13},\Phi)$. With N^*1, N^*2, and N^*4, compute the various E_c^r of (d) of Examples 16.2. Then compute the various graphs $\Gamma_c^r=(N^*c,\mathscr{E},I)$. Try to notice some interesting patterns.

Definitions 16.4. Let $\Gamma_i=(V_i,\mathscr{E}_i,I_i)$, $i=1,2$, be two graphs. The *cartesian product* of Γ_1 and Γ_2 is a graph $\Gamma_1\times\Gamma_2=(V_1\times V_2,\mathscr{E}_{12},I_{12})$, where $e\{(x,a),(y,b)\}\in I_{12}$ if and only if $x=y$ and $e\{a,b\}\in I_2$ or $e\{x,y\}\in I_1$ and $a=b$.

If a graph $\Gamma=(V,\mathscr{E},I)$ has $|V|$ rather small, one can often display the graph Γ on paper rather easily. Use $\bullet$ to represent a vertex, perhaps with its name placed by it like $\overset{x}{\bullet}$, and an edge between vertices $\overset{x}{\bullet}$ and $\overset{y}{\bullet}$ could be represented by something like a line segment, or a broken line segment, that is, something like $\overset{x}{\bullet}\!\!\longrightarrow\!\!\overset{y}{\bullet}$ or $\overset{x}{\bullet}\text{---}\overset{y}{\bullet}$.

Examples 16.5. Consider three graphs Γ_1, Γ_2, and Γ_3, where Γ_1 is the null graph with vertices $V_1=\{x,y\}$, where Γ_2 is the complete graph with vertices

$V_2 = \{p, q\}$, and Γ_3 is the complete graph with vertices $V_3 = \{a, b, c\}$. We can display Γ_1 as  ; we can display Γ_2 as  ; and we can display Γ_3 by

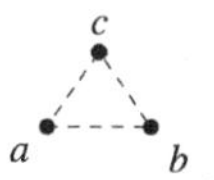

.

Compute $\Gamma_1 \times \Gamma_3$. It can be displayed as

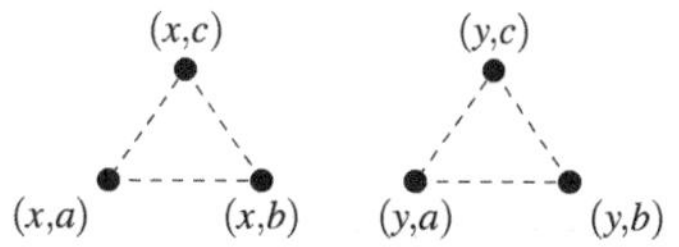

.

Compute $\Gamma_2 \times \Gamma_3$. It can be displayed as

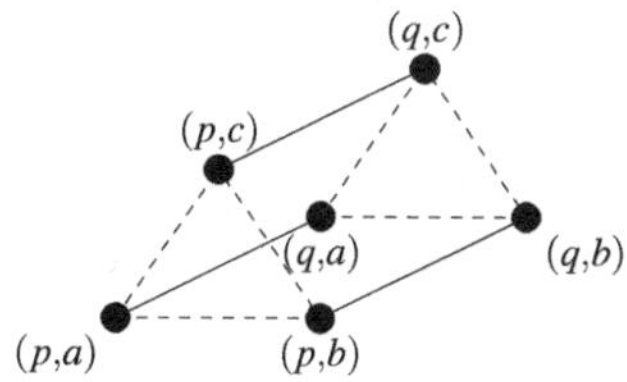

Finally, compute $\Gamma_1 \times (\Gamma_2 \times \Gamma_3)$. It can be displayed as

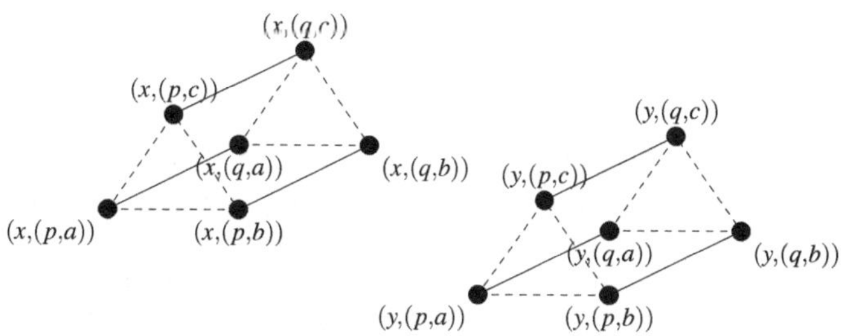

A disadvantage of displaying a graph for $\Gamma_1 \times (\Gamma_2 \times \Gamma_3)$ as we have done is

that the same graph can have very different appearances. For example,

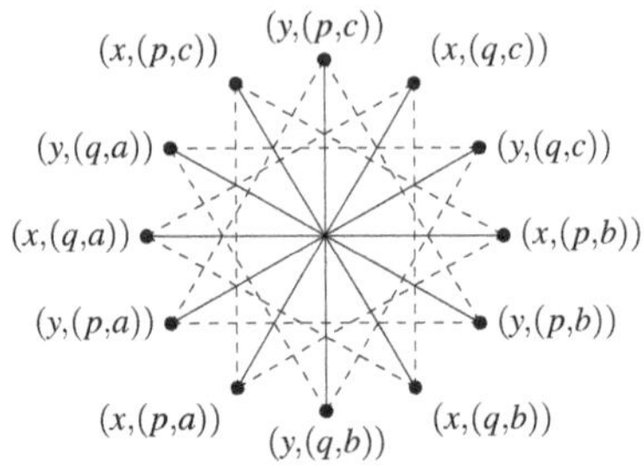

is also a display of the graph $\Gamma_1 \times (\Gamma_2 \times \Gamma_3)$.

Exercise 16.6. Let $(N, +, \cdot)$ be an IP-nearring constructed from the Ferrero pair $(\mathbb{Z}_{73}, \Phi)$ where $|\Phi| = 8$. Compute E_c^1 for the various values of c, and then construct the corresponding graphs Γ_c^1. Show that Γ_2^1, Γ_3^1, and Γ_{12}^1 have a cartesian product structure. Also, show that Γ_1^1 is a complete graph.

In §9, we constructed nets from IP-nearrings. With a finite IP-nearing $(N, +, \cdot)$, we obtained a $(v, k+2)$-net $(N \times N, \mathscr{L}_N)$, as proved in Theorem 9.2, where $k = |N^*a|$ for any nonzero $a \in N$, and $v = |N|$.

Theorem 16.7. *Let $(N, +, \cdot)$ be a finite IP-nearring with $v = |N|$ and $k = |N^*a|$ for any nonzero $a \in N$. Let $(N \times N, \mathscr{L}_N)$ be the $(v, k+2)$-net of Theorem 9.2, and suppose $k < v - 1$. From this net we construct a strongly regular graph Γ with valency $v_\Gamma = (k+2)(v-1)$ and parameters α and β of (16:8) given by $\alpha = (v-2) + (k+1)k$ and $\beta = (k+2)(k+1)$.*

Proof. First, we define $\Gamma = (N \times N, \mathscr{E}, I)$ as follows. If $x, y \in N \times N$, $x \neq y$, we say that $e\{x, y\} \in \mathscr{E}$ if and only if there is a "line" $L \in \mathscr{L}_N$ such that $x, y \in L$. So,

$$\mathscr{E} = \{\{x, y\} \mid x, y \in N \times N,\ x \neq y, \text{ and } \{x, y\} \subseteq L \text{ for some } L \in \mathscr{L}_N\}.$$

Since $\mathscr{L}_N \neq \varnothing$ and each $L \in \mathscr{L}_N$ has $|L| = |N| = v > 2$, we are assured that $\mathscr{E} \neq \varnothing$, and so Γ is not null. As long as $(N \times N, \mathscr{L}_N)$ is not an affine plane, there will be pairs of points $x, y \in N \times N$, $x \neq y$, with no line $M \in \mathscr{L}_N$ containing both x and y. So, if $(N \times N, \mathscr{L}_N)$ is not an affine plane, then Γ will not be complete. To be assured that $(N \times N, \mathscr{L}_N)$ is not an affine plane, we need only have $k = |N^* a| < v - 1$ for any nonzero $a \in N$, which we have, by hypothesis.

From the net $(N \times N, \mathscr{L}_N)$, we have a partition $\{\mathscr{L}(m_1), \ldots, \mathscr{L}(m_{k+2})\}$ of $\mathscr{L}_N$, where each $\mathscr{L}(m_i)$ is itself a partition of $N \times N$. So, for any fixed x in $N \times N$, x belongs to exactly one line L_i from each of these $\mathscr{L}(m_i)$, $1 \leq i \leq k+2$. In L_i, there are v points, so there are $v-1$ points y in L_i such that $e\{x,y\} \in \mathscr{E}$. If $e\{x,y\} \in \mathscr{E}$, then we cannot have $y \in L_i \cap L_j$ if $i \neq j$, by (b) of Definition 9.1, since two distinct points belong to at most one line of $\mathscr{L}_N$. So, Γ has valency $v_\Gamma = (k+2)(v-1)$, and, consequently, Γ is regular.

Suppose $e\{x,y\} \notin \mathscr{E}$. Let $\mathscr{L}_N(x) = \{L_1, \ldots, L_{k+2}\}$ be the $k+2$ lines of $\mathscr{L}_N$ on x, and let $\mathscr{L}_N(y) = \{M_1, \ldots, M_{k+2}\}$ be the $k+2$ lines of $\mathscr{L}_N$ on y, and choose the subscripts so that $L_i \cap M_i = \varnothing$ for each i, that is, L_i and M_i are parallel. This means $L_i, M_i \in \mathscr{L}(m_i)$ for each i, $1 \leq i \leq k+2$. If $i \neq j$, then $L_i \cap M_j = \{p_{ij}\}$ for some vertex $p_{ij} \in N \times N$. Hence, $p_{ij} \in V_\Gamma(x) \cap V_\Gamma(y)$.

If we should have $L_i \cap M_j = \{p_{ij}\} = \{p_{st}\} = L_s \cap M_t$, then $x, p_{ij} \in L_i \cap L_s$ and $y, p_{ij} \in M_j \cap M_t$. Consequently, $L_i = L_s$ and $M_j = M_t$, and so $i = s$ and $j = t$. Consequently, distinct pairs (i,j), $i \neq j$, yield distinct p_{ij}. This shows that $\beta = (k+2)(k+1)$, that is $|V_\Gamma(x) \cap V_\Gamma(y)| = (k+2)(k+1)$ if $e\{x,y\} \in \mathscr{E}$.

What happens when $e\{x,y\} \in \mathscr{E}$? Since $x \neq y$, and $e\{x,y\} \in \mathscr{E}$, we have $L_i = M_i$ for some i, $1 \leq i \leq k+2$, where $L_i \in \mathscr{L}_N(x) = \{L_1, L_2, \cdots, L_{k+2}\}$ and $M_i \in \mathscr{L}_N(y) = \{M_1, M_2, \ldots, M_{k+2}\}$. Without loss of generality, we take $M_i = M_1$ and $L_i = L_1$. Hence, $L_i \cap M_i = \varnothing$ for $2 \leq i \leq k+2$, and $L_i \cap M_j = \{p_{ij}\}$ for $i, j \in \{2, 3, \ldots, k+2\}$ and $i \neq j$. With $x, p_{ij} \in L_i$ and $y, p_{ij} \in M_j$ we have $p_{ij} \in V_\Gamma(x) \cap V_\Gamma(y)$, and, as shown above, we have $p_{ij} = p_{st}$ if and only if $(i,j) = (s,t)$ for $i, j, s, t \in \{2, 3, \ldots, k+2\}$. These p_{ij} account for $(k+1)k$ distinct vertices of $N \times N$ in $V_\Gamma(x) \cap V_\Gamma(y)$, and none of these are in $L_1 = M_1$. Of course, there are $v-2$ vertices in $L_1 \setminus \{x,y\}$ which are also in $V_\Gamma(x) \cap V_\Gamma(y)$. This accounts for all the points in $V_\Gamma(x) \cap V_\Gamma(y)$, so $|V_\Gamma(x) \cap V_\Gamma(y)| = (v-2) + (k+1)k = \alpha$. This completes the proof of Theorem 16.7. □

A strongly regular graph $\Gamma = (V, \mathscr{E}, I)$ is *connected* if, for any pair $\{x,y\} \subseteq V$, $x \neq y$, there is a sequence $(x_0, x_1, \ldots, x_n)$ with each $x_i \in V$, $x_0 = x$, $x_n = y$, and $e\{x_i, x_{i+1}\} \in \mathscr{E}$ for each i, $0 \leq i \leq n-1$.

Theorem 16.8. *For a finite IP-nearring $(N, +, \cdot)$, the graph $\Gamma = (N \times N, \mathscr{E}, I)$ of Theorem 16.7 is connected.*

Proof. If $x, y \in N$ and $x \neq y$, then either $e\{x,y\} \in \mathscr{E}$ or $e\{x,y\} \notin \mathscr{E}$. If $e\{x,y\} \in \mathscr{E}$, then $(x_0, x_1) = (x, y)$ is a necessary sequence. If $e\{x,y\} \notin \mathscr{E}$, we take any line $L \in \mathscr{L}_N$ on x and any $M \in \mathscr{L}_N$ on y so that $L \cap M \neq \varnothing$, that is $L \in \mathscr{L}(m_i)$, $M \in \mathscr{L}(m_j)$, and $\mathscr{L}(m_i) \cap \mathscr{L}(m_j) = \varnothing$. (See Definition 9.1.) Let $L \cap M = \{x_1\}$. Then $(x_0, x_1, x_2) = (x, x_1, y)$ is a necessary sequence. □

17 Applications to coding theory

Today, a great deal of information is transmitted from point A to point B in the form of 0s and 1s. A sequence $a_1 a_2 \ldots a_n$ of 0s and 1s represents a datum. If this sequence $a_1 a_2 \ldots a_n$ leaves A, one hopes that the same sequence $a_1 a_2 \ldots a_n$ will arrive at B. But, it may be that $a_1 a_2 \ldots a_n$ is transmitted from A and $b_1 b_2 \ldots b_n$ is received at B, and $a_i \neq b_i$ for some i. In transmitting a_i, there is the possibility that b_i is received, and $a_i \neq b_i$.

As an example, suppose 0010101 represents the letter x. If 0010101 is sent from A and 1010101 is received at B, then there is an error in transmission. If 1010101 represents the letter y, then the recipient at B must assume that the letter y was sent, even though, in truth, x was sent. In this example, only one small error was made, but yet false information was received, and it was not detected.

How could the receiver at B (1) know that false information had been received, and (2) correct the false information? An elementary example will quickly illustrate a possibility.

Suppose a communication system is designed to transmit exactly one of two values at a given time, a y for "yes" and an n for "no." Suppose 11110000 represents y and 00001111 represents n. If one wants to transmit a y, then one transmits 11110000. But what if 11110001 is received? Obviously an error has occurred, and if only one error has occurred then interpreting 11110001 at B as 11110000 will correct the error. So error detection and error correction can take place. If the communication system is highly reliable, then one is reasonably assured that a received 11110001 was meant to be a 11110000. But suppose 11001010 was received at B. The receiver at B cannot be confident of what was sent from A. Errors can be detected, but not necessarily corrected.

Certainly, there are 2^n distinct sequences $a_1a_2\ldots a_n$ of 0s and 1s. Perhaps this is significantly more than we need. For example, perhaps all we need are the 26 letters of the Roman alphabet, the 10 digits, 13 punctuation symbols, and one symbol for a blank. So, with a total of 50 symbols required, we could take $n = 6$, and with $2^6 = 64$, we have more than enough sequences $a_1a_2\ldots a_n$ to represent the 50 symbols needed for effective communication. However, with $n = 10$, we have 1024 such sequences $a_1a_2\ldots a_{10}$. In order to detect and correct errors, we will want to isolate 50 of these 1024 sequences as much as possible.

Exactly what do we mean when we say "we want to isolate" a sequence? Let $\mathbf{a} = a_1a_2\ldots a_n$ and $\mathbf{b} = b_1b_2\ldots b_n$ be two sequences of 0s and 1s of length n. Define

$$d(\mathbf{a},\mathbf{b}) = |\{i \mid 1 \le i \le n,\ a_i \ne b_i\}|. \tag{17:9}$$

So $d(\mathbf{a},\mathbf{b})$ counts the number of places where $\mathbf{a}$ differs from $\mathbf{b}$, and so it is a measure of how much $\mathbf{a}$ differs from $\mathbf{b}$. If $\mathbf{a}$ is transmitted from A and $\mathbf{b}$ is received at B, then $d(\mathbf{a},\mathbf{b})$ errors have occurred. Let $\mathbb{Z}_2^n$ denote all sequences

$a_1a_2\ldots a_n$ of 0s and 1s. Then d is a metric on $\mathbb{Z}_2^n$. That is, for all $\mathbf{a},\mathbf{b},\mathbf{c} \in Z_2^n$, (i) $d(\mathbf{a},\mathbf{b}) = d(\mathbf{b},\mathbf{a}) \geq 0$, (ii) $d(\mathbf{a},\mathbf{c}) \leq d(\mathbf{a},\mathbf{b}) + d(\mathbf{b},\mathbf{c})$, (iii) $d(\mathbf{a},\mathbf{a}) = 0$, and (iv) if $d(\mathbf{a},\mathbf{b}) = 0$, then $\mathbf{a} = \mathbf{b}$. To see that d is a metric on $\mathbb{Z}_2^n$ is immediate except for (ii). Suppose $\mathbf{a}$ and $\mathbf{b}$ differ at $i_1, i_2, \ldots, i_k$, so $d(\mathbf{a},\mathbf{b}) = k$. Suppose $\mathbf{b}$ and $\mathbf{c}$ differ at $j_1, j_2, \ldots, j_l$, so $d(\mathbf{b},\mathbf{c}) = l$. Also, suppose $\mathbf{a}$ and $\mathbf{c}$ differ at $s_1, s_2, \ldots, s_m$, so $d(\mathbf{a},\mathbf{c}) = m$. If $a_s \neq c_s$, then we cannot have both $a_s = b_s$ and $b_s = c_s$. So $s \in \{i_1, i_2, \ldots, i_k\}$ or $s \in \{j_1, j_2, \ldots, j_l\}$, or $s \in \{i_1, i_2, \ldots, i_k\} \cap \{j_1, j_2, \ldots, j_l\}$. Hence, $m \leq k + l$, or $d(\mathbf{a},\mathbf{c}) \leq d(\mathbf{a},\mathbf{b}) + d(\mathbf{b},\mathbf{c})$. If $\mathbf{a}_1, \mathbf{a}_2, \ldots, \mathbf{a}_M$ are M of the 2^n distinct sequences $a_1a_2\ldots a_n$ of 0s and 1s, then we want a positive r so that if $1 \leq d(\mathbf{b},\mathbf{a}_i) \leq r$, then $\mathbf{b} \notin \{\mathbf{a}_1, \ldots, \mathbf{a}_M\} \setminus \{\mathbf{a}_i\}$ for each i, $1 \leq i \leq M$. Then the $\mathbf{a}_1, \mathbf{a}_2, \ldots, \mathbf{a}_M$ are "isolated." We will return to this idea shortly.

Let $\mathscr{C}(n)$ denote a nonempty set of sequences $\mathbf{a} = a_1a_2\ldots a_n$ of 0s and 1s, where n is a positive integer measuring the *length* of the sequence. So $1 \leq |\mathscr{C}(n)| \leq 2^n$. If $\mathbf{a},\mathbf{b} \in \mathscr{C}(n)$ and $\mathbf{a} \neq \mathbf{b}$, then $1 \leq d(\mathbf{a},\mathbf{b}) \leq n$, and so

$$D = \min\{d(\mathbf{a},\mathbf{b}) \mid \mathbf{a},\mathbf{b} \in \mathscr{C}(n),\, \mathbf{a} \neq \mathbf{b}\}$$

exists, and we are assured that $1 \leq D \leq n$. If $M = |\mathscr{C}(n)|$, then we refer to $\mathscr{C}(n)$ as an *(n,M,D)-(binary) code*, and (n,M,D) is said to be the parameters of the code $\mathscr{C}(n)$. Our codes will be binary, in that each element of $\mathscr{C}(n)$ is a sequence of 0s and 1s. There is a function $w : \mathscr{C}(n) \to \{0,1,\ldots,n\}$ defined by

$$w(a_1a_2\ldots a_n) = |\{i \mid 1 \leq i \leq n,\, a_i = 1\}|.$$

So $w(a_1a_2\ldots a_n)$ is the *weight* of the *codeword* $\mathbf{a} = a_1a_2\ldots a_n \in \mathscr{C}(n)$.

Of all the possible codes $\mathscr{C}(n)$, some have advantages over others. When one has a finite tactical configuration $(N,\mathscr{B},\in)$, one can easily construct two codes, a *row code* $\mathscr{C}^A(v)$ and a *column code* $\mathscr{C}_A(b)$. Following the conventions in MacWilliams and Sloane [M & S], we define for a finite tactical configuration $(N,\mathscr{B},\in)$ an *incidence matrix* A, a $b \times v$ matrix of 0s and 1s. Let

$\mathscr{B} = \{B_1, B_2, \ldots, B_b\}$ and $N = \{x_1, x_2, \ldots, x_v\}$. Define $A = (a_{ij})$ where

$$a_{ij} = \begin{cases} 1, & \text{if } x_j \in B_i; \\ 0, & \text{otherwise.} \end{cases}$$

Let $\mathscr{C}^A(v)$ be the collection of the codewords $\mathbf{a}_1, \mathbf{a}_2, \ldots, \mathbf{a}_b$, where for each i, $\mathbf{a}_i = a_{i1}a_{i2}\ldots a_{iv}$, and let $\mathscr{C}_A(b)$ be the collection of the distinct codewords $\mathbf{b}_1, \mathbf{b}_2, \ldots \mathbf{b}_v$, where for each i, $\mathbf{b}_j = a_{1j}a_{2j}\ldots a_{bj}$. That is, $\mathscr{C}^A(v)$ consists of the b rows of A and $\mathscr{C}_A(b)$ consists of the distinct columns from the v columns of A. Then $\mathscr{C}^A(v)$ is a (v, b, D_v)-code for some D_v, and $\mathscr{C}_A(b)$ is a (b, v', D_b)-code for some D_b and for some v', $1 \le v' \le v$. Actually, any (n, M, D)-binary code $\mathscr{C}(n)$ has several incidence matrices A. Let

$$\mathscr{C}(n) = \{\mathbf{a}_1, \mathbf{a}_2, \ldots, \mathbf{a}_M\} \text{ with } \mathbf{a}_i = a_{i1}a_{i2}\ldots a_{in},$$

and then let $A = (a_{ij})$. Now $\mathscr{C}^A(n) = \mathscr{C}(n)$ and $\mathscr{C}_A(M)$ are as before. If one takes any $s \times t$ matrix $A = (a_{ij})$, where each $a_{ij} \in \{0, 1\}$, then let $\mathscr{C}^A(t)$ consist of the distinct rows of A and let $\mathscr{C}_A(s)$ consist of the distinct columns of A. So $\mathscr{C}^A(t)$ and $\mathscr{C}_A(s)$ are binary codes.

For a finite tactical configuration $(N, \mathscr{B}, \in)$, the codes $\mathscr{C}^A(v)$ and $\mathscr{C}_A(b)$ are nice in that they are constant weight codes. A code $\mathscr{C}(n)$ is a *constant weight code* if there is a number W so that $w(\mathbf{a}) = W$ for each $\mathbf{a} \in \mathscr{C}(n)$. Hence, $W = k$ for $\mathscr{C}^A(v)$ and $W = r$ for $\mathscr{C}_A(b)$. This is because each block $B \in \mathscr{B}$ has exactly k elements, and each $x \in N$ belongs to exactly r blocks.

Example 17.1. For the finite field $\mathbb{Z}_7$, we get $\Phi = \{2^\bullet, 4^\bullet, 1^\bullet\}$ and the Ferrero pair $(\mathbb{Z}_7, \Phi) = (N, \Phi)$. Then $N^*1 = \{2, 4, 1\}$ and $N^*3 = \{6, 5, 3\}$. Hence, the BIBD $(N, \mathscr{B}^*, \in) = (\mathbb{Z}_7, \mathscr{B}_3^*, \in)$ of §8 has $(v_1, \ldots, v_7) = (0, \ldots, 6)$ and the blocks of $\mathscr{B}^*$ are

$$\begin{array}{llll} B_1 = \{2,4,1\}, & B_5 = \{6,1,5\}, & B_8 = \{6,5,3\}, & B_{12} = \{3,2,0\}, \\ B_2 = \{3,5,2\}, & B_6 = \{0,2,6\}, & B_9 = \{0,6,4\}, & B_{13} = \{4,3,1\}, \\ B_3 = \{4,6,3\}, & B_7 = \{1,3,0\}, & B_{10} = \{1,0,5\}, & B_{14} = \{5,4,2\}. \\ B_4 = \{5,0,4\}, & & B_{11} = \{2,1,6\}, & \end{array}$$

We get the following incidence matrix A:

$$A = \begin{pmatrix} 0 & 0 & 1 & 1 & 0 & 1 & 0 \\ 0 & 0 & 0 & 1 & 1 & 0 & 1 \\ 1 & 0 & 0 & 0 & 1 & 1 & 0 \\ 0 & 1 & 0 & 0 & 0 & 1 & 1 \\ 1 & 0 & 1 & 0 & 0 & 0 & 1 \\ 1 & 1 & 0 & 1 & 0 & 0 & 0 \\ 0 & 0 & 0 & 1 & 0 & 1 & 1 \\ 1 & 0 & 0 & 0 & 1 & 0 & 1 \\ 1 & 1 & 0 & 0 & 0 & 1 & 0 \\ 0 & 1 & 1 & 0 & 0 & 0 & 1 \\ 1 & 0 & 1 & 1 & 0 & 0 & 0 \\ 0 & 1 & 0 & 1 & 1 & 0 & 0 \\ 0 & 0 & 1 & 0 & 1 & 1 & 0 \end{pmatrix}$$

The rows of A represent a BIBD with parameters

$$(v,b,r,k,\lambda) = (7,14,6,3,2).$$

The row code $\mathscr{C}^A(7)$ is a $(7,14,2)$-code with constant weight 3. The column code $\mathscr{C}_A(14)$ is a $(14,7,8)$-code with constant weight 6. For any two distinct columns $\mathbf{a}$ and $\mathbf{b}$ of A, we have $d(\mathbf{a},\mathbf{b}) = 2(r-\lambda) = 8$.

We have occasion to use the following proposition.

Proposition 17.2. *Let $\mathscr{C}(n)$ be a constant weight code. Then $D = 2\delta$, an even positive integer.*

Proof. For $\mathbf{a} = a_1a_2\ldots a_n$, $\mathbf{b} = b_1b_2\ldots b_n \in \mathscr{C}(n)$, $\mathbf{a} \neq \mathbf{b}$, suppose $1 = a_i \neq b_i$ for some i, $1 \leq i \leq n$. Then there is a j, $1 \leq j \leq n$, such that $a_j \neq b_j = 1$, since $w(\mathbf{a}) = w(\mathbf{b})$, and conversely. Hence, $\mathbf{a}$ differs from $\mathbf{b}$ in an even number of places, that is, $d(\mathbf{a},\mathbf{b})$ is even. This forces D to be even also. □

Let us return now to the idea of "isolation." For a code $\mathscr{C}(n)$, suppose $D \geq 2$. Let $\mathscr{C}(n) = \{\mathbf{a}_1, \mathbf{a}_2, \ldots, \mathbf{a}_M\}$. If $\mathbf{a} \in \mathscr{C}(n)$ is transmitted from A and $\mathbf{b}$ is received at B, then one tries to find an $\mathbf{a}_i$ so that $\mathbf{b} = \mathbf{a}_i$. If this search is successful, one assumes that $\mathbf{a}_i = \mathbf{a}$, that is, $\mathbf{a}_i$ was transmitted from A. Of course, this could be incorrect, but to be incorrect, at least D errors would have occurred in transmission. The next best thing is to find the $\mathbf{a}_i$ closest to $\mathbf{b}$. It is enough to have the existence of an $\mathbf{a}_i$ such that $d(\mathbf{a}_i, \mathbf{b}) < D/2$. For if $i \neq j$ and $d(\mathbf{a}_i, \mathbf{b}) = d(\mathbf{a}_j, \mathbf{b})$, then $D \leq d(\mathbf{a}_i, \mathbf{a}_j) \leq d(\mathbf{a}_i, \mathbf{b}) + d(\mathbf{b}, \mathbf{a}_j) = 2d(\mathbf{a}_i, \mathbf{b})$, and so $D/2 \leq d(\mathbf{a}_i, \mathbf{b})$. Of course, if we find an $\mathbf{a}_i$ such that $d(\mathbf{a}_i, \mathbf{b}) < D/2$, then we assume that $\mathbf{a}_i = \mathbf{a}$. This means that we can correct fewer than $D/2$ errors. If $\mathbf{b} \notin \mathscr{C}(n)$, then at least one error occurred during transmission. If fewer than D errors were made, then $\mathbf{b} \notin \mathscr{C}(n)$. Hence, we can detect $D-1$ or fewer errors. In summary,

Theorem 17.3. *Let $\mathscr{C}(n)$ be a binary code with $D \geq 2$. Then fewer than $D/2$ errors can be corrected, and fewer than D errors can be detected. One traditionally writes that up to $[(D-1)/2]$ errors can be corrected, and up to $D-1$ errors can be detected, where $[x]$ is the integer satisfying $[x] \leq x < [x]+1$.*

In our study, we frequently obtained a BIBD. In particular, every time we obtained a finite affine plane, a finite projective plane, a finite Möbius plane, or a finite IP-nearring, we also got a BIBD. Hence, the following theorem and its corollaries are applicable to the BIBDs obtained from our various sources. For example, whenever a finite planar nearring generates a BIBD, then the row code $\mathscr{C}^A(v)$ and the column code $\mathscr{C}_A(b)$ will have properties supplied by and its corollaries. But first, we determine the parameters of the BIBDs one obtains from finite affine planes and finite projective planes.

Three particular types of BIBDs are (a) finite affine planes, (b) finite projective planes, and (c) finite Möbius planes. We know the parameters

$$(v, b, r, k, \lambda) = (n^2+1, n(n^2+1), n(n+1), n+1, n+1)$$

of a finite Möbius plane in terms of its block size $n+1$, but we have never determined these for finite affine planes and finite projective planes. It will be useful to do it now.

Theorem 17.4. *Let the BIBD $(P, \mathscr{L})$ also be a finite affine plane, as in Definition 1.13, with $|L| = k$ for each $L \in \mathscr{L}$. Then the parameters of the BIBD $(P, \mathscr{L})$ are given by*

$$(v, b, r, k, \lambda) = (k^2, k^2 + k, k+1, k, 1).$$

Proof. For any $L \in \mathscr{L}$, let $p(L) = \{L = L_1, L_2, \ldots, L_t\}$ be the family of lines of $\mathscr{L}$ parallel to L. Since $p(L)$ forms a partition of P, we have $v = |P| = tk$ for $t = |p(L)|$. Take any $M \in \mathscr{L} \setminus p(L)$. Then $|M \cap L_i| = 1$ for each i, $1 \leq i \leq t$, and $M = \cup_{i=1}^{t}(M \cap L_i)$. Hence, $k = |M| = t$, from which we get $v = k^2$. From $vr = bk$, we get $k^2 r = bk$, and so $b = kr$. From $\lambda(v-1) = r(k-1)$ and $\lambda = 1$, we obtain $v - 1 = k^2 - 1 = r(k-1)$, and so $r = k+1$. Hence, $b = k(k+1) = k^2 + k$. □

Theorem 17.5. *Let the BIBD $(P, \mathscr{L})$ also be a finite projective plane, as in Definition 1.18, with $|L| = n+1$ for each $L \in \mathscr{L}$. Then the parameters of the BIBD $(P, \mathscr{L})$ are given by*

$$(v, b, r, k, \lambda) = (n^2+n+1, n^2+n+1, n+1, n+1, 1).$$

Proof. Take any $L \in \mathscr{L}$ and any $x \in P$ with $x \notin L$. There are r distinct lines $M_1, M_2, \ldots, M_r$ on x, each of which intersects L at exactly one point. Suppose $M_i \cap L = \{x_i\}$. If $x_i = x_j$, then $x, x_i \in M_i$ $x, x_i \in M_j$. Consequently, $M_i = M_j$. This follows since $\lambda = 1$. So, $|\{x_1, \ldots, x_r\}| = r$. If $y \in L \setminus \{x_1, \ldots, x_r\}$, then there is a unique line $M \in \mathscr{L}$ with $x, y \in M$. Hence, $M \in \{M_1, \ldots, M_r\}$, and so $M = M_s$ for some s, $1 \leq s \leq r$, and $M \cap L = M_s \cap L = \{x_s\}$. This forces $y \in \{x_1, \ldots, x_r\}$, a contradiction. Therefore, $L = \{x_1, \ldots, x_r\}$. This means that $k = r$.

With $k = n+1$, we now have $r = n+1$. From $vr = bk$ and $r = k$, we conclude that $v = b$. With $\lambda = 1$, $r = k$, and $k = n+1$, we obtain from $\lambda(v-1) = r(k-1)$ that $v - 1 = (n+1)n$. Thus, $b = v = n^2 + n + 1$. This completes the proof of Theorem 17.5. □

Theorem 17.6. *Let* $(N, \mathscr{B}, \in)$ *be a BIBD with incidence matrix A and parameters* (v, b, r, k, λ). *Let* $\mathscr{C}^A(v)$ *and* $\mathscr{C}_A(b)$ *denote the row code of A and the column code of A respectively. Let* $\mu = \max\{|B_i \cap B_j| \mid B_i, B_j \in \mathscr{B}, B_i \neq B_j\}$. *Then* D_v *for* $\mathscr{C}^A(v)$ *is* $2(k-\mu)$, *and* D_b *for* $\mathscr{C}_A(b)$ *is* $2(r-\lambda)$. *In fact,* $d(\boldsymbol{a}, \boldsymbol{b}) = 2(r-\lambda)$ *for any two distinct* $\boldsymbol{a}, \boldsymbol{b} \in \mathscr{C}_A(b)$.

Proof. Now B_i is used to make $\mathbf{a}_i$ in the incidence matrix and in $\mathscr{C}^A(v)$. Suppose $|B_i \cap B_j| = \mu_{ij}$. Then $d(\mathbf{a}_i, \mathbf{a}_j) = 2k - 2\mu_{ij}$. Hence, $D_v = 2k - 2\mu = 2(k-\mu)$.

Now $\mathbf{b}_j \in \mathscr{C}_A(b)$ corresponds to the jth column of the incidence matrix. This tells, by its 1s, exactly which $B_i \in \mathscr{B}$ contain $x_j \in N$, and there are exactly r such 1s in $\mathbf{b}_j$. Since $x_s, x_t \in N$, $x_s \neq x_t$, have exactly λ blocks $B_i \in \mathscr{B}$ with $x_s, x_t \in B_i$, there will be exactly λ identical positions in $\mathbf{b}_s$ and in $\mathbf{b}_t$ which are 1s. Hence, $d(\mathbf{b}_s, \mathbf{b}_t) = 2r - 2\lambda = 2(r-\lambda) = D_b$. □

An affine plane has $v = n^2$ points and $b = n^2 + n$ lines, so $r = n+1$, $k = n$, and $\mu = \lambda = 1$. Hence, we can easily prove

Corollary 17.7. *A finite affine plane* $(N, \mathscr{L}, \in)$ *with* $v = n^2$ *points and incidence matrix A has D for* $\mathscr{C}^A(n^2)$ *equal to* $2(k-1)$, *and D for* $\mathscr{C}_A(n^2+n)$ *equal to* $2(r-1)$. *In particular, for* $\mathscr{C}^A(n^2)$, *up to* $n-2$ *errors can be corrected and up to* $2n-3$ *errors can be detected. For* $\mathscr{C}_A(n^2+n)$, *up to* $n-1$ *errors can be corrected and up to* $2n-1$ *errors can be detected.*

Proof. Apply Theorem 17.3, Theorem 17.6, and the remarks made just before the statement of Corollary 17.7. □

Theorem 17.8. *A finite projective plane* $(N, \mathscr{L}, \in)$ *with* $|N| = v = n^2+n+1$ *points and incidence matrix A has D for* $\mathscr{C}^A(n^2+n+1)$ *equal to* $2(k-1) = 2n$, *and D for* $\mathscr{C}_A(n^2+n+1)$ *equal to* $2(r-1) = 2n$. *In particular, for* $\mathscr{C}^A(n^2+n+1)$, *up to* $n-1$ *errors can be corrected and up to* $2n-1$ *errors can be detected. The same is true for* $\mathscr{C}_A(n^2+n+1)$.

Proof. Here, the BIBD $(N, \mathscr{L}, \in)$ has parameters as in Theorem 17.5. So, apply Theorem 17.3 and Theorem 17.6. □

Note. As one extends a finite affine plane to a projective plane, as described in Theorem 3.10, we improve the error correcting and the error detecting capabilities. In doing so, however, we increase the length of each codeword by $n+1$ for the row code and increase the length of each codeword by 1 for the column code.

Corollary 17.9. *For a BIBD $(N, \mathscr{B}, \in)$ with parameters (v, b, r, k, λ) and μ of Theorem 17.6, let A be the incidence matrix. Then $\mathscr{C}^A(v)$ is a $(v, b, 2(k-\mu))$-code, and $\mathscr{C}_A(b)$ is a $(b, v, 2(r-\lambda))$-code.*

Proof. Apply Theorem 17.6 and remarks made before Proposition 17.2. But we need to say more about $\mathscr{C}_A(b)$. How can we be assured that $v = |\mathscr{C}_A(b)|$? That is, why are the columns of A distinct? If two columns are identical, say $\mathbf{b}_i = \mathbf{b}_j$, with $i \neq j$, then $x_i, x_j \in N$ are two distinct points which belong to exactly the same blocks. Hence, $\lambda = r$. From $\lambda(v-1) = r(k-1)$, we conclude that $v = k$. Hence, $b = 1$ and so $\mathscr{B} = \{N\}$. This makes $A = \begin{pmatrix} 1 & 1 & \cdots & 1 \end{pmatrix}$, which we really do not want. □

Corollary 17.10. *Let $(N, +, \cdot)$ be a finite circular planar nearring. Let A be the incidence matrix for the BIBD $(N, \mathscr{B}^*, \in)$. Then $\mathscr{C}^A(v)$ is a $(v, v(v-1)/k, 2(k-2))$-code, and $\mathscr{C}_A(b)$ is a $(v(v-1)/k, v, 2(v-k))$-code.*

Proof. Apply Corollary 17.9, Proposition 5.5, and the fact that circularity implies that the μ of Theorem 17.6 is 2. □

Corollary 17.11. *Let $(N, +, \cdot)$ be a finite planar nearring with BIBD $(N, \mathscr{B}, \in)$ with no basic blocks as subgroups. Let A be the incidence matrix for $(N, \mathscr{B}, \in)$. Then $\mathscr{C}^A(v)$ is a $(v, v(v-1)/(k-1), 2(k-\mu))$-code, and $\mathscr{C}_A(b)$ is a code with parameters $(v(v-1)/(k-1), v, 2k(v-k)/(k-1))$.*

Proof. Apply Corollary 17.9 and Theorem 15.1. □

The example in Example 17.1 gave us the column code $\mathscr{C}_A(14)$. It is a $(14,7,8)$-code with constant weight 6. Here, 7 elements of $\mathbb{Z}_2^{14}$ have been selected; they each have weight 6; and $d(\mathbf{a},\mathbf{b}) \geq 8$ for distinct $\mathbf{a},\mathbf{b} \in \mathscr{C}_A(14)$. Could you select a larger set from $\mathbb{Z}_2^{14}$, say $\mathscr{C}(14) \subseteq \mathbb{Z}_2^{14}$, where $|\mathscr{C}(14)| > 7$, where each $\mathbf{a} \in \mathscr{C}(14)$ has weight $w(\mathbf{a}) = 6$, and $d(\mathbf{a},\mathbf{b}) \geq 8$ for all distinct $\mathbf{a},\mathbf{b} \in \mathscr{C}(14)$?

Here is another way of looking at the problem we have just described. We have shown that $(\mathbb{Z}_2^{14}, d)$ is a metric space. For $\mathbf{a} \in \mathbb{Z}_2^{14}$, let

$$B(\mathbf{a},4) = \{\mathbf{b} \in \mathbb{Z}_2^{14} \mid d(\mathbf{a},\mathbf{b}) \leq 4\},$$

and think of $B(\mathbf{a},4)$ as the "circle" or "sphere" with center $\mathbf{a}$ and radius 4, together with its interior. Since $d(\mathbf{a},\mathbf{b}) \geq 8$ for distinct $\mathbf{a},\mathbf{b} \in \mathscr{C}_A(14)$, we think that the spheres, or circles, $B(\mathbf{a},4)$ and $B(\mathbf{b},4)$ are either disjoint or tangent, that is, they share no interior points. This should help to visualize the isolation we mentioned earlier. So, the code $\mathscr{C}_A(14)$ represents 7 spheres of radius 4 in $\mathbb{Z}_2^{14}$, and no two of these spheres share an interior point. These 7 spheres are packed into the metric space $\mathbb{Z}_2^{14}$.

How many spheres of radius 4 can be packed into $\mathbb{Z}_2^{14}$? Can you get more than 7? But our codewords each have weight 6. How many spheres of radius 4, having center at an $\mathbf{a} \in \mathbb{Z}_2^{14}$ with $w(\mathbf{a}) - 6$, can be packed into $\mathbb{Z}_2^{14}$? This is the essence of the question concerning $\mathscr{C}(14)$ above. In the language of coding theory, we are asking for the largest code in $\mathbb{Z}_2^{14}$ with constant weight 6. Johnson's bound, developed below, sometimes tells us how large such a code can be, and we shall see that we can actually obtain this bound at times.

Let n be a positive integer and consider the set Z_2^n. From Theorem 17.3, we learn that D controls the error correcting and error detecting capabilities of a binary code $\mathscr{C}(n)$. So, let us assume that a D, $1 \leq D \leq n$, is specified. Next choose a W, $1 \leq W \leq n$. For $1 \leq M \leq 2^n$, does there exist an (n,M,D)-code $\mathscr{C}(n)$ with constant weight W? Let $\mathscr{M}$ be the set of all M, $1 \leq M \leq 2^n$, such that there is an (n,M,D)-code $\mathscr{C}(n)$ with constant weight W. If $\mathscr{M} = \varnothing$, define $A(n,D,W) = 0$. If $\mathscr{M} \neq \varnothing$, define $A(n,D,W) = \max\mathscr{M}$. So, if $\mathscr{M} \neq \varnothing$, then there is a code $\mathscr{C}(n)$ with constant weight W and minimum distance D with exactly $A(n,D,W)$ codewords and there can be no larger code with constant weight W and minimum distance D in Z_2^n. An $(n,A(n,D,W),D)$-code $\mathscr{C}(n)$ is the best one can obtain with n, D, and W fixed, if one is looking to maximize the number of codewords. Just how large can $A(n,D,W)$ be? There is a bound, called *Johnson's bound* [M & S], which is attained by some of the codes constructed from BIBDs. So, in some sense, these codes are among the best. For a constant weight code, Proposition 17.2 assures us that $D = 2\delta$, an even integer, so $A(n,2t+1,W)$ is not positive for odd integers $2t+1$.

Theorem 17.12 (Johnson's bound). *If $W^2 - Wn + \delta n > 0$, then*

$$A(n,2\delta,W) \leq \frac{\delta n}{W^2 - Wn + \delta n}.$$

Proof. Let $\mathscr{C}(n)$ be an $(n,M,2\delta)$-binary code of constant weight W, where M is the Johnson's bound $A(n,2\delta,W)$. Let $\mathscr{C}(n) = \{\mathbf{a}_1, \mathbf{a}_2, \ldots, \mathbf{a}_M\}$, with code words $\mathbf{a}_i = a_{i1}a_{i2}\ldots a_{in}$, have incidence matrix $A = (a_{ij})$. For $\mathbf{a}_i, \mathbf{a}_j \in \mathscr{C}(n)$, the inner product $\mathbf{a}_i \cdot \mathbf{a}_j = \sum_{k=1}^{n} a_{ik}a_{jk}$. Notice that $\mathbf{a}_i \cdot \mathbf{a}_j = W$ if and only if $d(\mathbf{a}_i, \mathbf{a}_j) = 0$; $\mathbf{a}_i \cdot \mathbf{a}_j = W - 1$ if and only if $d(\mathbf{a}_i, \mathbf{a}_j) = 2$; $\ldots$; $\mathbf{a}_i \cdot \mathbf{a}_j = W - t$ if and only if $d(\mathbf{a}_i, \mathbf{a}_j) = 2t$. If $i \neq j$, then $d(\mathbf{a}_i, \mathbf{a}_j) = 2t$ cannot happen for $2t < 2\delta = D$. Hence, since $d(\mathbf{a}_i, \mathbf{a}_j) \geq D = 2\delta$ for $i \neq j$, we have for $i \neq j$, $\mathbf{a}_i \cdot \mathbf{a}_j \leq W - \delta$.

For each i, $1 \leq i \leq M$, we want to take all $\mathbf{a}_i \cdot \mathbf{a}_j$, $i \neq j$, and add the results. The sum is

$$S = \sum_{i=1}^{M} \sum_{\substack{j=1 \\ j \neq i}}^{M} \sum_{k=1}^{n} a_{ik}a_{jk}. \tag{17:10}$$

Since each $\mathbf{a}_i \cdot \mathbf{a}_j = \sum_{k=1}^{n} a_{ik}a_{jk} \leq W - \delta$, we have

$$S \leq M(M-1)(W-\delta). \tag{17:11}$$

We can rewrite S as

$$S = \sum_{k=1}^{n} \sum_{i=1}^{M} \sum_{\substack{j=1 \\ j \neq i}}^{M} a_{ik} a_{jk}. \tag{17:12}$$

Let s_k be the number of 1s in the kth column of A. Then the kth column contributes $s_k(s_k - 1)$ to the sum in (17:12). Hence

$$S = \sum_{k=1}^{n} s_k(s_k - 1) = \sum_{k=1}^{n} s_k^2 - \sum_{k=1}^{n} s_k. \tag{17:13}$$

But $\sum_{k=1}^{n} s_k = WM$, and so $S = \sum_{k=1}^{n} s_k^2 - WM$. Using the method of Lagrange multipliers from calculus, we wish to minimize $H(x_1, \ldots, x_n) = \sum_{k=1}^{n} x_k^2$ subject to $G(x_1, \ldots, x_n) = \sum_{k=1}^{n} x_k = WM$. We form

$$u(x_1, \ldots, x_n) = H(x_1, \ldots, x_n) + \lambda G(x_1, \ldots, x_n)$$

and solve the system

$$\begin{cases} \partial u / \partial x_i = 2x_i + \lambda = 0, \quad 1 \leq i \leq n, \\ \sum_{k=1}^{n} x_i = WM, \end{cases}$$

for λ and each x_i, $1 \leq i \leq n$. We get $x_i = -\lambda/2$ for each i, and so $\lambda = -2WM/n$. This makes each $x_i = WM/n$. Hence, $W^2M^2/n \leq \sum_{k=1}^{n} x_k^2$. From (17:13) and (17:11) we can write

$$\frac{W^2M^2}{n} - WM \leq S \leq M(M-1)(W - \delta), \tag{17:14}$$

which simplifies to

$$M(W^2 - Wn + \delta n) \leq \delta n, \tag{17:15}$$

from which our theorem follows. □

Our next theorem proves that every time we have a BIBD, then we get a column code $\mathscr{C}_A(b)$ which satisfies Johnson's bound! Planar nearrings give BIBDs in various ways, so planar nearrings are useful in constructing binary equal weight codes which reach Johnson's bound, hence they are among the best in this sense.

Theorem 17.13. *Let $(N, \mathscr{B}, \in)$ be a BIBD with parameters (v, b, r, k, λ), and let A be the incidence matrix for $(N, \mathscr{B}, \in)$. If*

$$\mathscr{B} = \{B_1, B_2, \ldots, B_b\} \neq \{N\},$$

then $\mathscr{C}_A(b)$ is a $(b, v, 2(r-\lambda))$-binary code with constant weight $W = r$, and

$$v = \frac{\delta n}{W^2 - Wn + \delta n} = A(b, 2(r-\lambda), r) = \frac{(r-\lambda)b}{r^2 - \lambda b}. \tag{17:16}$$

Proof. First apply Corollary 17.9 to recall that $\mathscr{C}_A(b)$ is a $(b, v, 2(r-\lambda))$-binary code. We noticed just before Proposition 17.2 that $W = r$. So (17:16) is all we have to prove. With $W = r$, $\delta = r - \lambda$, and $n = b$, we obtain that

$$A(v, 2(r-\lambda), r) \leq \delta n/(W^2 - Wn + \delta n) = (r-\lambda)b/(r^2 - \lambda b),$$

and certainly $|\mathscr{C}_A(b)| = v$. We will finish by showing $v = (r-\lambda)b/(r^2-\lambda b)$ and noting that $W^2 - Wn + \delta n = r^2 - \lambda b > 0$.

Recall that a BIBD's parameters satisfy $bk = vr$ and $\lambda(v-1) = r(k-1)$. So

$$b = vr - bk + b = vr - b(k-1).$$

Hence,

$$rb = vr^2 - br(k-1) = vr^2 - b\lambda(v-1) = vr^2 + \lambda b - \lambda bv.$$

So

$$rb - \lambda b = vr^2 - \lambda bv = (r^2 - \lambda b)v.$$

Now $v > 0$, and certainly $\lambda \leq r \leq b$. If $\lambda = b$, then $\lambda = r$, and for $\lambda = r$, we have $v = k$, and so $\mathscr{B} = \{N\}$. Consequently, $\lambda < r$, and so $rb - \lambda b > 0$, which gives us that $r^2 - \lambda b > 0$. But $W^2 - Wn + \delta n = r^2 - \lambda b > 0$, and this completes our proof. □

Notice that Theorem 17.13 not only tells us that $\mathscr{C}_A(b)$ is maximal but it gives an explicit formula for $A(v, 2(r-\lambda), r)$ in terms of the parameters of the BIBD. It is a bonus whenever an explicit formula for an $A(n, 2\delta, W)$ is known.

Theorem 17.14. *If $2\delta \geq 4$, then*

$$A(n,2\delta,W) \leq [\frac{n}{W}A(n-1,2\delta,W-1)] \leq \frac{n}{W}A(n-1,2\delta,W-1).$$

Proof. Let $\mathscr{C}(n)$ be an $(n,A(n,2\delta,W),2\delta)$-code with constant weight W and incidence matrix A. Hence, $\mathscr{C}^A(n) = \mathscr{C}(n)$. Let

$$M = A(n,2\delta,W) = |\mathscr{C}(n)| \quad \text{and} \quad \mathscr{C}^A(n) = \{\mathbf{a}_1, \mathbf{a}_2, \ldots, \mathbf{a}_M\}.$$

First note that A has exactly $WA(n,2\delta,W)$ entries which are 1. We now count these 1s another way and get an upper bound for them.

Take the kth column of A. Let M_k denote the number of 1s in the kth column. Take the $\mathbf{a}_i = a_{i1}a_{i2}\ldots a_{in}$ with $a_{ik} = 1$, eliminate a_{ik}, and let $\mathbf{b}_i$ be the resulting codeword in $\mathbb{Z}_2^{n-1}$ of length $n-1$ and weight $W-1$. Altogether, these codewords $\mathbf{b}_i$ form a code $\mathscr{C}_k^A(n-1)$ and $\mathscr{C}_k^A(n-1)$ is an $(n-1, M_k, 2\delta_k)$-code for some $\delta_k \geq \delta$ since $d(\mathbf{b}_i, \mathbf{b}_j) \geq 2\delta_k \geq 2\delta$, if $i \neq j$. Hence,

$$M_k = |\mathscr{C}_k^A(n-1)| \leq A(n-1, 2\delta_k, W-1).$$

Let $\delta' = \min\{\delta_1, \delta_2, \ldots, \delta_n\}$. As a result of a little lemma in the next paragraph, we will have each

$$M_k \leq A(n-1, 2\delta_k, W-1) \leq A(n-1, 2\delta', W-1).$$

Also, the number of 1s in A is $\sum_{k=1}^{n} M_k$. Hence,

$$WA(n,2\delta,W) = \sum_{k=1}^{n} M_k \leq \sum_{k=1}^{n} A(n-1,2\delta',W-1) = nA(n-1,2\delta',W-1),$$

from which we infer that

$$A(n,2\delta,W) \leq (n/W)A(n-1,2\delta',W-1).$$

We will be through when we have $A(n-1,2\delta',W-1) \leq A(n-1,2\delta,W-1)$. What we will show is the lemma that if we keep n and W constant, then

$A(n,2\delta,W)$ is a decreasing function of $D=2\delta$, that is, if $D'=2\delta'\le 2\delta=D$, then $A(n,D',W)\ge A(n,D,W)$. Let $\mathscr{C}(n)=\{\mathbf{a}_1,\mathbf{a}_2,\ldots,\mathbf{a}_M\}$ be a constant weight binary code with $D=2\delta$, and, without loss of generality, we will insist that $d(\mathbf{a}_1,\mathbf{a}_2)=2\delta$. We will replace $\mathbf{a}_1$ with an $\mathbf{a}_1'$. Choose s and t where $a_{1s}=0$ and $a_{2s}=1$, and where $a_{1t}=1$ and $a_{2t}=0$. Set $a_{1s}'=1$, $a_{1t}'=0$, and $a_{1i}'=a_{1i}$ if $i\notin\{s,t\}$. Let $\mathbf{a}_1'=a_{11}'a_{12}'\ldots a_{1n}'$. Now $d(\mathbf{a}_1',\mathbf{a}_2)=2(\delta-1)=d(\mathbf{a}_1,\mathbf{a}_2)-2$. What about $d(\mathbf{a}_1',\mathbf{a}_i)$ for $i\ge 2$? There are four cases for $a_{is},a_{it}\in\{0,1\}$. If $a_{is}=a_{it}$, then $d(\mathbf{a}_1',\mathbf{a}_i)=d(\mathbf{a}_1,\mathbf{a}_i)$. If $a_{is}\ne a_{it}$, then $d(\mathbf{a}_1',\mathbf{a}_i)\in\{d(\mathbf{a}_1,\mathbf{a}_i)\pm 2\}$. So $\mathscr{C}'(n)=\{\mathbf{a}_1',\mathbf{a}_2,\ldots,\mathbf{a}_M\}$ has minimal distance $D'=2\delta'=2(\delta-1)=D-2$. As long as $D=2\delta\ge 4$, one can replace $\mathscr{C}(n)$ by $\mathscr{C}'(n)$, and $|\mathscr{C}'(n)|=|\mathscr{C}(n)|$, and the minimal distance of $\mathscr{C}'(n)$ is $D'=2\delta'=2(\delta-1)=D-2$, where D is the minimal distance of $\mathscr{C}(n)$. Hence, if $M=A(n,2\delta,W)$, then $M\le A(n,2(\delta-1),W)$. A direct application now assures us that we really do have

$$A(n-1,2\delta',W-1)\le A(n-1,2\delta,W-1),$$

as required, and this completes the proof of Theorem 17.14. □

Lemma 17.15. $\delta\le W$.

Proof. If codewords $\mathbf{a}$ and $\mathbf{b}$, $\mathbf{a}\ne\mathbf{b}$, have 1s in exactly τ common positions, then $2(W-\tau)=d(\mathbf{a},\mathbf{b})\ge D=2\delta$. Hence, $2W\ge 2\delta+2\tau$ and $W\ge\delta+\tau$. Since $\tau\ge 0$, we obtain $W\ge\delta$. □

Corollary 17.16. $A(n,2\delta,\delta)=[n/\delta]$.

Proof. If $W=\delta$ and $D=2\delta$, then for $\mathbf{a}\ne\mathbf{b}$, the τ in the proof of Lemma 17.15 is 0. That is, no two distinct codewords $\mathbf{a}$ and $\mathbf{b}$ have a 1 in a common position. Hence, $0\le n-\delta A(n,2\delta,\delta)<\delta$, and so $0\le(n/\delta)-A(n,2\delta,\delta)<1$. □

Corollary 17.17.

$$\begin{aligned}A(n,2\delta,W)&\le\left[\frac{n}{W}\left[\frac{n-1}{W-1}\left[\cdots\left[\frac{n-W+\delta}{\delta}\right]\cdots\right]\right]\right]\\&\le\frac{n(n-1)\cdots(n-W+\delta+1)}{W(W-1)\cdots(\delta+1)}\left[\frac{n-W+\delta}{\delta}\right].\end{aligned}$$

Proof. From Theorem 17.14, we have

$$A(n,2\delta,W) \le [(n/W)A(n-1,2\delta,W-1)] \le (n/W)A(n-1,2\delta,W-1).$$

Applying Theorem 17.14 to $A(n-1,2\delta,W-1)$, we obtain

$$\begin{aligned} A(n,2\delta,W) &\le [(n/W)[((n-1)/(W-1))A(n-2,2\delta,W-2)]] \\ &\le (n(n-1)/W(W-1))A(n-2,2\delta,W-2). \end{aligned}$$

Continuing we obtain

$$\begin{aligned} A(n,2\delta,W) &\le \left[\frac{n}{W}\left[\frac{n-1}{W-1}\left[\cdots\left[\frac{n-W+\delta+1}{\delta+1}A(n-W+\delta,2\delta,\delta)\right]\cdots\right]\right]\right] \\ &\le \frac{n(n-1)\cdots(n-W+\delta+1)}{W(W-1)\cdots(\delta+1)}A(n-W+\delta,2\delta,\delta). \end{aligned}$$

Now apply Corollary 17.16 to get the desired results. □

Since any finite affine plane $(N,\mathscr{L},\in)$ is a BIBD with the parameters $(v,b,r,k,\lambda) = (m^2,m^2+m,m+1,m,1)$ for some m, from Theorem 17.13 we know that the column code $\mathscr{C}_A(m^2+m)$ of the incidence matrix A for $(N,\mathscr{L},\in)$ is maximal in the sense that $A(m^2+m,2m,m+1) = m^2 = |\mathscr{C}_A(m^2+m)|$. But the row code $\mathscr{C}^A(m^2)$ of A is also maximal in the sense that $A(m^2,2(m-1),m) = m^2+m$.

Theorem 17.18. *Let $(N,\mathscr{L},\in)$ be a finite affine plane, hence a BIBD with parameters $(v,b,r,k,\lambda) = (m^2,m^2+m,m+1,m,1)$ for some m. Let A be the incidence matrix for $(N,\mathscr{L},\in)$. Then*

$$A(m^2,2(m-1),m) = |\mathscr{C}^A(m^2)| = m^2+m.$$

Proof. We have, from Corollary 17.17 and Corollary 17.16,

$$\begin{aligned} m^2+m = |\mathscr{C}^A(m^2)| &\le A(m^2,2(m-1),m) \\ &\le (m^2/m)A(m^2-1,2(m-1),m-1) \\ &= m[(m^2-1)/(m-1)] = m(m+1) = m^2+m. \end{aligned}$$

Hence, $m^2+m = |\mathscr{C}^A(m^2)| = A(m^2+m,2(m-1),m+1)$. □

Again, we get an explicit formula for $A(m^2+m, 2(m-1), m+1)$ where m is the number of points in a line of an affine plane.

All the affine planes which we get from finite planar nearrings give optimal $\mathscr{C}^A(m^2)$ and $\mathscr{C}_A(m^2+m)$. Actually, Theorem 17.18 should be the corollary of the more general

Theorem 17.19. *Let $(N, \mathscr{B}, \in)$ be a BIBD with parameters (v, b, r, k, λ) with $\lambda = 1$. Let A be the incidence matrix for $(N, \mathscr{B}, \in)$. Then*

$$b = |\mathscr{B}| = |\mathscr{C}^A(v)| = A(v, 2(k-1), k).$$

Proof. Again, with $\lambda = 1$, we have $D = 2(k-1)$ and $W = k$. Now apply Corollary 17.17 and Corollary 17.16 and recall that

$$bk = vr \text{ and } \lambda(v-1) = v-1 = r(k-1),$$

so

$$\begin{aligned} b = |\mathscr{B}| &= |\mathscr{C}^A(v)| \leq A(v, 2(k-1), k) \\ &\leq (v/k)A(v-1, 2(k-1), k-1) = (v/k)[(v-1)/(k-1)] \\ &= vr/k = bk/k = b. \end{aligned}$$

Hence, $b = |\mathscr{B}| = |\mathscr{C}^A(v)| = A(v, 2(k-1), k)$, as was to be shown. □

All the affine spaces constructed in §7 from finite planar nearrings are BIBD with $\lambda = 1$, and so Theorem 17.19 applies to each of these.

Johnson's bound was not applicable for the proof of Theorem 17.18 since $W^2 - Wn + \delta n = 0$. For Theorem 17.19, $W^2 - Wn + \delta n = k^2 - v$, and so we have no immediate assurance that $W^2 - Wn + \delta n > 0$. The techniques of the proofs of Theorem 17.18 and Theorem 17.19 are also applicable for finite Möbius planes.

Start with a BIBD $(N, \mathscr{B}, \in)$ having parameters

$$(v, b, r, k, \lambda) = (n^2+1, n(n^2+1), n(n+1), n+1, n+1)$$

with the additional property that for every three distinct points $x, y, z \in N$, there is a unique $B \in \mathscr{B}$ with $x, y, z \in B$. Hence, the μ of Theorem 17.6 is 2. Such a BIBD, as you should well know from Definitions 1.24, is also called a *Möbius plane*. Let A be the incidence matrix for $(N, \mathscr{B}, \in)$, so $\mathscr{C}^A(n^2+1)$ is an $(n^2+1, n(n^2+1), 2(n-1))$-code with constant weight $W = n+1$. Then

$$\begin{aligned} n(n^2+1) = b &\leq A(n^2+1, 2(n-1), n+1) \\ &\leq ((n^2+1)/(n+1))A(n^2, 2(n-1), n) \\ &\leq ((n^2+1)n^2/(n+1)n)A(n^2-1, 2(n-1), n-1) \\ &= (n^2+1)n^2(n^2-1)/(n+1)n(n-1) = n(n^2+1). \end{aligned}$$

So we have

Theorem 17.20. *Let a BIBD $(N, \mathscr{B}, \in)$ be a Möbius plane with $v = n^2+1$. Let A be the incidence matrix for $(N, \mathscr{B}, \in)$. Then*

$$|\mathscr{C}^A(v)| = A(n^2+1, 2(n-1), n+1) = n(n^2+1).$$

Again, Johnson's bound is not applicable since the denominator of Johnson's bound is $(n+1)^2 - (n+1)(n^2+1) + (n-1)(n^2+1) = -(n+1)^2 < 0$.

Can there exist a BIBD $(N, \mathscr{B}, \in)$ with the parameters of a Möbius plane but which is not a Möbius plane? In other words, do we need the extra condition that every three distinct points belong to a unique block $B \in \mathscr{B}$? In [L], Heinz Lüneburg, a contemporary geometer, gives examples of such BIBD that are not Möbius planes.

We have pointed out that the denominator for Johnson's bound is not positive for the row codes $\mathscr{C}^A(v)$ of the BIBDs of Theorem 17.18 and Theorem 17.20. The same is true for the row codes $\mathscr{C}^A(v)$ of the BIBDs $(N, \mathscr{B}^*, \in)$ for finite circular planar nearrings $(N, +, \cdot)$.

Proposition 17.21. *Let $(N, +, \cdot)$ be a finite circular planar nearring of order v with BIBD $(N, \mathscr{B}^*, \in)$. Let A be the incidence matrix for $(N, \mathscr{B}^*, \in)$ and $\mathscr{C}^A(v)$ the* *corresponding row code which has constant weight* $W = k$ *and distance* $D = 2\delta = 2(k-\mu) = 2(k-2)$. *Then* $W^2 - Wv + \delta v < 0$.

Proof. Since $k = |B|$ for each $B \in \mathscr{B}^*$ and k divides $v-1$, there is an integer t such that $v = kt+1$. Now $W^2 - Wv + \delta v = k^2 - 2(kt+1)$. So the answer to when $W^2 - Wv + \delta v > 0$ is equivalent to when $k^2 - 2tk - 2 > 0$, which is, of course, when $k < t - \sqrt{t^2+2} < 0$ or $t + \sqrt{t^2+2} < k$. Since $0 < k$, we concentrate on the latter. With $t = (v-1)/k$, we then seek to answer the question when is $(v-1)/k + \sqrt{((v-1)/k)^2+2} < k$? This will happen exactly when $\sqrt{2v} < k$.

We also have $k \leq (3+\sqrt{4v-7})/2$ since $(N,+,\cdot)$ is circular. So we will need $2\sqrt{2v} < 3+\sqrt{4v-7}$, which means $2 < v < 8$. But $v-1 = kt$ and $1 \notin \{k,t\}$, so $v \in \{5,7\}$. With $v = 5$, then $k = t = 2$, but

$$k = 2 > 2+\sqrt{4+2} = t+\sqrt{t^2+2}$$

is not valid. With $v = 7$ and $k = 2$, we obtain $t = 3$, and

$$k = 2 > 3+\sqrt{9+2} = t+\sqrt{t^2+2}$$

is not valid. Finally, for $v = 7$ and $k = 3$, we have $t = 2$. But then

$$k = 3 > 2+\sqrt{4+2} > 4$$

is not valid. We have now exhausted all the possibilities. □

Problem 17.22. Do we have $b = v(v-1)/k = A(v,2(k-2),k)$ for the row code $\mathscr{C}^A(v)$ of any BIBD $(N,\mathscr{B}^*,\in)$ from a finite circular planar nearring? If so, which ones?

Even though Proposition 17.21 is a negative result about the row code $\mathscr{C}^A(v)$ of a BIBD $(N,\mathscr{B}^*,\in)$ of a finite circular planar nearring, we know that the codes $\mathscr{C}^A(v)$ can correct up to $k-3$ errors and k can be arbitrarily large. Also, $b = |\mathscr{C}^A(v)| = v(v-1)/k$ can be very large. We can construct a new code from $\mathscr{C}^A(v)$, and, in doing so, even improve on its error correcting and error detecting capabilities. The price we pay is in the number of code words. But if $|\mathscr{C}^A(v)|$ is sufficiently large, perhaps we can pay the price.

Let $(N, \mathscr{B}^*, \in)$ be the BIBD of a finite circular planar nearring $(N, +, \cdot)$ of order v with incidence matrix A. Start with the row code $\mathscr{C}^A(v)$. Take an arbitrary $B_1 \in \mathscr{B}^*$ and list all the blocks $B \in \mathscr{B}^*$ which intersect B_1 in two or more places. That is, we include B_1 in this list. For $x, y \in B_1$, there are $\lambda - 1 = k - 2$ blocks $B \in \mathscr{B}^*$ with $B \neq B_1$ and $|B \cap B_1| = 2$. For $\{u, v\} \neq \{x, y\}$, $u, v \in B_1$, we again obtain $\lambda - 1 = k - 2$ new blocks $C \in \mathscr{B}^*$ with $C \neq B_1$ and $|C \cap B_1| = 2$. Let

$$\mathscr{B}_1^* = \{B \in \mathscr{B}^* \mid |B \cap B_1| = 2\} \cup \{B_1\}.$$

Then $|\mathscr{B}_1^*| = \binom{k}{2}(k-2) + 1$. So $\mathscr{B}_1^*$ consists of a list of $\binom{k}{2}(k-2) + 1$ blocks which intersect B_1 twice, together with B_1 itself.

Next, take $B_2 \in \mathscr{B}^* \setminus \mathscr{B}_1^*$ and repeat the procedure. That is, let

$$\mathscr{B}_2^* = \{B \in \mathscr{B}^* \setminus \mathscr{B}_1^* \mid |B \cap B_2| = 2\} \cup \{B_2\}.$$

Obviously, $|\mathscr{B}_2^*| \leq \binom{k}{2}(k-2) + 1$. The third step is to take $B_3 \in \mathscr{B}^* \setminus (\mathscr{B}_1^* \cup \mathscr{B}_2^*)$ and repeat the procedure. Let

$$\mathscr{B}_3^* = \{B \in \mathscr{B}^* \setminus (\mathscr{B}_1^* \cup \mathscr{B}_2^*) \mid |B \cap B_3| = 2\} \cup \{B_3\}.$$

In general, let

$$B_u \in \mathscr{B}^* \setminus (\mathscr{B}_1^* \cup \cdots \cup \mathscr{B}_{u-1}^*)$$

and let

$$\mathscr{B}_u^* = \{B \in \mathscr{B}^* \setminus (\mathscr{B}_1^* \cup \cdots \cup \mathscr{B}_{u-1}^*) \mid |B \cap B_u| = 2\} \cup \{B_u\}.$$

Since $|\mathscr{B}^*| < \infty$, we eventually come to an end. When we do, we let

$$\mathscr{C} = \{B_1, B_2, \ldots, B_u\}$$

where u is the smallest positive integer such that $\mathscr{B}^* = \mathscr{B}_1^* \cup \cdots \cup \mathscr{B}_u^*$. Let $\mathscr{C}(v)$ be the row code corresponding to the incidence matrix constructed from $\mathscr{C}$.

Let us call the procedure described above to construct a code $\mathscr{C}(v)$ from the $\mathscr{C}^A(v)$ for a finite circular planar nearring the *FHP-procedure* since it came from [FHP]. We can obtain upper and lower bounds for a $|\mathscr{C}(v)|$. First note that if $\mathscr{C} = \{B_1, B_2, \ldots, B_u\}$ and $\mathscr{C}(v) = \{\mathbf{b}_1, \mathbf{b}_2, \ldots, \mathbf{b}_u\}$, then $d(\mathbf{b}_i, \mathbf{b}_j) \geq 2(k-1)$, if $i \neq j$, since $|B_i \cap B_j| \leq 1$. Hence, for $\mathscr{C}(v)$, we have $D = 2\delta \geq 2(k-1)$, if $i \neq j$. So a $\mathscr{C}(v)$ can correct at least one more error than can $\mathscr{C}^A(v)$. From Theorem 8.3, Theorem 17.14, and Corollary 17.16, we obtain

$$v(v-1)/k = b = |\mathscr{C}^A(v)| \leq v(v-1)(v-2)/k(k-1)(k-2)$$

and

$$|\mathscr{C}(v)| \leq v(v-1)/k(k-1).$$

Hence,

$$v(v-1)/k\left(\binom{k}{2}(k-2)+1\right) \leq |\mathscr{C}(v)| \leq v(v-1)/k(k-1).$$

So we have

Theorem 17.23. *For a finite circular planar nearring of order v with circles having k points each, let $\mathscr{C}(v)$ be a code produced by the FHP-procedure. Then*

$$\frac{v(v-1)}{k\left(\binom{k}{2}(k-2)+1\right)} \leq |\mathscr{C}(v)| \leq \frac{v(v-1)}{k(k-1)},$$

and $\mathscr{C}(v)$ can correct at least one more error than $\mathscr{C}^A(v)$ can.

Problem 17.24. Is the u for the FHP-procedure an invariant of the finite circular planar nearring? That is, does u depend upon the choices of $B_1, B_2, \ldots, B_u$? If not, is there an algorithm which will maximize u?

In [FHP], the question is raised as to whether the denominator of Johnson's bound for the row code of a BIBD, $W^2 - Wn + \delta N$, is always nonpositive. We can give here only one example where $W^2 - Wn + \delta n > 0$ and, in doing so, give infinitely many more examples where $W^2 - Wn + \delta n < 0$.

In Proposition 5.8 and Theorem 5.9, let $q = p^n$. Then for any circular planar nearring $(N, +, \cdot)$ constructed from a Ferrero pair (K, Φ) of Theorem 5.9, we obtain a BIBD $(N, \mathscr{B}^*, \in)$ with parameters

$$(v, b, r, k, \lambda) = (q^2, q^2(q-1), q^2 - 1, q+1, q),$$

and since $(N, +, \cdot)$ is circular, the μ of Theorem 17.6 is 2. So for $W^2 - Wn + \delta n$, we have $W = q+1$, $n = q^2$, and $\delta = q-1$. Asking when $W^2 - Wn + \delta n > 0$ translates into asking when $-q^2 + 2q + 1 > 0$. Since $q > 0$, the answer is when $0 < q < 1 + \sqrt{2}$. Hence, $q = 2$ is our only possibility. For $q = 2$, then $W = 3$, $n = r$, and $\delta = 1$ make $W^2 - Wn + \delta n = 1 > 0$. Hence, $b = 4$, but $A(n, 2\delta, W) = A(4, 2, 3) \leq (1 \cdot 4)/1 = 4$, so again we have $b = |\mathscr{C}^A(4)| = A(4, 2, 3)$.

With this one example of a $\mathscr{C}^A(v)$ with $W^2 - Wn + \delta n > 0$, there is added intrigue to the following problem suggested in [FHP].

Problem 17.25. Does there exist a relatively large family of BIBD $(N, \mathscr{B}, \in)$ whose incidence matrix A yields a row code $\mathscr{C}^A(v) = \mathscr{C}^A(n)$ with positive denominator $W^2 - Wn + \delta n$ for Johnson's bound?

We have seen several times already that $|\mathscr{C}^A(n)| = A(n, 2\delta, W)$ even though $W^2 - Wn + \delta n \leq 0$. One naturally raises the question

Problem 17.26. For a BIBD $(N, \mathscr{B}, \in)$, with $n = v = |N|$, when is $|\mathscr{C}^A(n)| = A(n, 2\delta, W)$?

All the codes we have dealt with so far are constant weight codes. On page 530 of [M & S], it is mentioned that "constant weight codes have a number of practical applications." Then numerous references are given. This brings up

Problem 17.27. For the constant weight codes arising from finite planar nearrings, which have practical applications and why?

And then in [M & S], on page 530, we also see a "Research Problem" which we shall borrow and list here with appropriate modifications.

Problem 17.28. "Find good methods of encoding and decoding constant weight codes" arising from finite planar nearrings.

In Chapter 21 of [M & S], one is led to believe that association schemes are useful for coding theory, but, in this chapter, a different definition of an association scheme is given. Now we have met with association schemes in §13. The two definitions of association schemes are not quite the same, but they may in fact be equivalent, and certainly there is common ground. So we have

Problem 17.29. Do any of the association schemes produced in §13 have any nice applications to coding theory?

Exercise 17.30. Look up the definition in [M & S] of an association scheme. Is it equivalent to the one included in these notes?

Exercise 17.31. Is the conclusion of Theorem 17.20 valid for all BIBD with parameters $(v,b,r,k,\lambda) = (n^2+1, n(n^2+1), n(n+1), n+1, n+1)$?

18 Finite affine planes and finite projective planes are BIBDs

With Corollary 17.9, we see that any BIBD with $\lambda = 1$ gives a row code with the number of codewords $b = A(v, 2(k-1), k)$ as well as a column code from Theorem 17.13 with the number of codewords

$$v = A(b, 2(r-1), r) = \delta n/(w^2 - wn + \delta n).$$

Such BIBDs include the finite affine planes and the finite projective planes which we have constructed from finite fields. In Definition 1.13 and in Definition 1.18, we demanded that a finite affine plane and a finite projective plane first be a BIBD. Later, in Definitions 3.1 and in Definition 3.9, we made no such requirements. This is analogous to what was also done for Möbius planes. In Definitions 1.24, we required that a finite Möbius plane be a BIBD. However, in Definitions 3.20, we made no such requirement. But, with Theorem 3.28, we reconciled the differences showing that a finite Möbius plane, according to Definitions 3.20, is also a BIBD and a Möbius plane according to Definitions 1.24. We have never done the corresponding work for finite affine planes nor for finite projective planes. It is the purpose of this section to bring these two points into harmony.

Since the idea of a BIBD is a central theme of this work, we chose to introduce finite affine planes, finite projective planes, and finite Möbius planes within the framework of a BIBD. We also wanted to limit §1 to introducing the principle ideas. It would have been distracting in §1 to give the definition as in Definitions 3.1, Definition 3.9, and Definitions 3.20, and then to prove that, in the finite case, each is a BIBD. However, here at the end of this chapter, after having seen that BIBDs with $\lambda = 1$ are particularly important, it seems appropriate to show that all finite affine planes and all finite projective planes are also BIBDs.

Theorem 18.1. *Let $(P, \mathscr{L})$ be an affine plane with $|P| < \infty$. Then $(P, \mathscr{L})$ is a BIBD with parameters $(v, b, r, k, \lambda) = (k^2, k(k+1), k+1, k, 1)$ for some positive integer k.*

Proof. We will first define the family $\mathscr{A}(L)$ of lines parallel to a given line L. Each such family $\mathscr{A}(L)$ will form a partition of P. All such $\mathscr{A}(L)$ will form a partition of $\mathscr{L}$. Then we shall show the existence of three distinct lines $L, M, N \in \mathscr{L}$ such that if $X, Y \in \{L, M, N\}$ and $X \neq Y$, then $|X \cap Y| = 1$. From this, we will be able to deduce that $|A| = |\mathscr{A}(B)|$ for arbitrary $A, B \in \mathscr{L}$. This gives the existence of a positive integer k with $|A| = k$ for each $A \in \mathscr{L}$. With $|\mathscr{A}(L)| = k$, we will get $v = |P| = k^2$. Next, we will show that an arbitrary $x \in P$ belongs to exactly $k+1$ distinct lines. We begin with a $\lambda = 1$ from the definition of an affine plane, so then we will have a BIBD and can compute that $b = k(k+1)$ from the equation $vr = bk$. Let's begin!

Choose any $L \in \mathscr{L}$, and we can do this since $\mathscr{L} \neq \varnothing$. Let

$$\mathscr{A}(L) = \{M \in \mathscr{L} \mid M = L \text{ or } M \cap L = \varnothing\}.$$

If $x \in P \setminus L$, then there is a unique $M \in \mathscr{L}$ such that $x \in M$ and $L \cap M = \varnothing$. Hence, $P = \cup\mathscr{A}(L)$. If $M_1, M_2 \in \mathscr{A}(L)$ and $M_1 \neq M_2$, then we need to see that $M_1 \cap M_2 = \varnothing$. First, if $L \in \{M_1, M_2\}$, then we have $M_1 \cap M_2 = \varnothing$ from the definition of $\mathscr{A}(L)$ and the fact that $M_1, M_2 \in \mathscr{A}(L)$. So, we can be assured that $L \notin \{M_1, M_2\}$. If $u \in M_1 \cap M_2$, then $u \notin L$, and there is a unique $M \in \mathscr{L}$ with $u \in M$ and $M \cap L = \varnothing$. But $u \in M_1 \cap M_2$ and each $M_i \cap L = \varnothing$. What can we do but conclude that $M_1 \cap M_2 = \varnothing$ if $M_1, M_2 \in \mathscr{A}(L)$ and $M_1 \neq M_2$? This means that any such $\mathscr{A}(L)$ is a partition of P. It also means that $\mathscr{A}(L) = \mathscr{A}(T)$ for any $T \in \mathscr{A}(L)$.

Certainly $L \in \mathscr{L}(L)$ for each $L \in \mathscr{L}$, and so $\mathscr{L} = \cup_{L \in \mathscr{L}} \mathscr{A}(L)$. Suppose that $\mathscr{A}(L)$ and $\mathscr{A}(M)$ are not the same. We want to show that $\mathscr{A}(L) \cap \mathscr{A}(M) = \varnothing$. If $T \in \mathscr{A}(L) \cap \mathscr{A}(M)$, then $\mathscr{A}(T) = \mathscr{A}(L)$ and $\mathscr{A}(T) = \mathscr{A}(M)$. So $\mathscr{A}(L) \neq \mathscr{A}(M)$ implies $\mathscr{A}(L) \cap \mathscr{A}(M) = \varnothing$. This means that the collection of all $\mathscr{A}(L)$, $L \in \mathscr{L}$, form a partition of $\mathscr{L}$.

Since $\mathscr{L} \neq \varnothing$, there is an $L \in \mathscr{L}$, and we also have that $|L| \geq 2$ and the existence of a $y \in P \setminus L$. With $x_1, x_2 \in L$ and $x_1 \neq x_2$, we have lines $M, N \in \mathscr{L}$ with $y, x_1 \in M$ and $y, x_2 \in N$ since $y \neq x_1$ and $y \neq x_2$. Certainly $M \neq L$ and $N \neq L$. If $M = N$, then $x_1, x_2 \in M$, $x_1, x_2 \in L$ and $x_1 \neq x_2$ imply that $L = M$, since two distinct points belong to exactly one line L. Hence, $M \neq N$, and so L, M, and N are three distinct lines in $\mathscr{L}$. Hence, $|\mathscr{L}| \geq 3$.

Take any $L \in \mathscr{L}$ and any $M \in \mathscr{L} \setminus \mathscr{A}(L)$. We will see that $|M| = |\mathscr{A}(L)|$.

If $\mathscr{A}(L) = \{L = L_1, L_2, \ldots, L_t\}$ and $|\mathscr{A}(L)| = t$, then $|M \cap L_i) = 1$ for each i, $1 \leq i \leq t$. This is because $M \notin \mathscr{A}(L)$, so we cannot have $M \cap L_i = \varnothing$ since $\mathscr{A}(L) = \mathscr{A}(L_i)$ for each i, $1 \leq i \leq t$. From the definition of an affine plane, we cannot have $|M \cap L_i| \geq 2$ for any i, $1 \leq i \leq t$. Hence, each $|M \cap L_i| = 1$, $1 \leq i \leq t$. With $M = M \cap P = M \cap (L_1 \cup L_2 \cup \cdots \cup L_t) = (M \cap L_1) \cup \cdots \cup (M \cap L_t)$, and with $(M \cap L_i) \cap (M \cap L_j) = \varnothing$ for $i \neq j$, we get $|M| = \sum_{i=1}^{t} |M \cap L_i| = t$. This assures us that $|M| = |\mathscr{A}(L)|$.

With three distinct lines $L, M, N \in \mathscr{L}$, we have $|L| = |\mathscr{A}(M)|$, $|L| = |\mathscr{A}(N)|$ and $|M| = |\mathscr{A}(N)|$. This gives us $|L| = |M|$ and $|\mathscr{A}(M)| = |\mathscr{A}(N)|$. So, any two lines $L, M \in \mathscr{L}$ have the same cardinality, i.e., $|L| = |M|$, and for any two lines $M, N \in \mathscr{L}$, we have $|\mathscr{A}(M)| = |\mathscr{A}(N)|$. Moreover, for any $A, B \in \mathscr{L}$, we have $|A| = |\mathscr{A}(B)|$. Let this common value be k, since we would have $|A| = k$ for each $A \in \mathscr{L}$ and we are developing a proof that $(P, \mathscr{L})$ is a BIBD. With $|L| = |\mathscr{A}(L)|$, $\mathscr{A}(L) = \{L = L_1, L_2, \ldots, L_k\}$, and $L_i \cap L_j = \varnothing$ for $i \neq j$, we have $P = \cup \mathscr{A}(L)$ and so $v = |P| = \sum_{i=1}^{k} |L_i| = k^2$.

Choose an arbitrary $x \in P$. There is a line $L \in \mathscr{L}$ with $|L| \geq 2$. If $x \in L$, then there is a $y \in P \setminus L$ and a unique line $M \in \mathscr{L}$ with $y \in M$ and $L \cap M = \varnothing$. Hence, $x \notin M$. If $x \notin L$, then let $M = L$. One way or another, we get the existence of a line $M \in \mathscr{L}$ with $x \notin M$. So, there is a line $A \in \mathscr{L}$ with $x \in A$ and $A \cap M = \varnothing$. This gives one line A with $x \in A$. Let $M = \{m_1, \ldots, m_k\}$. For each m_i, there is a unique line $A_i \in \mathscr{L}$ with $x, m_i \in A_i$. Now we have the lines $A, A_1, A_2, \ldots, A_k$ each containing x, and each $A_i \neq A$. If $i \neq j$ and $A_i = A_j$, then $m_i, m_j \in A_i = A_j$ and $m_i \neq m_j$. So, m_i and m_j both belong to both M and $A_i = A_j$. But $x \in A_i = A_j$ and $x \notin M$, so $M \neq A_i = A_j$. This contradictions allows us to conclude that the A_i, $1 \leq i \leq k$, are distinct. So, we now have $k+1$ distinct lines containing x. If there is another $T \in \mathscr{L}$ with $x \in T$, then either $T \cap M = \varnothing$ or $|T \cap M| = 1$. If $T \cap M = \varnothing$, then T and A are distinct lines containing x and $T \cap M = A \cap M = \varnothing$, which cannot be. Also, if $|T \cap M| = 1$, then $T \cap M = \{m_i\}$ for some i. Since $m_i, x \in T \cap A_i$, we again have a contradiction. We must then conclude that A, A_1, ..., A_k are exactly the $k+1$ distinct lines of $\mathscr{L}$ containing x. Since $x \in P$ is arbitrary, we have that each point in P belongs to exactly $r = k+1$ distinct lines in $\mathscr{L}$.

From the definition of an affine plane, each pair of distinct points of P belong to exactly one line $B \in \mathscr{L}$, and so $\lambda = 1$ and now we have that $(P, \mathscr{L})$ is a BIBD. From $vr = bk$, we compute $b = k(k+1)$. This completes the proof of Theorem 18.1. □

There is a very close relationship between affine planes and projective planes. One can always obtain one from the other. Here, we shall demonstrate how one obtains an affine plane from a projective plane. One simply chooses an arbitrary line of the projective plane and removes its influence from the structure. What is left is an affine plane.

For more detail, let $(P, \mathscr{L})$ be any projective plane, and since $\mathscr{L} \neq \varnothing$, we select any $A \in \mathscr{L}$. Let $\mathscr{B} = \mathscr{L} \setminus \{A\}$, the remaining lines of the projective plane $(P, \mathscr{L})$. For each $B \in \mathscr{B}$, $|B \cap A| = 1$. Let $B' = B \setminus A$, so we are removing from B the unique point which A and B have in common. Let $\mathscr{L}' = \{B' = B \setminus A \mid B \in \mathscr{B}\}$ and let $P' = P \setminus A$. In $(P', \mathscr{L}')$, we have removed A's influence from $(P, \mathscr{L})$.

Theorem 18.2. *$(P', \mathscr{L}')$ is an affine plane.*

Proof. From (d) of Definition 3.9, there is an $x \in P$ with $x \notin A$. Hence, $P' \neq \varnothing$. Since $|A| \geq 3$, there are at least three distinct points $a_1, a_2, a_3 \in A$, and the a_i, together with x, determine three distinct lines $A_1, A_2, A_3 \in \mathscr{L}$, with $x, a_i \in A_i$, $i = 1,2,3$. Each $A_i' = A_i \setminus A \in \mathscr{L}'$, so $\mathscr{L}' \neq \varnothing$. For $L' = L \setminus A \in \mathscr{L}'$, we have $|L| \geq 3$, and $|L \cap A| = 1$, so $|L'| \geq 2$. This gives (c) of Definitions 3.1. For $L' = L \setminus A \in \mathscr{L}'$, there is a point $x \in P$ with $x \notin L$. If $x \in P' = P \setminus A$, then we have (d) of Definitions 3.1. Otherwise, $x \in A$. Since $|L| \geq 3$, there are at least two distinct points $y_1, y_2 \in L \setminus A = L'$, and there is a unique line $M \in \mathscr{L}$ with $x, y_1 \in M$. With $|M| \geq 3$ and $M \cap A = \{x\}$, we have the existence of a point

$t \in M \setminus (L \cup A)$, and so $t \in P'$ and $t \notin L'$. Now we have (d) of Definitions 3.1. It remains to show (a) and (b) of Definitions 3.1.

If we take two distinct points $x, y \in P'$, then $x, y \in P$, and so there is a unique $L \in \mathscr{L}$ with $x, y \in L$. Since $x, y \in P$, we have $\{x, y\} \cap A = \varnothing$, and so $L \in \mathscr{B}$ and $L' \in \mathscr{L}'$ with $x, y \in \mathscr{L}'$. Obviously, there can be no other $M' \in \mathscr{L}'$ with $x, y \in M'$, so (a) of Definitions 3.1 is satisfied.

For (b) of Definitions 3.1, take any $L' = L \setminus A \in \mathscr{L}'$ and any $x \in P' \setminus L'$. Let $L \cap A = \{a\}$, and so there is a unique $M \in \mathscr{L}$ with $x, a \in M$, and so $M \cap A = \{a\}$, which assures us that $x \in M'$ and that $L' \cap M' = \varnothing$. This gives us existence for (b). If $N' = N \setminus A$ is another such line, $N' \in \mathscr{L}'$, with $x \in N'$ and $L' \cap N' = \varnothing$, then $N \cap A = \{b\} \neq \{a\}$, since M is the only line of $\mathscr{L}$ with $a, x \in M$. So $b \notin L \cap N$, but yet $|L \cap N| = 1$, and so there is a $y \in L \cap N$ and $y \notin A$. Hence, $y \in P'$ and $y \in L' \cap N'$, contrary to $L' \cap N' = \varnothing$. This gives uniqueness of M' and so (b) of Definitions 3.1 is valid for $(P', \mathscr{L}')$.

Having shown that (a), (b), (c), and (d) of Definitions 3.1 are valid for $(P', \mathscr{L}')$ and that $P' \neq \varnothing$, and that $\mathscr{L}' \neq \varnothing$, we can certainly conclude that $(P', \mathscr{L}')$ is an affine plane. □

Exercise 18.3. Let $(P', \mathscr{L}')$ be an affine plane. Describe how one can construct a projective plane $(P, \mathscr{L})$ from $(P', \mathscr{L}')$.

Theorem 18.4. *A projective plane $(P, \mathscr{L})$ with $|P| < \infty$ is a BIBD with parameters*

$$(v, b, r, k, \lambda) = (n^2 + n + 1, n^2 + n + 1, n + 1, n + 1, 1)$$

for some positive integer n.

Proof. Take any $A \in \mathscr{L}$ and let $|A| = n + 1$, so $n \geq 2$. We construct the affine plane $(P', \mathscr{L}')$ as was done for Theorem 18.2. Then $(P', \mathscr{L}')$ is a BIBD with parameters $(n^2, n(n+1), n+1, n, 1)$. We already have that $v = |P| = n^2 + n + 1$, $b = |\mathscr{L}| = n^2 + n + 1$, $k = |L| = |L'| + 1 = n + 1$, and $\lambda = 1$. Hence we need only to show that each $x \in P$ belongs to exactly $n + 1$ distinct lines in $\mathscr{L}$. We get this immediately for any $x \in P'$. If $x \in P \setminus P' = A$, then A is one of the lines on x. If $\mathscr{B}(x) = \{L \in \mathscr{L} \mid L \cap A = \{x\}\}$, then $\mathscr{B}'(x) = \{L' = L \setminus A \mid L \in \mathscr{B}(x)\}$ is exactly $\mathscr{A}(L')$ of the proof of Theorem 18.1 for any $L' \in \mathscr{B}'(x)$. Therefore, $|\mathscr{B}(x)| = |\mathscr{B}'(x)| = |\mathscr{A}(L')| = n$. So, even an $x \in A$ has $n+1$ lines in $\mathscr{L}$ containing it. Thus, $r = n+1$, and we have that $(P, \mathscr{L})$ is a BIBD with parameters as claimed. □

Chapter IV

Planar nearrings and geometry

Some of the nicer applications of planar nearrings have been in the area of geometry. It is here that the internal structure of a planar nearring is most dramatic. It is here that one finds an abundance of symmetry which completely disappears in rings. This is remarkable since a ring is significantly more symmetric than a ring, at least on the surface, that is, as one examines the definition and as one studies the algebraic structure. This symmetry is most apparent in circular planar nearrings.

One should be able to recognize that our applications to geometry are very similar to those of the previous chapter, that is, applications to combinatorics. It is hoped that the reader will be able to see that our applications in this chapter are closer, conceptually, with traditional geometric ideas.

19 Circular planar nearrings

In §1, nine ideas of interest were introduced each via a **Problem**. Except for Problem 1.7, we have addressed each of these ideas for a solution. For

Problem 1.1, we can use Theorem 15.2, Example 7.6, and **7.3** (the Ferrero Planar Nearring Factory). To solve Problem 1.12, we let $F = \mathbb{Z}_5$, and then construct $(F \times F, \mathscr{L}_F)$ as in Theorem 13.4. Then rename the points in $F \times F = \{(0,0),(0,1),\ldots,(0,4),(1,0),\ldots,(1,4),\ldots,(4,0),\ldots,(4,4)\}$ with the points in $\mathbb{Z}_{25} = \{0,1,\ldots,24\}$. A solution to Problem 1.17 is somewhat similar. Starting with $(F \times F, \mathscr{L}_F)$ as shown, we construct the projective plane as taught in Theorem 3.10. As points, we have $P' = (F \times F) \cup I = (F \times F) \cup \{0,1,2,3,4,\infty\}$. Rename these 31 points with points of $\mathbb{Z}_{31} = \{0,1,\ldots,30\}$. This gives a solution to Problem 1.17. Similarly, for Problem 1.23, we must make a quadratic extension K of $F = \mathbb{Z}_5$. Then apply Theorem 3.25. If we rename the points of K with the elements of $\mathbb{Z}_{25}$, we will have a solution to Problem 1.23. A solution to Problem 1.30 is provided with an application of Theorem 3.30 to $F = \mathbb{Z}_7$.

In §2, we developed Theorem 2.13 and Theorem 2.14, which can be used to solve Problem 1.38 and Problem 1.39. Finally, Problem 1.40 has a solution as provided by Theorem 2.15, and again by Theorem 8.1.

To solve Problem 1.7, we can use Theorem 8.3, Example 7.6, and **7.3**, so we have enough tools to solve all the **Problems** in §1. But, we have not really discussed circular BIBDs, except for Möbius planes. One might suspect that, of the ideas introduced in §1, we have saved the best for last. Perhaps the reader will agree after studying this chapter.

19.1 The bound for circularity

We have shown in Theorem 8.4 that if N is a finite circular IP-nearring, and the BIBD $(N, \mathscr{B}^*)$ has parameters v, b, r, k and λ, then $k \leq (3+\sqrt{4v-7})/2$. We want to show that this upper bound for k, in terms of v, is as good as one can get, in general. We will get a circular BIBD from a field K with $|K| = q^2 = v$, and circles B with $|B| = [(3+\sqrt{4q^2-7})/2]$, where $[x]$ is the greatest integer function.

This next result is a very special case of a very important theorem of algebra, but we only need what we present here.

Proposition 19.1. *Let F be a field with $p = x^2 + ax + b \in F[x]$. If $p(m) = 0$, then $p = (x-m)(x-n)$ for $n = -m-a$, and so $p(\alpha) \neq 0$ if $\alpha \in F \setminus \{m,n\}$.*

Proof. With $0 = p(m) = m^2 + am + b$, we let $n = -m-a$, and so

$$p(n) = n^2 + an + b = (-m-a)^2 + a(-m-a) + b = 0.$$

Also,

$$(x-m)(x-n) = x^2 - (m+n)x + mn = x^2 + ax + (-m^2 - ma) = x^2 + ax + b.$$

If $\alpha \in F \setminus \{m,n\}$, then $p(\alpha) = (\alpha - m)(\alpha - n) \neq 0$. □

So, a polynomial $p \in F[x]$ of degree 2 has at most two distinct roots in the field F.

Theorem 19.2. *Let K be a quadratic extension of a field F. Then there is a field automorphism σ of K of order 2 such that $\sigma(a) = a$ if and only if $a \in F$.*

Proof. We can take $K = F[x]/(p)$ for some irreducible $p \in F[x]$, and then we can think of the elements of K as being of the form $a + bx$, with $a, b \in F$. We first assume that the characteristic of F is 2. In this case, we define $\sigma(a+bx) = a + b(1+x) = (a+b) + bx$. Recalling that $u + u = 0$ for each $u \in K$, one gets that $\sigma^2 = 1_F$, $\sigma \neq 1_F$, and that σ is an automorphism of the field K where $\sigma(a) = a$ if and only if $a \in F$. If the characteristic of F is $\neq 2$, we define $\sigma(a+bx) = a - bx$. One easily sees that $\sigma^2 = 1_F$, $\sigma \neq 1_F$, and that σ is an automorphism of K where $\sigma(a) = a$ if and only if $a \in F$. □

Exercise 19.3. Show that σ, as defined in the proof of Theorem 19.2, really is a field automorphism, both when the characteristic of F is 2, and when it is not 2. Also, show that $\sigma(a) = a$ if and only if $a \in F$ for both cases.

Proposition 19.4. *Let K and L be finite fields of characteristic p with $L \subset K$, K of order p^{2n}, and L of order p^n, so K is of dimension 2 over L. Let the automorphism group of K over L be $\mathrm{Aut}_L K = \{1, \sigma\}$. The norm function of K over L is the function $N : K^* \to L^*$ defined by $N(x) = x\sigma(x)$. Then N is a group homomorphism whose kernel has order $p^n + 1$.*

Proof. For $x, y \in K^*$, $N(xy) = xy\sigma(xy) = xy\sigma(x)\sigma(y) = N(x)N(y)$. Now $|L^*| = p^n - 1$, so if the image of N is L^*, we have, since $|K^*| = p^{2n} - 1$, that $|\ker N| = p^n + 1$. So we proceed to show that the image of N is L^*.

For an $x\sigma(x) = N(x)$, $\sigma(x\sigma(x)) = \sigma(x)\sigma^2(x) = x\sigma(x)$, so $x\sigma(x) \in L^*$, the fixed field of σ. Hence, $N(K^*) \subseteq L^*$. For $H = \{x^2 \mid x \in L^*\}$, we have $|L^*/H| \leq 2$, and if $x^2 \in H$, $x \in L$, then $x^2 = x \cdot \sigma(x)$, so H is in the image of N. We have $H \subseteq N(K^*) \subseteq L^*$, and since $|L^*/H| \leq 2$, either $N(K^*) = H$ or $N(K^*) = L^*$. If $p = 2$, then $H = L^*$, and so $N(K^*) = L^*$, and we have our desired result.

Assume $p \neq 2$. Then $K = L(\omega) = L[X]/(X^2 + \omega^2)$, where $\omega^2 \in L^* \setminus H$ and $\omega \notin L$. So $z \in K$ if and only if $z = x + \omega y$ for some $x, y \in L$. Applying N, we obtain $N(z) = z\sigma(z) = (x + \omega y)(x - \omega y) = x^2 - \omega^2 y^2$. So $N(K^*)$ contains all $x^2 - \omega^2 y^2 \neq 0$ and, in particular, all $-\omega^2 y^2 \neq 0$ and, specifically, $-\omega^2$. If $N(K^*)$ contains one more element than H, then $N(K^*) = L^*$. So, if $-\omega^2 \notin H$, then $N(K^*) = L^*$.

If $-\omega^2 \in H$, and $H = \{h_1, h_2, \ldots, h_t\}$, then

$$H = \{h_1, h_2, \ldots, h_t\} = \{-\omega^2 h_1, -\omega^2 h_2, \ldots, -\omega^2 h_t\}.$$

If $N(K^*) = H$, then each $x^2 - \omega^2 y^2 \in H$ for any $x, y \in L$, as long as $\{x, y\} \neq \{0\}$. But $H = \{x^2 \mid x \in L^*\} = \{h_1, h_2, \ldots, h_t\}$, so $\{x^2 - \omega^2 y^2 \mid x, y \in L,\ \{x, y\} \neq \{0\}\} = \{h_i + h_j \mid h_i, h_j \in H\}$. In particular, this means that $N(K^*) = H$ is closed with respect to +, making it a subgroup of L^+, putting $0 \in H$, which cannot be. So the only honourable thing to do is to conclude that $N(K^*) = L^*$. As we said earlier, this forces $|\ker N| = p^n + 1$. □

We note that the subgroup $\ker N$ of K^* in the above proof is the unique subgroup of K^* of order p^n+1 since K^*, as the multiplicative group of a finite field, is cyclic.

Recall that for any multiplicative subgroup S of K^* with $|S| \geq 2$, we have a Ferrero pair (K, Φ) where $\Phi = \{a^\bullet \mid a \in S\}$. We usually identify a and $a^\bullet$, and regard Φ as a subgroup of K^*. This way, we will also simply the notation for a block $\Phi(a)+b$ for $a, b \in K$, $a \neq 0$, by writing it as $\Phi a + b$.

Theorem 19.5. *For K, F, σ, and p as in Proposition 19.4, let $\Phi = \ker N$. If $(K,+,*)$ is any planar nearring constructed from the Ferrero pair (K,Φ), then $(K, \mathscr{B}^*, \in)$ is circular.*

Proof. We want to show that two distinct circles can have *at most* two points in common. Take two arbitrary distinct circles K^*a+b and K^*c+d. Let $x \in (K^*a+b) \cap (K^*c+d)$. Our goal is to prove that x can have at most two values.

The proof offered here is in terms of field theory. Now $K^*a+b = \Phi a+b$ and $K^*c+d = \Phi c+d$, where $\Phi \triangleleft K^*$. So we are letting $x \in (\Phi a+b) \cap (\Phi c+d)$. This makes $x = ma+b$, and $x = nc+d$ for some $m, n \in \Phi$. We can then write $x-d = ma+(b-d)$ and $x-d = nc$. Being in a field, we have $(x-d)c^{-1} = mac^{-1}+(b-d)c^{-1}$ and $(x-d)c^{-1} = n$, where $m, n \in \Phi = \ker N$. This puts $(x-d)c^{-1} \in [\Phi ac^{-1}+(b-d)c^{-1}] \cap \Phi$.

Now we will show $x \in [\Phi a+b] \cap \Phi$, where $\Phi a+b \neq \Phi$, can have at most two values. Such an x has $x = ma+b$, and $x = n$ for some $m, n \in \Phi$, or, equivalently, $x-b = ma$, and $x = n$. Applying the homomorphism N we obtain

$$\begin{cases} N(x-b) = N(a), \\ N(x) = 1. \end{cases}$$

From the proof of Proposition 19.4, we know that $x = x_1+\omega x_2$ and $b = b_1+\omega b_2$, with the x_i, $b_i \in L$, and

$$N(x-b) = (x_1-b_1)^2 - \omega^2(x_2-b_2)^2 = N(a) \in H = \{r^2 \mid r \in L^*\}.$$

(If $\omega^2 = -1$, then this is the traditional equation for a circle of radius r, where $r^2 = N(a)$, and centre (b_1, b_2).) Also, $N(x) = x_1^2 - \omega^2 x_2^2 = 1$. Expanding, we obtain

$$\begin{cases} x_1^2 - 2b_1x_1 + b_1^2 - \omega^2 x_2^2 + 2\omega^2 b_2 x_2 - \omega^2 b_2^2 = N(a), \\ x_1^2 - \omega^2 x_2^2 = 1. \end{cases}$$

So $x = x_1 + \omega x_2$ must satisfy these equations simultaneously. Hence

$$-2b_1x_1 + b_1^2 + 2\omega^2 b_2 x_2 - \omega^2 b_2^2 = N(a) - 1,$$

which can be written in the form

$$x_2 = Ax_1 + B. \tag{19:1}$$

Substituting into $x_1^2 - \omega^2 x_2^2 = 1$ yields

$$\begin{aligned} 1 &= x_1^2 - \omega^2 (Ax_1 + B)^2 \\ &= x_1^2 - \omega^2 (A^2 x_1^2 + 2ABx_1 + B^2) \\ &= (1 - \omega^2 A^2) x_1^2 - 2AB\omega^2 x_1 - \omega^2 B^2, \end{aligned}$$

a polynomial of degree at most 2 with coefficients in a field. So, there are at most two solutions for x_1. For each solution, (19:1) gives exactly one solution for x_2. Thus, there are at most two solutions for x. □

In particular, we have

Corollary 19.6. *Let $p \geq 3$ be a prime number. Let $K = F_{p^2}$, the field of order p^2, and Φ the subgroup of K^* of order $p+1$. Then (K, Φ) is a circular Ferrero pair.*

Proposition 19.7. *Let (N, Φ) be a circular Ferreor pair. If $\Psi \leq \Phi$ with $|\Psi| \geq 2$, then (N, Ψ) is also a circular Ferrero pair.*

Proposition 19.8. *If $v = n^2$ and $n > 1$, then $[(3 + \sqrt{4v - 7})/2] = n + 1$.*

Proof. Certainly, $\sqrt{n^2-7/4}<n$, and since $(n-1)^2=n^2-2n+1<n^2-7/4$ for $n>1$, we obtain $n-1<\sqrt{n^2-7/4}$. So $n-1<\sqrt{n^2-7/4}<n$. Adding $3/2$, we have

$$n+1/2<(3+\sqrt{4n^2-7})/2<n+3/2.$$

This puts $[(3+\sqrt{4n^2-7}/2]\in\{n,n+1\}$. If $[(3+\sqrt{4n^2-7})/2]=n$, then

$$n\le(3+\sqrt{4n^2-7})/2<n+1.$$

Multiplying by 2 and subtracting 3 would yield

$$0<2n-3\le\sqrt{4n^2-7}<2n-1.$$

Upon squaring and subtracting the common term of $4n^2$ we get

$$-12n+9\le-7<-4n+1.$$

From $-7<-4n+1$ we learn that $4n<8$, or $n<2$. But $n\ge2$, so we must have the other alternative, namely, $[(3+\sqrt{4n^2-7}/2]=n+1$. □

Thus, we have a circular BIBD from a field with $v=q^2$ elements, and circles B with $q+1$ elements as promised. On the other hand, it is easy to see that if Ψ is a subgroup of Φ with $|\Psi|\ge2$, then $(K,\mathscr{B}^*_\Psi,\in)$ is also circular. Combine this fact and Theorem 19.5, we arrive at another conclusion. Given an integer $k\ge4$, there are infinite many primes in the arithmetical sequence $km-1$, $m=1,2,\ldots$, by Dirichlet's Theorem. Pick any prime p in this sequence, then k is a divisor of $p+1$. Thus by Theorem 19.5, the BIBD $(K,\mathscr{B}^*,\in)$ obtained from the Ferrero pair (K,Φ), K the field of order p^2 and Φ the subgroup of K^* order $p+1$, is circular, and so $(K,\mathscr{B}^*_\Psi,\in)$ is circular with block size k if Ψ is the subgroup of Φ of order k. Therefore, *we can have a circular BIBD for any specified block size.*

The following theorem is actually a rewriting of Theorem 8.2 using the terminology of Ferroer pairs.

Proposition 19.9. *Let (N,Φ) be a Ferrero pair with N finite. Then $\Phi a+b=\Phi c+d$ if and only if $b=d$ and $\Phi a=\Phi c$.*

Using the language of Ferreor pair, we can generalize Proposition 19.9 a little bit.

Proposition 19.10. *Let (N,Φ) be a Ferrero pair where N is abelian and Φ is of finite order k. If $kx\neq 0$ for all $x\in N\setminus\{0\}$, then $\Phi a+b=\Phi c+d$ if and only if $b=d$ and $\Phi a=\Phi c$.*

Proof. Assume that $\Phi a=\Phi b+c$ where $a,b,c\in N$ with a and b nonzero. We need only to show that $c=0$. Let $\psi\in\Phi\setminus\{1\}$. Then $\psi(a)+\cdots+\psi^s(a)=0$ where $s>1$ is the order of ψ [Gr, Lemma 10.1.1]. Let $\Psi=\langle\psi\rangle$, the cyclic subgroup of Φ generated by ψ. For every element $\phi\in\Phi$, we have

$$\phi\psi(a)+\ldots\phi\psi^s(a)=\phi(\psi(a)+\cdots+\psi^s(a))=\phi(0)=0.$$

Since Φ is the union of cosets of Ψ, we see that

$$\sum_{\phi\in\Phi}\phi(a)=\sum_{i=1}^{t}\sum_{j=1}^{s}\phi_i\psi^j(a)=\sum_{i=1}^{t}\phi_i\big(\psi(a)+\cdots+\psi^j(a)\big)=0$$

where $\{\phi_1,\ldots,\phi_t\}\subseteq\Phi$ is the set of a complete list of coset representatives of Ψ in Φ. Similarly, we have $\sum_{\phi\in\Phi}\phi(b)=0$.

Now,

$$0=\sum_{\phi\in\Phi}\phi(a)=\sum_{\phi\in\Phi}(\phi(b)+c)=\left(\sum_{\phi\in\Phi}\phi(b)\right)+kc,$$

where k is the order of Φ. By the assumption, $c=0$. This proves the result. □

Corollary 19.11. *Let (F,Φ) be a Ferrero pair where F is a field and Φ a finite subgroup of F^*. Then for any $a,c\in F^*$ and $b,d\in F$, $\Phi a+b=\Phi c+d$ if and only if $b=d$ and $\Phi a=\Phi c$.*

19.2 Characterizations of circular planar nearrings

Next, we turn to the question that *given a Ferrero pair* (N,Φ), *is it possible to tell whether* $(N,\mathscr{B}^*,\in)$ *constructed from* (N,Φ) *is circular or not without actually constructing it?* That is, we would like to seek for some characterizations of circularity of BIBDs constructed from finite Ferrero pairs which resorts to no constructing of the blocks. For example, the bound given in the previous discussion is not sufficient. The question seems to be hopeless for an answer if we were dealing with the most general BIBDs as they may come up using very different methods. Surprisingly enough, with Ferrero pairs behind, there is indeed some way to tell whether a BIBD is circular or not by looking at the Ferrero pair used for it!

Although there is a characterization for general situation, what we shall present here is only for BIBDs constructed from finite fields as it fits better the scope of this book.

First of all, we look closely at the circularity in the BIBDs constructed from fields with finite Φ. Later, this will provide us a computational basis.

For a Ferrero pair (K,Φ) where K is a field and Φ a finite subgroup of K^* of order k, we set $\widetilde{\Phi}=\{\frac{1-\phi}{1-\psi}\mid \phi,\psi\in\Phi\setminus\{1\},\phi\neq\psi\}$. Certainly, we have $|\widetilde{\Phi}|\leq(k-1)(k-2)$.

Proposition 19.12. *Let* (K,Φ) *be a Ferrero pair where* Φ *is a finite subgroup of* K^* *and let* $(K,\mathscr{B}^*,\in)$ *be the BIBD constructed from* (K,Φ). *Then* $(K,\mathscr{B}^*,\in)$ *is not circular if and only if* $|\widetilde{\Phi}|<(k-1)(k-2)$.

Proof. Suppose first that $(K,\mathscr{B}^*,\in)$ is not circular. Then there are $a,b,c\in K^*$ such that $\Phi a\cap(\Phi b+c)$ contains at least three distinct points. Thus for some $\phi_i,\psi_i\in\Phi$, $i=1,2,3$, we have

$$\phi_1 a=\psi_1 b+c,\quad \phi_2 a=\psi_2 b+c,\quad \phi_3 a=\psi_3 b+c.$$

From these identities, we get

$$(\phi_1 - \phi_2)a = (\psi_1 - \psi_2)b \quad \text{and} \quad (\phi_1 - \phi_3)a = (\psi_1 - \psi_3)b$$

and so

$$\frac{1 - \phi_1^{-1}\phi_2}{1 - \phi_1^{-1}\phi_3} = \frac{1 - \psi_1^{-1}\psi_2}{1 - \psi_1^{-1}\psi_3}.$$

Here, we have $\phi_1^{-1}\phi_2 \in \Phi \setminus \{1\}$, $\psi_1^{-1}\psi_2 \in \Phi \setminus \{1\}$, $\phi_1^{-1}\phi_2 \neq \phi_1^{-1}\phi_3$, and $\psi_1^{-1}\psi_2 \neq \psi_1^{-1}\psi_3$. If $\phi_1^{-1}\phi_2 = \psi_1^{-1}\psi_2$, then $1 - \phi_1^{-1}\phi_2 = 1 - \psi_1^{-1}\psi_2$, and so $\psi_1(\phi_1 - \phi_2) = \phi_1(\psi_1 - \psi_2)$. But then

$$\psi_1(\psi_1 - \psi_2)b = \psi_1(\phi_1 - \phi_2)a = \phi_1(\psi_1 - \psi_2)a$$

which would yield $\psi_1 b = \phi_1 a$; consequently, $c = 0$, a contradiction. Therefore $\phi_1^{-1}\phi_2 \neq \psi_1^{-1}\psi_2$. Similarly, $\phi_1^{-1}\phi_3 \neq \psi_1^{-1}\psi_3$. What we have just seen is that if $(K, \mathscr{B}^*, \in)$ is not circular, then $|\widetilde{\Phi}| < (k-1)(k-2)$.

Conversely, assume that $|\widetilde{\Phi}| < (k-1)(k-2)$. Then there are $\phi_1, \phi_2, \psi_1, \psi_2 \in \Phi \setminus \{1\}$ with $\phi_1 \neq \phi_2$, $\psi_1 \neq \psi_2$, $\phi_1 \neq \psi_1$, and $\frac{1-\phi_1}{1-\phi_2} = \frac{1-\psi_1}{1-\psi_2}$ $(= u$, say). Then we have $1 - \phi_1 = u - \phi_2 u$ and $1 - \psi_1 = u - \psi_2 u$. Therefore, in $\Phi \cap (\Phi u + (1-u))$, we find

$$\phi_1 = \phi_2 u + (1-u), \quad \psi_1 = \psi_2 u + (1-u), \quad \text{and} \quad 1 = 1 \cdot u + (1-u),$$

which are distinct. Then $(K, \mathscr{B}^*, \in)$ cannot be circular. □

Rremark 19.13. Now, take K to be a finite extension field of the finite field L and let Φ be a subgroup of L^*. Then Φ is also a subgroup of K^*, and the blocks of the BIBD $(L, \mathscr{B}_1^*, \in)$ constructed from (L, Φ) are also blocks of the BIBD $(K, \mathscr{B}_2^*, \in)$ constructed from (K, Φ). Therefore, if $(L, \mathscr{B}_1^*, \in)$ is not circular, then $(K, \mathscr{B}_2^*, \in)$ cannot be circular either. Conversely, if $(K, \mathscr{B}_2^*, \in)$ is not circular, then $|\widetilde{\Phi}| < (k-1)(k-2)$ by the Proposition 19.12. But then $(L, \mathscr{B}_1^*, \in)$ cannot be circular since Φ is the same in L as in K. Therefore, we have that $(L, \mathscr{B}_1^*, \in)$ is circular if and only if $(K, \mathscr{B}_2^*, \in)$ is circular.

Let $k > 1$ be a positive integer. Since a BIBD of block size two or three is certainly circular, we assume that $k \geq 4$. We now want to construct two sets $\mathscr{P}_k$ and $\mathscr{P}_k'$ of prime numbers having the following properties. Take a prime p. Then either there is a natural number $n \in \mathbb{N}$ such that $k|(p^n - 1)$ or there is not. If no such n exists, then we put $p \in \mathscr{P}_k'$. If there indeed is such an n, then there is a minimal one. We still call it n. Then there is a unique subgroup Φ of the field $F = F_{p^n}^*$ of order k. Now, consider the BIBD $(F, \mathscr{B}^*, \in)$ constructed from the pair (F, Φ). Either it is circular or it is not. If it is not circular, then we put $p \in \mathscr{P}_k$.

Now, assume that p is a prime in $\mathscr{P}_k$. Let $F = F_{p^s}$ be a finite field having a multiplicative subgroup Φ of order k. We now argue that the BIBD constructed from (F, Φ) is not circular. To see this, let n be the minimal positive integer such that $k \mid (p^n - 1)$, and let $L = F_{p^n}$. As $s \geq n$, we can write $s = nt + r$ where $r, t \in \mathbb{Z}$ and $0 \leq r < n$. Then from

$$p^s - 1 = (p^n - 1)(p^{s-n} + p^{s-2n} + \cdots + p^{s-tn}) + p^r - 1$$

we infer that $k|(p^\nu - 1)$. By the minimality of n, we see that $t = 0$, i.e. $n|s$. This means that F is an extension field of L. Then by Remark 19.13, we know that the BIBD constructed from (F, Φ) cannot be circular.

We have shown the following characterization.

Theorem 19.14. *Let $F = F_{p^n}$ be a finite field, where p is a prime and n a positive integer. Let Φ be a multiplicative subgroup of F of order $k \geq 4$. Then the BIBD constructed from (F, Φ) is circular if and only if $p \notin \mathscr{P}_k$, where p is the characteristic of F.*

Even though we don't know what $\mathscr{P}_k$ is, this is still a very nice characterization for circular BIBDs that can be constructed from finite field. But we know more! First of of, we know that for each $k \geq 4$, the set $\mathscr{P}_k^* = \mathscr{P}_k' \cup \mathscr{P}_k$ cannot consist of all prime numbers. This is because that if $p = mk - 1$ is a prime for some $m \in \mathbb{N}$, which always exists, then $k|(p+1)$ and $k|(p^2-1)$ (hence p is not contained in $\mathscr{P}'_k$), and we know that (F_{p^2}, Φ) where Φ is the multiplicative subgroup of F_{p^2} of order $p+1$ is circular. Take the subgroup Ψ of Φ of order k, then the BIBD constructed from (F_{p^2}, Ψ) is also circular. This means that $p \notin \mathscr{P}_k$.

Theorem 19.15. *Each of the sets $\mathscr{P}_k$ is finite.*

The proof of this nice result will be carried out with a constructive method. Thus, one may write a program on a computer to compute rather efficiently the sets $\mathscr{P}_k$.

The steps are as follows.

1. A link between circularity and the non-singularity of a finite set of two-by-two matrices is established.

2. Each of these matrices has a corresponding matrix over the field $\mathbb{Q}(z)$, where $z \in \mathbb{C}$ is a primitive k^{th} root of unity. These matrices are non-singular over $\mathbb{Q}(z)$ and the determinants are of the form $P(z)$ for some special polynomials $P(x) \in \mathbb{Z}[x]$.

3. Taking norms of these $P(z)$ over $\mathbb{Q}$ we obtain nonzero integers. If the characteristic of F is a divisor some of these integers, then the corresponding matrices over F are singular, and hence the BIBD in consideration cannot be circular; otherwise, it is circular.

Let F be a field and Φ a finite subgroup of F^* with $|\Phi| = k \geq 4$. Since Φ is cyclic, $\Phi = \langle \zeta \rangle$ where ζ is a primitive k^{th} root of unity. For $a, b, c, d \in \Phi \setminus \{1\}$ with $(a, b) \neq (c, d)$ and $(a, c) \neq (b, d)$ define the mapping $\tau = \tau_{a,b,c,d} : F^2 \to F^2$ by the rule

$$(x, y) \mapsto ((a-1)x - (b-1)y, (c-1)x - (d-1)y) \quad \text{for all } x, y \in F. \tag{19:2}$$

If τ is not bijective, then there are $(x,y) \neq (x',y')$ in F^2 such that $\tau(x,y) = \tau(x',y')$. Put $s = x - x'$ and $t = y - y'$. It is direct to check that $\tau(s,t) = (0,0)$. Thus, $(a-1)s = (b-1)t$ and $(c-1)s = (d-1)t$. Setting $z = s - t$. Then we have

$$s = t + z, \quad as = bt + z, \quad \text{and} \quad cs = dt + z,$$

and so

$$|\Phi s \cap (\Phi t + z)| \geq 3.$$

To see that (F,Φ) is not circular, we have to check that $\Phi s \neq \Phi t + z$. If this is not the case, then $z = 0$ by Corollary 19.11, and so $s = t$. But then from $as = bt$, or equivalently, $a^{-1}bt = s$, we infer that $a^{-1}b = 1$, i.e. $a = b$, which is a contradiction. Therefore, $\Phi s \neq \Phi t + z$ and (F,Φ) is not circular.

Conversely, suppose that (F,Φ) is not circular. Then we can find $s,t,z \in F^*$ such that $|(\Phi s) \cap (\Phi t + z)| \geq 3$. Thus, there are $\phi_i, \psi_i \in \Phi$, $i = 1,2,3$, such that $\phi_i \neq \phi_j$ and $\psi_i \neq \psi_j$ if $i \neq j$, and

$$\phi_i s = \psi_i t + z, \quad i = 1,2,3.$$

From these we derive

$$(\phi_2\phi_1^{-1} - 1)\phi_1 s = (\phi_2 - \phi_1)s = (\psi_2 - \psi_1)t = (\psi_2\psi_1^{-1} - 1)\psi_1 t,$$
$$(\phi_3\phi_1^{-1} - 1)\phi_1 s = (\phi_3 - \phi_1)s = (\psi_3 - \psi_1)t = (\psi_3\psi_1^{-1} - 1)\psi_1 t.$$

Putting $a = \phi_2\phi_1^{-1}$, $b = \psi_2\psi_1^{-1}$, $c = \phi_3\phi_1^{-1}$, $d = \psi_3\psi_1^{-1}$, $x = \phi_1 s$ and $y = \psi_1 t$, we see that $(a,b) \neq (c,d)$, $(a,c) \neq (b,d)$, and

$$\tau_{a,b,c,d}(s,t) = ((a-1)x - (b-1)y, (c-1)x - (d-1)y) = (0,0).$$

Since $(x,y) \neq (0,0)$ and certainly $\tau_{a,b,c,d}(0,0) = (0,0)$, we conclude that $\tau_{a,b,c,d}$ is not bijective.

Summarizing, we have

Proposition 19.16. *Let F be a field and Φ a finite subgroup of F^* of order ≥ 4. Then (F,Φ) is circular if and only if $\tau_{a,b,c,d}$ is bijective for all $a,b,c,d \in \Phi\setminus\{1\}$ with $(a,b) \neq (c,d)$ and $(a,c) \neq (b,d)$.*

Rremark 19.17. The mapping τ can be represented as a matrix $M = M_{a,b,c,d} = \left(\begin{smallmatrix} a-1 & b-1 \\ c-1 & d-1 \end{smallmatrix}\right)$, and so τ is bijective if and only if M is invertible, and equivalently,

$$(a-1)(d-1)-(b-1)(c-1) \neq 0.$$

Writing a,b,c,d in terms of powers of ζ, we see that $M = M_{i,j,s,t} = \left(\begin{smallmatrix} \zeta^i-1 & \zeta^j-1 \\ \zeta^s-1 & \zeta^t-1 \end{smallmatrix}\right)$ for some $i,j,s,t \in \{1,\ldots,k-1\}$ with $(i,j) \neq (s,t)$ and $(i,s) \neq (j,t)$, and M is invertible if and only if the element

$$\det M_{i,j,s,t} = (\zeta^i-1)(\zeta^t-1)-(\zeta^j-1)(\zeta^s-1)$$

is nonzero.

Now, for positive integers i, j, s, and t, we define a polynomial $\omega_{i,j,s,t}(x)$ in $\mathbb{Z}[x]$ by

$$\begin{aligned}\omega_{i,j,s,t}(x) &= (x^i-1)(x^t-1)-(x^j-1)(x^s-1)\\ &= x^{i+t}-x^i-x^t-x^{j+s}+x^j+x^s.\end{aligned}$$

Also, for $k \geq 3$, we set

$$\Omega_k = \{\omega_{i,j,s,t}(x) \mid i,j,s,t \in \{1,\ldots,k-1\},\ (i,j) \neq (s,t) \text{ and } (i,s) \neq (j,t)\}.$$

Since the coefficients of the polynomials in Ω_k are ± 1, we can always view Ω_k as a subset of $\mathbb{Z}_p[x]$ for any prime p. We can then use Remark 19.17 to reformulate Proposition 19.16 as

Theorem 19.18. *Let F be a field and $\Phi = \langle \mu \rangle$ a finite multiplicative subgroup of F of order k, $k \geq 2$. Then (F,Φ) is circular if and only if $\omega_{i,j,s,t}(\mu) \neq 0$ for all $\omega_{i,j,s,t} \in \Omega_k$.*

Now we consider in $\mathbb{C}$ and put $z = e^{2\pi \mathrm{i}/k}$. If Φ is the multiplicative subgroup $\{1,z,\ldots,z^{k-1}\}$ of $\mathbb{C}^*$, then certainly $(\mathbb{C},\Phi)$ is circular. Hence,

Corollary 19.19. $\omega_{i,j,s,t}(z) \neq 0$ *for all* $\omega_{i,j,s,t} \in \Omega_k$.

Since $z^k = 1$, $\mathbb{Q}(z) = \mathbb{Q}[z]$, the subset of $\mathbb{C}$ consisting the elements of the form

$$a_0 + a_1 z + \cdots + a_{k-1} z^{k-1}, \quad a_i \in \mathbb{Q}, 0 \leq i \leq k-1.$$

Also, with $z^k = 1$, we see that if $P(x) \in \mathbb{Q}[x]$, a polynomial with coefficients from $\mathbb{Q}$, then $P(z) \in \mathbb{Q}[z]$. In $\mathbb{Q}[z]$, there is a ring $\mathbb{Z}[z] = \{P(z) \mid P(x) \in \mathbb{Z}[x]\}$. For each element $a \in \mathbb{Q}[z]$, we can define a linear map $\tilde{a}$ on $\mathbb{Q}[z]$ by left multiplication, i.e. $\tilde{a}(y) = ay$ for all $y \in \mathbb{Q}[z]$. As $\{1, z, \ldots, z^{k-1}\}$ as a basis of $\mathbb{Q}[z]$, for $1 \leq i \leq k$, there are $a_{1i}, \ldots, a_{ki} \in \mathbb{Q}$ such that $\tilde{a}(z^{i-1}) = \sum_{j=1}^{k} a_{ji} z^{j-1}$. That is $\tilde{a}$ can be represented as a $k \times k$ matrix

$$T_a = \begin{pmatrix} a_{11} & a_{12} & \cdots & a_{1k} \\ a_{21} & a_{22} & \cdots & a_{2k} \\ \vdots & \vdots & \ddots & \vdots \\ a_{k1} & a_{k2} & \cdots & a_{kk} \end{pmatrix}.$$

The determinate of T_a is called the regular norm of a in $\mathbb{Q}[z]$ over $\mathbb{Q}$, and is denoted by $N(a)$. It is clear that $N(a)$ is a rational number and $N(a) \neq 0$ implies that $a \neq 0$. Moreover, if $a \neq 0$, then $\tilde{a}^{-1} = \widetilde{a^{-1}}$, and so $N(a) \neq 0$. Next, if $P(x) = a_0 + a_1 x + \cdots + a_{m-1} x^{m-1} + x^m \in \mathbb{Q}[x]$ with $m = kq + r$, $0 \leq q$ and $0 \leq r < k$, then $P(z) = Q(z)$, where $Q(x) = b_0 + b_1 x + \cdots + b_{k-1} x^{k-1} \in \mathbb{Q}[x]$ with b_i's given by the rule (1) if $q = 0$, then $b_i = a_i$ for $i \leq r$ and $b_i = 0$ for $r < i < k$, and (2) if $q > 0$, then $b_i = a_{0k+i} + \cdots + a_{kq+i}$ for $i \leq r$ and $b_i = a_{0k+i} + \cdots + a_{k(q-1)+i}$ for $r < i < k$. Hence we have

$$T_a = \begin{pmatrix} b_0 & b_{k-1} & \cdots & b_1 \\ b_1 & b_0 & \cdots & b_2 \\ \vdots & \vdots & \ddots & \vdots \\ b_{k-1} & b_{k-2} & \cdots & b_0 \end{pmatrix},$$

and so $N(a)$ is a sum of products of the b_i's, and consequently, a sum of products of the a_i's. In particular, if $P(x)$ has integer coefficients, then $N(a)$

is an integer. Therefore, for i, j, s, t as in Corollary 19.19, the norms $N_{i,j,s,t} = N(\omega_{i,j,s,t}(z))$ are nonzero integers.

Now, we turn to $F = F_{p^s}$, a finite filed of order p^s where p is a prime not dividing k, and $s \geq 0$ an integer such that $k \mid (p^s - 1)$. We denote by ζ the primitive k^{th} root of unity in F and let Φ the multiplicative subgroup of F^* generated by ζ. By Proposition 19.12, (F, Φ) is not circular if and only if $\omega_{i,j,s,t}(\zeta) = 0$ for some $\omega_{i,j,s,t}(x) \in \Omega_k$.

Inside F, there is a subfield $K = \mathbb{Z}_p(\zeta)$ generated by $\mathbb{Z}_p$ and ζ. Just like $\mathbb{Q}(\zeta) = \mathbb{Q}[\zeta]$, we have $\mathbb{Z}_p(\zeta) = \mathbb{Z}_p[\zeta]$, i.e. each element of $\mathbb{Z}_p(\zeta)$ is of the form

$$a_0 + a_1\zeta + \cdots + a_{k-1}\zeta^{k-1}$$

with $a_i \in \mathbb{Z}_p$, $0 \leq i < k$.

For each element $a \in \mathbb{Z}_p(\zeta)$, the linear map on $\mathbb{Z}_p(\zeta)$ induced by the left multiplication by a also has a matrix representation L_a with respective to the basis $\{1, \zeta, \zeta^2, \ldots, \zeta^{k-1}\}$. The regular norm of a in $\mathbb{Z}_p(\zeta)$ over $\mathbb{Z}_p$, denoted as $N_p(a)$, is just the determinate of L_a. Just like the case in $\mathbb{Q}(\zeta)$, we have $N_p(a) \in \mathbb{Z}_p$ and that $N_p(a) = 0$ if and only if $a = 0$.

Take a polynomial $\omega_{i,j,s,t}(x) \in \Omega_k$. We notice that the (i,j)-entry l_{ij} of the matrix $L_{\omega_{i,j,s,t}(\zeta)}$ is congruent to the (i,j)-entry t_{ij} of the matrix $T_{\omega_{i,j,s,t}(e^{2\pi\mathbf{i}/k})}$ modulo p. Consequently,

$$N_p(\omega_{i,j,s,t}(\zeta)) \equiv N(\omega_{i,j,s,t}(e^{2\pi\mathbf{i}/k})) \pmod{p},$$

and so, (F, Φ) is not circular if and only if p is a divisor of $N(\omega_{i,j,s,t}(e^{2\pi\mathbf{i}/k}))$ for some $\omega_{i,j,s,t}(x) \in \Omega_k$.

Summering, we have shown that

Theorem 19.20. *Let $k \geq 3$. Then*

$$\mathscr{P}_k = \{p \mid p \text{ is a prime divisor of } N(\omega_{i,j,s,t}(e^{2\pi\mathbf{i}/k})) \text{ for some } \omega_{i,j,s,t}(x) \in \Omega_k\}.$$

Since there are only finitely many polynomials in Ω_k, this proves Theorem 19.15, namely, $\mathscr{P}_k$ is a finite set. Moreover, it gives an explicit description of $\mathscr{P}_k$ so that one can construct the members of $\mathscr{P}_k$.

20 Double planar nearrings

As §6 ended, the reader's intellectual curiosity was stimulated and immediately abandoned as he was informed that $(\mathbb{C}, +, *, \circ)$ is the prototype of a double planar nearring. Actually, the study of double planar nearrings has really just begun, but early investigations promise a rich harvest of interesting and curious geometric relationships. Anyone with an appreciation of classical Euclidean geometry should also appreciate how some of the classical geometric ideas can be transferred to any field, finite or infinite. Let's begin with the definition, and then learn how to construct infinitely many double planar nearrings.

Definition 20.1. A *double nearring* $(N, +, *, \circ)$ is an ordered quadruple where each of the ordered triples $(N, +, *)$ and $(N, +, \cdot)$ is a nearring, and where $*$ and $\circ$ are each left distributive over the other. That is,

$$a * (b \circ c) = (a * b) \circ (a * c) \tag{20:3}$$

and

$$a \circ (b * c) = (a \circ b) * (a \circ c) \tag{20:4}$$

for all $a, b, c \in N$. If each of the nearrings $(N, +, *)$ and $(N, +, \circ)$ is planar, then $(N, +, *, \circ)$ is a *double planar nearring*.

That $(\mathbb{C}, +, *, \circ)$ is an example of a double planar nearring follows from the following theorem.

Theorem 20.2. *Let $(F,+,\cdot)$ be a division ring where the multiplicative group $(F^*,\cdot)$, $F^* = F\setminus\{0\}$, is a direct product of subgroups Φ and Γ. If $a = a_\Phi a_\Gamma$, $a_\Phi \in \Phi$, $a_\Gamma \in \Gamma$, is the unique factorization of a defined by the direct product $F^* = \Phi\Gamma$, and if $*$ and $\circ$ are defined by*

$$a * b = \begin{cases} 0, & \text{if } a = 0; \\ a_\Phi \cdot b, & \text{if } a \neq 0, \end{cases}$$

and

$$a \circ b = \begin{cases} 0, & \text{if } a = 0; \\ a_\Gamma \cdot b, & \text{if } a \neq 0, \end{cases}$$

then $(F,+,*,\circ)$ is a double nearring. If Φ and Γ are each nontrivial, then $(F,+,*,\circ)$ is a double planar nearring.

Proof. First, we shall see that $*$ is associative. Of course, this will show that $\circ$ is also associative. Let $a,b,c \in F$. If $a = 0$, then

$$a*(b*c) = 0, \text{ and } (a*b)*c = 0*c = 0.$$

If $b = 0$, then

$$a*(b*c) = a*0, \text{ and } (a*b)*c = 0*c = 0.$$

When $a,b \in F^*$, they have unique factorizations $a = a_\Phi a_\Gamma$ and $b = b_\Phi b_\Gamma$. Then

$$a*(b*c) = a_\Phi\cdot(b_\Phi\cdot c) = a_\Phi\cdot b_\Phi\cdot c$$

and

$$(a*b)*c = (a_\Phi\cdot b)*c = (a_\Phi\cdot b_\Phi\cdot b_\Gamma)*c = (a_\Phi\cdot b_\Phi)\cdot c = a_\Phi\cdot b_\Phi\cdot c.$$

Hence, $*$ and $\circ$ are associative.

Next, we shall see that $*$ is left distributive over $+$. For $a,b,c \in F$, we first consider the case where $a = 0$. Then

$$a*(b+c) = 0 = 0+0 = (a*b)+(a*c).$$

When $a = a_\Phi a_\Gamma \neq 0$, then

$$a * (b+c) = a_\Phi \cdot (b+c) = (a_\Phi \cdot b) + (a_\Phi \cdot c) = (a*b) + (a*c).$$

So, $*$ and $\circ$ are each left distributive over $+$.

As we show that $*$ is left distributive over $\circ$, we will also, in fact, be showing that $\circ$ is left distributive over $*$. Having done this, we'll be able to conclude that $(F,+,*,\circ)$ is a double planar nearring. So, for $a,b,c \in F$, we first let $a = 0$. Then

$$a * (b \circ c) = 0 = 0 \circ 0 = (a*b) \circ (a*c).$$

With $b = 0$, we have

$$a * (b \circ c) = a * 0 = 0 = 0 \circ (a*c) = (a*b) \circ (a*c).$$

So, with $a = a_\Phi a_\Gamma$, $b = b_\Phi b_\Gamma \in F^*$, we have

$$a * (b \circ c) = a_\Phi \cdot (b_\Gamma \cdot c)$$

and

$$(a*b) \circ (a*c) = (a_\Phi \cdot b_\Phi \cdot b_\Gamma) \circ (a_\Phi \cdot c) = b_\Gamma \cdot (a_\Phi \cdot c) = a_\Phi \cdot (b_\Gamma \cdot c).$$

(Here we are using Theorem A4.45, which assures us that the elements of Φ commute with the elements of Γ.) So, $(F,+,*,\circ)$ is indeed a double nearring.

We are now free to focus on planarity. Again, what we prove about Φ will also be true about Γ. Since Φ is nontrivial, we know that $=_m$ for $*$ has $|F/{=_m}| \geq 3$, and for $a,b \in F^*$, $a =_m b$ if and only if $a_\Phi = b_\Phi$. If $a \in F^*$, then $a \neq_m 0$. Consider an equation

$$a * x = b * x + c$$

with $a \neq_m b$. This equation becomes $a_\Phi \cdot x = b_\Phi \cdot x + c$, if $0 \notin \{a,b\}$, which has the unique solution $(-b_\Phi + a_\Phi)^{-1} \cdot c$, since $a_\Phi \neq b_\Phi$. If $0 \in \{a,b\}$, then one easily sees the equation also has a unique solution. So, $(F,+,*)$, and consequently, $(F,+,\circ)$, are each planar, and this completes the proof of the theorem. □

We studied the division ring of quaternions $(Q,+,\cdot)$ in §4. The quaternions give us an interesting place to look at a particular double planar nearring $(Q,+,*,\cdot)$. Let

$$\Phi = \{r \in Q \mid r > 0\}$$

and

$$\Gamma = \{a+bi+cj+dk \mid a^2+b^2+c^2+d^2 = 1\}.$$

Then certainly $Q^* = \Phi\Gamma$ and $\Phi \cap \Gamma = \{1\}$. Easily, Φ is a normal subgroup of Q^*, and from Theorem 4.8, one quickly gets that Γ is also a normal subgroup of Q^*. So, Q^* is the direct product of Φ and Γ, and Theorem 20.2 applies.

This is a generalization of the $(\mathbb{C},+,*,\cdot)$ at the end of §6. Here,

$$\Phi = \{r \in \mathbb{C} \mid r > 0\}$$

and

$$\Gamma = \{a+bi \mid a^2+b^2 = 1\}.$$

Then $\mathbb{C}^* = \Phi\Gamma$ and $\Phi \cap \Gamma = \{1\}$. Since $(\mathbb{C}^*,\cdot)$ is abelian, Φ and Γ are each normal. So, $\mathbb{C}^*$ is the direct product of Φ and Γ.

Let $(F,+,\cdot)$ be any finite field. Then $|F| = q$ for some positive integer q. It can be shown that $q = p^n$ for some prime p and some positive integer n, but we will not need this fact here. We do know that each $(\mathbb{Z}_p,+,\cdot)$, p a prime, is a field, and for any finite field F with $|F| = q$, there is a quadratic extension K of F and $|K| = q^2$. This is sufficient for our needs.

Any finite field $(F,+,\cdot)$ has for its multiplicative group $(F^*,\cdot)$, a cyclic group of order $q-1$. If s divides $q-1$, then F^* has a cyclic subgroup Φ of order s. Suppose $q-1 = st$, where s and t are relative prime factors of $q-1$. Let Φ and Γ be the subgroups of F^* of orders s and t, respectively. Certainly Φ and Γ are normal since F^* is abelian. Also, since $|\Phi \cap \Gamma|$ divides each of

$|\Phi|$ and $|\Gamma|$, we have that $\Phi \cap \Gamma = \{1\}$. If $\Phi\Gamma = \{\phi\gamma \mid \phi \in \Phi,\ \gamma \in \Gamma\}$, then $\Phi\Gamma$ is a subgroup of F^*. Since $|\Phi\Gamma| = |\Phi| \cdot |\Gamma| = st = q-1$, we have $F^* = \Phi\Gamma$. Hence, F^* is the direct product of Φ and Γ. The reader is now invited to experiment by applying Theorem 20.2 to various finite fields $(F,+,\cdot)$, where $|F^*| = q-1 = st$, with s and t relatively prime, and to actually find subgroups Φ and Γ of orders s and t, respectively. It may be useful to keep a notebook of your observations.

21 Rays, angles, segments, triangles, spheres, and circles

Using the $(\mathbb{C},+,*,\circ)$ and $(Q,+,*,\circ)$ of §20 as models, we can usually get nice geometric relationships from a double planar nearring, hence from most fields, finite and infinite. For $(\mathbb{C},+,*,\circ)$, we saw early in §6 that

$$\mathscr{B} = \{\mathbb{C} * a + b \mid a,b \in \mathbb{C},\ a \neq 0\}$$

is the family of all rays in the classical Euclidean plane $\mathbb{C}$, and that

$$\mathscr{B}^* = \{\mathbb{C}^* \circ a + b \mid a,b \in \mathbb{C},\ a \neq 0\}$$

is the family of all circles in the classical Euclidean plane $\mathbb{C}$. Similarly, if we consider

$$\mathscr{B} = \{Q * a + b \mid a,b \in Q,\ a \neq 0\},$$

we have the family of all rays in classical four-dimensional Euclidean space, and if we consider

$$\mathscr{B}^* = \{Q^* \circ a + b \mid a,b \in Q,\ a \neq 0\},$$

we have the family of all spheres in classical four-dimensional Euclidean space. Note that $\mathbb{C}^* \circ a + b$ is the circle in $\mathbb{C}$ with center b and radius $|a|$, and that $Q^* \circ a + b$ is the sphere in Q with center b and radius $|a|$. In addition, $\mathbb{C} * a + b$ and $Q * a + b$ would each be the ray with initial point b and passing through the point $a + b$. How do we carry these ideas over to other double planar nearrings?

For a double planar nearring $(F, +, *, \circ)$, we also have that $(F, +, \circ, *)$ is a double planar nearring. If we want $F * a$ to be the ray from 0 through a, then we would want $F * (-a)$ to be the ray from 0 through $-a$. Usually, we would want $F * a \neq F * (-1)$ for all $a \in F^*$. In fact, we would want $F * a \cap F * (-1) = \{0\}$ for each $a \in F^*$. It could be, as in our examples $(\mathbb{C}, +, *, \circ)$ and $(Q, +, *, \circ)$, that $F * a \neq F * (-a)$ and $F * a \cap F * (-a) = \{0\}$ for each $a \in F^*$, and $F^* \circ a = F^* \circ (-a)$ for each $a \in F^*$. (Or, it could be that $F \circ a \neq F \circ (-a)$ and $F \circ a \cap F \circ (-a) = \{0\}$ for each $a \in F^*$, and $F^* * a = F^* * (-a)$ for each $a \in F^*$. In this case, replace $*$ with $\cdot$ and $\cdot$ with $*$.)

Definition 21.1. Let $(F, +, \cdot, \circ)$ be a double planar nearring. Suppose that for each $a \in F^*$, $F * a \neq F * (-a)$ and $F * a \cap F * (-a) = \{0\}$. Then for $a \in F^*$ and $b \in F$, we call $F * a + b$ *the ray from b through $a + b$*. In addition, suppose $F^* \circ a = F^* \circ (-a)$ for all $a \in F^*$. Then, for $a \in F^*$ and $b \in F$, we call $F^* \circ a + b$ *the circle with center b and radius a.*

Observations and definitions 21.2. Let $(F, +, \cdot)$ be a division ring having a factorization of its multiplicative group F^* into a direct product of nontrivial subgroups Φ and Γ. Suppose $-1 \notin \Phi$ but yet $-1 \in \Gamma$. Then the corresponding double planar nearring $(F, +, *, \circ)$ has the property that $F * a \neq F * (-a)$ and $F * a \cap F * (-a) = \{0\}$ for every $a \in F^*$. In this case, we will say that the double planar nearring $(F, +, *, \circ)$ *has rays*. (This phrase, then, implies that $(F, +, *, \circ)$ is constructed from a division ring $(F, +, \cdot)$ as described here and in §20.) The rays are the elements of the family $\mathscr{R} = \{F * a + b \mid a, b \in F,\ a \neq 0\}$. The *ray from b through $a + b$* is $F * a + b$, which we shall also denote by $\overrightarrow{b, a + b}$.

If $a \in F^*$ has factorization $a = a_\Phi a_\Gamma$, we introduce the notation $|a| = a_\Phi$

and $\theta(a) = a_\Gamma$. So $a = |a|\theta(a)$, and we say that $|a|$ is the *length* or *absolute value* of a, and that $\theta(a)$ is the *angle* of $\overrightarrow{0,a} = F * a$.

The following theorems shows that our angles and lengths have some familiar properties.

Theorem 21.3. *Suppose a double planar nearring $(F,+,*,\circ)$ has rays. Then for $a,b \in F^*$, we have:*

(a) $|ab| = |a||b|$;

(b) $|-a| = |a|$;

(c) $\theta(ab) = \theta(a)\theta(b)$;

(d) $\theta(-a) = -\theta(a)$;

(e) $||a|| = |a|$ *and* $\theta(|a|) = 1$;

(f) $\theta(\theta(a)) = \theta(a)$ *and* $|\theta(a)| = 1$.

Proof. For $a = a_\Phi a_\Gamma = |a|\theta(a)$ and $b = b_\Phi b_\Gamma = |b|\theta(b)$, we have

$$ab = a_\Phi a_\Gamma b_\Phi b_\Gamma = a_\Phi b_\Phi a_\Gamma b_\Gamma = |a||b|\theta(a)\theta(b),$$

so $|ab| = |a||b|$ and $\theta(ab) = \theta(a)\theta(b)$. Hence, (a) and (c) follow from the uniqueness of factorization of ab relative to the direct product $F^* = \Phi\Gamma$. Also, since

$$-a = (-1)a_\Phi a_\Gamma = a_\Phi((-1)a_\Gamma) = a_\Phi(-a_\Gamma) = |a|(-\theta(a)),$$

we have $\theta(-a) = -\theta(a)$ and $|-a| = |a|$, and so (b) and (d) also are consequences of uniqueness.

Since $|a| = |a| \cdot 1$ and $|a| = ||a||\theta(|a|)$, we get $||a|| = |a|$ and $\theta(|a|) = 1$. Similarly, $\theta(a) = 1 \cdot \theta(a)$ and $\theta(a) = |\theta(a)|\theta(\theta(a))$, and so $|\theta(a)| = 1$ and $\theta(\theta(a)) = \theta(a)$. □

Theorem 21.4. *Suppose a double planar nearring $(F,+,*,\circ)$ has rays. Then $F * a = F * b$ if and only if $\theta(a) = \theta(b)$. Similarly, $F^* \circ a = F^* \circ b$ if and only if $|a| = |b|$. In fact, $F * a = F \cdot \theta(a)$ and $F^* \circ a = F^* \cdot |a|$.*

Proof. Since

$$\begin{aligned} F*a &= \{f*a \mid f\in F\} = \{|f|\theta(f)|a|\theta(a)\mid f\in F\} \\ &= \{(|fa|\theta(f))\theta(a)\mid f\in F\} = F\cdot\theta(a), \end{aligned}$$

and

$$\begin{aligned} F^*\circ a &= \{f\circ a\mid f\in F^*\} = \{|f|\theta(f)|a|\theta(a)\mid f\in F^*\} \\ &= \{(|f|\theta(fa))|a|\mid f\in F^*\} = F^*\circ|a|, \end{aligned}$$

the proof becomes very easy. □

Withe Theorem 21.4, we are prepared to define the angle of a ray $\overrightarrow{0,a}$ and the angle from ray $\overrightarrow{c,c+a}$ to the ray $\overrightarrow{c,c+b}$.

Definitions 21.5. Suppose a double planar nearring $(F,+,*,\circ)$ has rays. For $a\in F^*$, the *angle of the ray* $\overrightarrow{0,a}=F*a$ is $\theta(a)$. If $b\in F^*$, the *angle from the ray* $\overrightarrow{0,a}$ *to the ray* $\overrightarrow{0,b}$ is $\theta(b)\theta(a)^{-1}$. Let $c\in F$. The *angle from the ray* $\overrightarrow{c,c+a}$ *to the ray* $\overrightarrow{c,c+b}$ is also $\theta(b)\theta(a)^{-1}$.

In the classical way of doing things, any nonzero complex number a can be written $a=re^{i\mu}$ for $r=|a|$ and a unique μ, $0\le\mu<2\pi$. Here, $r\in\Phi$ and $e^{i\mu}\in\Gamma$, as described in §20. Traditionally, we would say that μ is the argument of a, or the angle of a, but here we are saying that $\theta(a)=e^{i\mu}$ is the angle of a. Traditionally, we may say that the angle from the ray $\overrightarrow{c,a+c}$ to the ray $\overrightarrow{c,b+c}$ is μ, but here, we are saying that the angle from the ray $\overrightarrow{c,a+c}$ to the ray $\overrightarrow{c,b+c}$ is $e^{i\mu}$. The elements of Γ measure the size of angles in this work.

Definition 21.6. Suppose $(F,+,*,\circ)$ has rays. If $a,b,c\in F$ are three distinct points, then the *angle* $\angle abc$ is defined to be $\theta(a-b)\theta(c-b)^{-1}$.

Note: In Definition 21.6, we did not tell the reader that $(F,+,*,\circ)$ is a double planar nearring. From here on, we'll assume the reader will infer this from the notation and the phrase "has rays."

Angles in our sense have many familiar properties.

Theorem 21.7. *Suppose $(F,+,*,\circ)$ has rays. Let $a,b,c \in F$ be any three distinct points. Then*

(a) *for any $x \in F$, $\angle(a+x)(b+x)(c+x) = \angle abc$;*
(b) *$\angle(au)(bu)(cu) = \angle abc$ for any $u \in F^*$.*

Proof. First,

$$\begin{aligned}\angle(a+x)(b+x)(c+x) &= \theta[(a+x)-(b+x)]\theta[(c+x)-(b+x)]^{-1}\\ &= \theta(a-b)\theta(c-b)^{-1}\\ &= \angle abc.\end{aligned}$$

Second,

$$\begin{aligned}\angle(au)(bu)(cu) &= \theta[au-bu]\theta[cu-bu]^{-1} = \theta[(a-b)u]\theta[(c-b)u]^{-1}\\ &= \theta(a-b)\theta(u)[\theta(c-b)\theta(u)]^{-1}\\ &= \theta(a-b)\theta(u)\theta(u)^{-1}\theta(c-b)^{-1}\\ &= \theta(a-b)\theta(c-b)^{-1}\\ &= \angle abc.\end{aligned}$$

□

Definitions 21.8. Suppose $(F,+,*,\circ)$ has rays. For any $x \in F$, define $T_x : F \to F$ by $T_x(a) = a+x$. Then T_x is a *translation*, and we say that T_x is the *translation by x*. For $u \in F^*$, define $M_u : F \to F$ by $M_u(a) = au$. If $u \in \Phi$, then M_u is a *dilatation* and M_u is the *dilatation by u*. (We think of a dilatation as a stretching or shrinking while preserving both directions and angles.) If $u \in \Gamma$, then M_u is a *rotation* and M_u is the *rotation by u*. (We would want a rotation to preserve the length of a point a and to preserve the angle $\angle abc$ defined by three distinct points $a,b,c \in F$.)

Definitions 21.9. Suppose $(F,+,*,\circ)$ has rays. For $a,b \in F$, $a \neq b$, we define the *segment $\overline{a,b}$ with endpoints a and b*, or simply the *segment $\overline{a,b}$*, for short, as

$$\overline{a,b} = [F*(a-b)+b] \cap [F*(b-a)+a].$$

So, $\overline{b,a} = [F*(a-b)+b] \cap [F*(b-a)+a] = \overline{a,b}$. The *length of* $\overline{a,b}$ is defined as $|b-a|$. So, the length of $\overline{b,a}$ is $|a-b| = |-(b-a)| = |b-a|$, which is the length of $\overline{a,b}$. With this reassurance, we also write $|\overline{a,b}| = |b-a|$ for the length of the segment $\overline{a,b}$.

Theorem 21.10. *Suppose $(F,+,*,\circ)$ has rays. For $a,b,x \in F$, $a \neq b$, and $u \in \Gamma$, we have*

(a) $|\overline{a+x,b+x}| = |\overline{a,b}|$;

(b) $|\overline{ax,bx}| = |x||\overline{a,b}|$, *if* $x \neq 0$;

(c) $|\overline{au,bu}| = |\overline{a,b}|$.

That is, the length of a segment is preserved by translations and rotations. The map M_x maps the segment $\overline{a,b}$ of length $|\overline{a,b}|$ onto a segment $\overline{ax,bx}$ of length $|x|$ times $|\overline{a,b}|$, the length of $\overline{a,b}$.

Proof. We have $|\overline{a+x,b+x}| = |(b+x)-(a+x)| = |b-a| = |\overline{a,b}|$, and if $x \neq 0$, then $ax \neq bx$, and so $|\overline{ax,bx}| = |bx-ax| = |(b-a)x| = |b-a||x| = |x||\overline{a,b}|$. If $u \in \Gamma$, then $|u| = 1$, and so $|\overline{au,bu}| = |\overline{a,b}|$.

Again, with $x \neq 0$, consider that

$$\begin{aligned}\overline{ax,bx} &= [F*(ax-bx)+bx] \cap [F*(bx-ax)+ax] \\ &= [F*(a-b)x+bx] \cap [F*(b-a)x+ax] \\ &= [(F*(a-b)+b)x] \cap [(F*(b-a)+a)x] = \overline{a,b}\cdot x.\end{aligned}$$

This completes the proof of Theorem 21.10. □

With the theory that we have developed, we are ready to introduce the idea of a triangle and its measurements.

Definitions 21.11. Suppose $(F,+,*,\circ)$ has rays, and consider three distinct points $a,b,c \in F$. Suppose $\{-1,1\} \cap \{\angle abc, \angle cab, \angle bca\} = \varnothing$. Then we say that the three point set $\{a,b,c\}$ is *triangular*. Suppose $\{a,b,c\}$ is triangular. We define the *triangle* $\triangle abc$ as an ordered pair (P,O), where

$$P = \overline{a,b} \cup \overline{b,c} \cup \overline{c,a},$$

a set, and is called the *points* of $\triangle abc$, and

$$O = (a,b,c),$$

a permutation, and is called the *orientation* of $\triangle abc$. (Here, (a,b,c) denotes the map $\pi : \{a,b,c\} \to \{a,b,c\}$ where $\pi(a) = b$, $\pi(b) = c$, and $\pi(c) = a$.)

Proposition 21.12. *Suppose that $(F,+,*,\circ)$ has rays and that $\{a,b,c\}$ is triangular. Then $\triangle abc = \triangle cab = \triangle bca$.*

Proof. Easily, $\overline{a,b} \cup \overline{b,c} \cup \overline{c,a} = \overline{c,a} \cup \overline{a,b} \cup \overline{b,c} = \overline{b,c} \cup \overline{c,a} \cup \overline{a,b}$, and, as permutations, $(a,b,c) = (c,a,b) = (b,c,a)$. □

Proposition 21.13. *Suppose $(F,+,*,\circ)$ has rays and that $\{a,b,c\}$ is triangular. Then $\triangle abc$ and $\triangle acb$ have the same set of points, but their orientations are inverses of each other.*

Proof. Certainly,

$$\overline{a,b} \cup \overline{b,c} \cup \overline{c,a} = \overline{a,c} \cup \overline{c,b} \cup \overline{b,a}.$$

Also,

$$(a,b,c) \circ (a,c,b) = (a,c,b) \circ (a,b,c)$$

is the identity map on $\{a,b,c\}$. □

Definitions 21.14. For a triangle $\triangle abc$, we define its *measurements* as the set of ordered pairs

$$M(\triangle abc) = \{(\angle abc, |\overline{a,c}|), (\angle cab, |\overline{c,b}|), (\angle bca, |\overline{b,a}|)\}.$$

Of course, $\angle abc$ is an *angular measurement* , and $|\overline{a,c}|$ is a *lateral measurement* of the opposite side $\overline{a,c}$, so

$$AM(\triangle abc) = \{\angle abc, \angle cab, \angle bca\}$$

is the set of *angular measurements* of $\triangle abc$, and

$$LM(\triangle abc) = \{|\overline{a,c}|, |\overline{c,b}|, |\overline{b,a}|\}$$

is the set of *lateral measurements* of $\triangle abc$. The *alternate measurements* of $\triangle abc$ is the set

$$M^a(\triangle abc) = \{((\angle abc)^{-1}, |\overline{a,c}|), ((\angle cab)^{-1}, |\overline{c,b}|), ((\angle bca)^{-1}, |\overline{b,a}|)\}.$$

We say that the triangle $\triangle abc$ is *equilateral* if $|LM(\triangle abc)| = 1$, and is *equiangular* if $|AM(\triangle abc)| = 1$. We say that the $\triangle abc$ is *isosceles* if $|LM(\triangle abc)| \leq 2$, and is a *right triangle* if $AM(\triangle abc)$ has an element of order 4. In fact, $\angle bca$ is a *right angle* if and only if $(\angle bca)^2 = -1$.

Proposition 21.15. *The measurements of $\triangle acb$ equals the alternate measurements of $\triangle abc$, i.e., $M(\triangle acb) = M^a(\triangle abc)$.*

Proof. We have

$$M(\triangle acb) = \{(\angle acb, |\overline{a,b}|), (\angle bac, |\overline{b,c}|), (\angle cba, |\overline{c,a}|)\}$$

and

$$M^a(\triangle abc) = \{((\angle abc)^{-1}, |\overline{a,c}|), ((\angle cab)^{-1}, |\overline{c,b}|), ((\angle bca)^{-1}, |\overline{b,a}|)\}.$$

Certainly, $|\overline{a,b}| = |\overline{b,a}|$, $|\overline{b,c}| = |\overline{c,b}|$ and $|\overline{c,a}| = |\overline{a,c}|$. Also,

$$(\angle xyz)^{-1} = [\theta(x-y)\theta(z-y)^{-1}]^{-1} = \theta(z-y)\theta(x-y)^{-1} = \angle zyx.$$

Hence, $(\angle abc)^{-1} = \angle cba$, and so $(\angle cba, |\overline{c,a}|) = ((\angle abc)^{-1}, |\overline{a,c}|)$, etc. □

So, $\triangle abc$ and $\triangle acb$ are closely related, differing only in the way their angles are measured, and this is a consequence of their orientations being inverses of one another.

Definition 21.16. Two triangles $\triangle abc$ and $\triangle xyz$ are *congruent*, and we write $\triangle abc \cong \triangle xyz$, if $M(\triangle abc) = M(\triangle xyz)$ or $M(\triangle abc) = M^a(\triangle xyz)$. (If $M(\triangle abc) = M^a(\triangle xyz)$, then $M(\triangle abc) = M(\triangle xzy)$.)

Theorem 21.17. *Given $\triangle abc$. Let $x = a + b - c$. Then $\triangle abc \cong \triangle axb$.*

Proof. We have

$$M(\triangle abc) = \{(\angle abc, |\overline{a,c}|), (\angle cab, |\overline{c,b}|), (\angle bca, |\overline{b,a}|)\}$$

and

$$M(\triangle axb) = \{(\angle axb, |\overline{a,b}|), (\angle bax, |\overline{b,x}|), (\angle xba, |\overline{x,a}|)\}.$$

Also,

$$\begin{aligned}
\angle axb &= \theta(a-x)\theta(b-x)^{-1} \\
&= \theta(a-(a+b-c))\theta(b-(a+b-c))^{-1} \\
&= \theta(c-b)\theta(c-a)^{-1} \\
&= \theta[(b-c)(-1)]\theta[(a-c)(-1)]^{-1} \\
&= \theta(a-c)\theta(-1)[\theta(a-c)\theta(-1)]^{-1} \\
&= \theta(b-c)\theta(a-c)^{-1} \\
&= \angle bac,
\end{aligned}$$

and $|\overline{a,b}| = |\overline{b,a}|$. Hence,

$$(\angle axb, |\overline{a,b}|) = (\angle bca, |\overline{b,a}|).$$

Similarly,

$$(\angle bax, |\overline{b,x}|) = (\angle abc, |\overline{a,c}|) \quad \text{and} \quad (\angle xba, |\overline{x,a}|) = (\angle cab, |\overline{c,b}|).$$

Hence, $M(\triangle abc) = M(\triangle axb)$ and so $\triangle abc \cong \triangle axb$. □

When we discuss the angles of a $\triangle abc$ in the classical Euclidean plane, i.e., the plane $\mathbb{C}$ of complex numbers, we usually think of the angular measurements as measuring how far two rays from a common point are separated. For example, in a $30^\circ - 60^\circ - 90^\circ$ right triangle, we may have $\angle abc = 30^\circ$ or $\pi/6$ radians, in the traditional sense. We don't really ever say that $\angle abc$ equals -30°, or -330°.

In school, we learned that the sum of the angles of a triangle $\triangle abc$ is 180°, or π radians. Consider any triangle $\triangle abc$ in the complex plane. Let $\angle abc = \beta$, $\angle bca = \gamma$, and $\angle cab = \alpha$, in the traditional sense. Then $\alpha + \beta + \gamma = \pi$, if we use radians. Now, let's translate that to our way of measuring angles. We have $\angle abc = e^{i\beta}$, $\angle bca = e^{i\gamma}$, and $\angle cab = e^{i\alpha}$. Also, $\angle abc \cdot \angle bca \cdot \angle cab = e^{i\beta} \cdot e^{i\gamma} \cdot e^{i\alpha} = e^{i(\alpha+\beta+\gamma)} = e^{i\pi} = -1$. So, the statement that "the sum of the angles of a triangle is 180° or π radians" gets translated to "the product of the angles of a triangle is -1." It is nice to know that this result holds universally for any triangle in F whenever $(F, +, *, \circ)$ has rays.

Theorem 21.18. *Suppose $(F, +, *, \circ)$ has rays. Let $\triangle abc$ be any triangle in F. Then the product of its angles is -1. That is*

$$\angle abc \cdot \angle bca \cdot \angle cab = -1.$$

Proof. We have done most of the work and have chosen the notation so that the proof is very easy. To be sure,

$$\begin{aligned}
&\angle abc \cdot \angle bca \cdot \angle cab \\
&= \theta(a-b)\theta(c-b)^{-1} \cdot \theta(b-c)\theta(a-c)^{-1} \cdot \theta(c-a)\theta(b-a)^{-1} \\
&= \theta(a-b)\theta(c-b)^{-1}\theta((c-b)(-1))\theta(a-c)^{-1}\theta((a-c)(-1))\theta((-1)(a-b))^{-1} \\
&= \theta(a-b)\theta(c-b)^{-1}\theta(c-b)\theta(-1)\theta(a-c)^{-1}\theta(a-c)\theta(-1)[\theta(-1)\theta(a-b)]^{-1} \\
&= \theta(a-b)\theta(-1)\theta(-1)\theta(a-b)^{-1}\theta(-1) \\
&= -1,
\end{aligned}$$

since $-1 \in \Gamma$ and so $\theta(-1) = -1$. □

In $\mathbb{C}$, there is the following classical geometric property. If $\overline{a,b}$ is a diameter of a circle S, and if $x \in S \setminus \{a,b\}$, then $\angle axb$ is a right angle. This result carries over to the quaternions Q as follows. If $\overline{a,b}$ is a diameter of any sphere S in Q, and if $x \in S \setminus \{a,b\}$, then $\angle axb$ is a right angle. We will see that angles in Q also have direction.

Of course, for the division ring $(Q,+,\cdot)$, we have $Q^* = \Phi\Gamma$ as described in §20. With Γ as our angles, we shall need to know what the right angles are. In Definitions 21.14, we noted that $\gamma \in \Gamma$ is a right angle if and only if $\gamma^2 = -1$. Let $\gamma = a + bi + cj + dk \in \Gamma$ as described in §4. From

$$-1 = \gamma^2 = (a^2 - b^2 - c^2 - d^2) + 2abi + 2acj + 2adk,$$

we learn that $ab = ac = ad = 0$. Hence, $a = 0$ or $b = c = d = 0$. If $b = c = d = 0$, then $-1 = a^2 - b^2 - c^2 - d^2 = a^2$, which cannot be, since $a \in \mathbb{R}$. So, $a = 0$. Hence $\gamma = bi + cj + dk$ where $b^2 + c^2 + d^2 = 1$. This is a nice generalization of right angles as they appear in $\mathbb{C}$, where the right angles are simply i and $-i$ on the unit circle. So, we have

Proposition 21.19. *For the double planar nearring $(Q,+,*,\circ)$, as described in §20, an angle $\gamma \in \Gamma$ is a right angle if and only if $\gamma = bi + cj + dk$ where $b^2 + c^2 + d^2 = 1$.*

With Proposition 21.19, from any nonzero $x = ai + bj + ck$, we can make a right angle by computing $u = (a^2 + b^2 + c^2)^{-1/2}x$. Then $u \in \Gamma$ is a right angle.

Now take any sphere $Q^* \circ a + x$, $a \neq 0$. A *diameter* of $Q^* \circ a + x$ is any segment $\overline{-a+x, a+x}$. Note that we can have $Q^* \circ a + x = Q^* \circ b + x$ for any $b \in Q^* \circ a$, and so for such a $b \in Q^* \circ a$, $\overline{-b+x, b+x}$ is also a diameter. Also, $b = \gamma \cdot a$ for some $\gamma \in \Gamma$, if $b \in Q^* \circ a$, since $b = q \circ a = (|q|\theta(q)) \circ a = \theta(q) \cdot a$, and $\theta(q) \in \Gamma$ for some $q \in Q^*$. So, fix a diameter $\overline{-a+x, a+x}$ of the sphere $Q^* \circ a + x$ and consider any other point $b + x = \gamma \cdot a + x \in (Q^* \circ a + x) \setminus \{-a+x, a+x\}$. We want to show that the angle $\mu = \angle(-a+x)(\gamma\cdot a+x)(a+x)$ is a right angle. Now $\mu = \angle(-a)(\gamma\cdot a)a$ by Theorem 21.7, and so

$$\begin{aligned}\mu &= \theta(-a-\gamma\cdot a)\theta(a-\gamma\cdot a)^{-1} = \theta[(-1-\gamma)a]\theta[(1-\gamma)a]^{-1}\\ &= \theta(-1-\gamma)\theta(a)\theta(a)^{-1}\theta(1-\gamma)^{-1} = \theta(-1-\gamma)\theta(1-\gamma)^{-1}\\ &= \theta(-1)\theta(1+\gamma)\theta(1-\gamma)^{-1} = (-1)\theta((1+\gamma)/(1-\gamma)).\end{aligned}$$

Hence, $\mu^2 = \theta[((1+\gamma)/(1-\gamma))^2]$.

Suppose $\gamma = a+bi+cj+dk$ where $a^2+b^2+c^2+d^2=1$, i.e., $\gamma\in\Gamma$. Then for $v = bi+cj+dk$, we have

$$\begin{aligned}\frac{1+\gamma}{1-\gamma} &= \frac{(1+\gamma)\overline{(1-\gamma)}}{|1-\gamma|^2} = \frac{[(1+a)+v][(1-a)+v]}{1-2a+a^2+b^2+c^2+d^2}\\ &= \frac{(1+a)(1-a)+2v+v^2}{2-2a} = \frac{1-a^2+2v-b^2-c^2-d^2}{2(1-a)}\\ &= \frac{2v}{2(1-a)} = \frac{v}{1-a}\end{aligned}$$

where $\bar{z}$ is defined in §4. Hence,

$$\left(\frac{1+\gamma}{1-\gamma}\right)^2 = \frac{v^2}{(1-1)^2} = \frac{-(b^2+c^2+d^2)}{(1-a)^2}.$$

Consequently, since $(b^2+c^2+d^2)/(1-a)^2 \in \Phi$, we have

$$\theta\left[\left(\frac{1+\gamma}{1-\gamma}\right)^2\right] = \theta\left[\frac{-(b^2+c^2+d^2)}{(1-a)^2}\right] = -1.$$

In summary, we have the following pleasing result.

Theorem 21.20. *Let $\overline{-a+x, a+x}$ be any diameter of an arbitrary sphere $Q^*\circ a+x$, $a\neq 0$, in $(Q,+,*,\circ)$. If $b\in(Q^*\circ a+x)\setminus\{-a+x, a+x\}$, then $\angle(-a+x)b(a+x)$ is a right angle.*

Think about the right angles $\gamma\in\Gamma$. They have direction, since they are of the form $\gamma = bi+cj+dk$ where $b^2+c^2+d^2=1$. There are infinitely many right angles in Γ.

22 The group of units of the rings of integers $(\mathbb{Z}_{p^2},+,\cdot)$

Consider the ring of integers $(\mathbb{Z}_{p^2},+,\cdot)$, where p is an odd prime. We want to determine $\mathscr{U}(\mathbb{Z}_{p^2},\cdot,1)$, the group of units of the monoid $(\mathbb{Z}_{p^2},\cdot,1)$.

Suppose $u \in \mathbb{Z}_{p^2}$. Recall that the set $\mathbb{Z}_{p^2} = \{0,1,2,\ldots,p^2-1\}$, a subset of the set of integers $\mathbb{Z} = \{0,\pm1,\pm2,\ldots\}$. Suppose also that u and p^2 are relatively prime, i.e., $(u,p^2)=1$. Then there are unique integers $s,t \in \mathbb{Z}$ so that in the ring $(\mathbb{Z},+,\cdot)$, $su+tp^2=1$. Let $s = mp^2+s'$, with $m,s' \in \mathbb{Z}$, and $0<s'<p^2$. Then $1 = su+tp^2 = (mp^2+s')u+tp^2 = s'u+(mu+t)p^2$, and $s' \in \mathbb{Z}_{p^2}$. With $u,s' \in \mathbb{Z}_{p^2}{}^* = \{1,2,\ldots,p^2-1\}$, we have $s'u=1$ in the ring $\mathbb{Z}_{p^2}$, since $s'u = -(mu+t)p^2+1$ in the ring $\mathbb{Z}$. Hence, $s'=u^{-1}$ in $\mathscr{U}(\mathbb{Z}_{p^2},\cdot,1)$, and we have $u \in \mathscr{U}(\mathbb{Z}_{p^2},\cdot,1)$. Hence, $\mathbb{Z}_{p^2} \setminus \{0,p,2p,\ldots,(p-1)p\} \subseteq \mathscr{U}(\mathbb{Z}_{p^2},\cdot,1)$. If $kp, lp \in \{0,p,2p,\ldots,(p-1)p\}$, then in the ring $\mathbb{Z}_{p^2}$, $kp\cdot lp = 0$, and so $(\mathbb{Z}_{p^2},\cdot,1)$ and $\{0,p,2p,\ldots,(p-1)p\}$ has empty intersection. This completes the proof of

Theorem 22.1. *In the ring of integers $(\mathbb{Z}_{p^2},+,\cdot)$, the group of units $\mathscr{U}(\mathbb{Z}_{p^2},\cdot,1) = \mathbb{Z}_{p^2} \setminus \{0,p,2p,\ldots,(p-1)p\}$ and has order $p(p-1)$.*

Proposition 22.2. *Let $\Gamma = \{1,p+1,2p+1,\ldots,(p-1)p+1\}$. Then Γ is a subgroup of $\mathscr{U}(\mathbb{Z}_{p^2},\cdot,1)$ and is isomorphic to a cyclic group of order p.*

Proof. With $kp+1, lp+1 \in \Gamma$, we have $(kp+1)(lp+1) = klp^2+(k+l)p+1$ in $\mathbb{Z}$, and so $(kp+1)(lp+1) = (k+l)p+1 \in \Gamma$. The map $\phi : \mathbb{Z}_p \to \Gamma$ defined by $\phi(k) = kp+1$ is then an isomorphism from the cyclic group $(\mathbb{Z}_p,+)$ onto Γ. □

Our next goal is to get a subgroup Φ of $\mathscr{U}(\mathbb{Z}_{p^2},\cdot,1)$ of order $p-1$. With $\mathbb{Z}_p = \{0,1,2,\ldots,p-1\}$, $\mathbb{Z}_{p^2} = \{0,1,2,\ldots,p^2-1\}$, and $\mathbb{Z} = \{0,\pm1,\pm2,\ldots\}$, we have $\mathbb{Z}_p \subset \mathbb{Z}_{p^2} \subset \mathbb{Z}$. However, as rings $(\mathbb{Z}_p,+,\cdot)$, $(\mathbb{Z}_{p^2},+,\cdot)$, and $(\mathbb{Z},+,\cdot)$, we do not have that $(\mathbb{Z}_p,+,\cdot)$ is a subring of either $(\mathbb{Z}_{p^2},+,\cdot)$ or $(\mathbb{Z},+,\cdot)$, and $(\mathbb{Z}_{p^2},+,\cdot)$ is not a subring of $(\mathbb{Z},+,\cdot)$. Nevertheless, elements $k,l \in \mathbb{Z}_p$ can be multiplied (a) in $\mathbb{Z}_p$, (b) in $\mathbb{Z}_{p^2}$, and (c) in $\mathbb{Z}$, and we may get the same, or, perhaps, different answers.

As a field, $(\mathbb{Z}_p,+,\cdot)$ has its multiplicative group $\mathbb{Z}_p^*$ isomorphic to any cyclic group of order $p-1$. In particular, there is an element $u \in \mathbb{Z}_p^*$ so that $\langle u\rangle = \{u,u^2,\ldots,u^{p-1}\} = \mathbb{Z}_p^*$. Next, consider the same u, but now as an element of $\mathbb{Z}$. In $\mathbb{Z}$, compute u, u^2, …, and u^{p-1}, and let

$$u^i = m_i p^2 + u_i,$$

where $m_i, u_i \in \mathbb{Z}$ and $0 \le u_i < p^2$. So $u = u_1$. Let $\phi : \mathbb{Z} \to \mathbb{Z}_{p^2}$ be the ring epimorphisms defined by $\phi(t) = [t]_{p^2} \in \mathbb{Z}_{p^2}$, as introduced in (b) of Examples A2.21. Then $\phi(u^i) = \phi(u)^i = u_1^i$, and $\phi(u^i) = u_i$, so $u_1^i = u_i$ in $\mathbb{Z}_{p^2}$. Since $u^{p-1} = 1$ in $\mathbb{Z}_p$, we have $u^{p-1} = mp+1$ for some $m \in \mathbb{Z}$. So, in $\mathbb{Z}_{p^2}$, $u_1^{p-1} \in \Gamma$. If $u_1^{p-1} = kp+1$, $k \neq 0$, then $\langle u_1\rangle = \mathscr{U}(\mathbb{Z}_{p^2},\cdot,1)$, and so $\Phi = \langle u_1^p\rangle$ is a cyclic subgroup of order $p-1$. If $u_1^{p-1} = 1$, let $\Phi = \{u_1,u_1^2,\ldots,u_1^{p-1} = 1\}$ in $\mathbb{Z}_{p^2}$. If $u_1^k = 1$ in $\mathbb{Z}_{p^2}$ for some k, $1<k<p-1$, then $u_1 = np^2+1 = (np)p+1$ for some $n \in \mathbb{Z}$, and so $u_1^k = u^k = 1$ in Z_p^*. But $u^k \neq 1$ in $\mathbb{Z}_p^*$ for $1<k<p-1$, and, therefore $u_1^k = 1$ cannot be true for such k, $1<k<p-1$. Again, Φ is a cyclic subgroup of order $p-1$.

We can now state with confidence

Proposition 22.3. *The group of units $\mathscr{U}(\mathbb{Z}_{p^2},\cdot,1)$ has a cyclic subgroup Φ of order $p-1$.*

With Proposition 22.2 and Proposition 22.3, we will prove

Theorem 22.4. *The group of units $\mathscr{U}(\mathbb{Z}_{p^2},\cdot,1)$ is the internal direct product of the Γ of* Proposition 22.2 *and the Φ of Proposition 22.3. Moreover, $\mathscr{U}(\mathbb{Z}_{p^2},\cdot,1)$ is isomorphic to the external direct product of groups $\mathbb{Z}_p \times \mathbb{Z}_{p-1}$.*

Proof. From Theorem A4.16, we get that $\Gamma \cap \Phi = \{1\}$. Let $H = \Gamma\Phi$, the internal direct product of Γ and Φ, and a subgroup of $\mathscr{U}(\mathbb{Z}_{p^2},\cdot,1)$. Since Γ is isomorphic to $(\mathbb{Z}_p,+)$ and Φ is isomorphic to $(\mathbb{Z}_{p-1},+)$, we have that H is isomorphic to $\mathbb{Z}_p \times \mathbb{Z}_{p-1}$. Hence, $|H| = |\mathbb{Z}_p \times \mathbb{Z}_{p-1}| = p(p-1) = |\mathscr{U}(\mathbb{Z}_{p^2},\cdot,1)|$. Therefore, $H = \mathscr{U}(\mathbb{Z}_{p^2},\cdot,1)$. □

Theorem 22.5. *Each* $(\mathbb{Z}_{p^2},\Phi)$ *is a Ferrero pair.*

Proof. We identify each $\phi \in \Phi$ with $\phi^\bullet : \mathbb{Z}_{p^2} \to \mathbb{Z}_{p^2}$ defined by $\phi^\bullet(x) = \phi x$ for all $x \in \mathbb{Z}_{p^2}$. So, we think of Φ as a group of automorphisms of $(\mathbb{Z}_{p^2},+)$. If $\phi \in \Phi$, $x \in \mathbb{Z}_{p^2}$, and $\phi^\bullet(x) = x$, then $\phi x = x$. Hence, $(-\phi+1)x = (-\phi+1)0 = 0$. But $-\phi+1 \in \mathscr{U}(\mathbb{Z}_{p^2},\cdot,1)$, if $\phi \neq 1$. Hence, if $\phi \neq 1$, then $x = 0$, and so Φ is a group of fixed point free automorphisms of $(\mathbb{Z}_{p^2},+)$. This, along with the fact that $|\Phi| \neq 1$, is exactly what is needed to conclude that $(\mathbb{Z}_{p^2},\Phi)$ is a Ferrero pair. □

23 Affine configurations

The affine plane in Example 1.14 was constructed from an IP-nearring in a rather curious way. The IP-nearring used was that of Example 7.9, and was also a motivating force for Theorem 13.7, and was used as well for Example 13.3. One must consult [C] to see how it was used in Example 13.3, however, since techniques used required more knowledge of group theory than we have presented in §A4.

The affine plane in Example 1.14 is one of the three obtained as a substructure of $(\mathbb{Z}_9,\mathscr{B},\in)$, where for $N = \mathbb{Z}_9$, we have $\mathscr{B} = \{Na+b \mid a,b \in N,\ a \neq 0\}$. Let $\mathscr{L}_0$ be the lines of Example 1.14. Then $\mathscr{L}_0 \subset \mathscr{B}$, and note that $\{K_1,K_2,K_3\} = N/N3$, the cosets of the subgroup $N3$ of $N = \mathbb{Z}_9$. Also, notice that

$$\{L_1,L_2,L_3\} = \{N1+0, N1+3, N1+6\},$$

$$\{M_1,M_2,M_3\} = \{N2+0, N2+3, N2+6\},$$

and

$$\{N_1,N_2,N_3\} = \{N4+0, N4+3, N4+6\}.$$

So, the elements of the subgroup $N3$ are used to select blocks $Na+x$ for each "basic block" $Na \neq N3$.

Next, in place of the coset $N3$, use the elements of the coset $N3+1$ to select the blocks for each basic block $Na \neq N3$. So, let

$$\mathscr{L}_1 = N/N3 \cup \{Na+x \mid Na \in \mathscr{B},\ Na \neq N3,\ x \in N3+1\}.$$

Since $\mathscr{L}_1 = \{L+1 \mid L \in \mathscr{L}_0\}$, one has immediately that $(N, \mathscr{L}_1, \in)$ is also an affine plane. That is, the map $1^+ : N \to N$, defined by $1^+(x) = x+1$, defines a bijective transformation which maps the lines in $\mathscr{L}_0$ onto the lines in $\mathscr{L}_1$. Notice that $\mathscr{L}_0 \cap \mathscr{L}_1 = N/N3$, the cosets of the subgroup $N3$, i.e., the line $N3$ and all those lines parallel to $N3$.

Finally, use the elements of the coset $N3+2$ to select the blocks for each basic block $Na \neq N3$. Then

$$\mathscr{L}_2 = N/N3 \cup \{Na+x \mid Na \in \mathscr{B},\ Na \neq N3,\ x \in N3+2\}.$$

Since $\mathscr{L}_2 = \{L+2 \mid L \in \mathscr{L}_0\}$, one gets that $(N, \mathscr{L}_2, \in)$ is an affine plane. Also, $2^+ : N \to N$, defined by $2^+(x) = x+2$, defines a bijective translation which maps the lines in $\mathscr{L}_0$ onto the lines in $\mathscr{L}_2$.

We have three affine planes $(N, \mathscr{L}_i, \in)$, $i = 0,1,2$, and if $i, j \in \{0,1,2\}$, $i \neq j$, then $\mathscr{L}_i \cap \mathscr{L}_j = N/N3$. Notice that $(N, N/N3, \in)$ is a net, and that $N/N3$ consists of the line $N3$ and all the lines parallel to $N3$.

This example, discovered by Helmut Karzel in 1979, provides motivation for the ideas of an affine configuration.

Definition 23.1. Let $(N, \mathscr{T}, \in)$ be a tactical configuration. Suppose $\mathscr{T} = \cup_{i \in I} \mathscr{L}_i$ for some index set I, that each $(N, \mathscr{L}_i, \in)$ is an affine plane, and that there is a net $(N, \mathscr{P}, \in)$, with $\mathscr{P} \subset \mathscr{T}$, such that if $i, j \in I$, $i \neq j$, then $\mathscr{L}_i \cap \mathscr{L}_j = \mathscr{P}$. Then $(N, \mathscr{T}, \in)$ is an *affine configuration*, and $\mathscr{P}$ is the *core* of this affine configuration. Since $(N, \mathscr{P}, \in)$ is a net, $\mathscr{P}$ has a partition $\mathbf{P} = \{\mathscr{P}_i \mid i \in T\}$ for some index set T, where each $\mathscr{P}_i$ is itself a partition of N. Such a $\mathscr{P}_i$ is the family of all lines parallel to and including any $L \in \mathscr{P}_i$, and is called the *pencil* of lines for L. We say that $(N, \mathscr{T}, \in)$ is an *affine configuration with* $|T|$ *pencils.*

Of course, the nearring $(\mathbb{Z}_9, +, *)$ from Example 7.9 yields an affine configuration $(\mathbb{Z}_9, \mathscr{B}, \in)$ with one pencil. Our first goal is to generalize this example.

With a Ferrero pair $(\mathbb{Z}_{p^2}, \Phi)$ of Theorem 22.5, constructed from the ring of integers $(\mathbb{Z}_{p^2}, +, \cdot)$, we can construct an IP-nearring $(\mathbb{Z}_{p^2}, +, *)$. Let $N = \mathbb{Z}_{p^2}$ and write $\Phi = \{\phi_1, \phi_2, \ldots, \phi_{p-1}\}$. Then $N * p = \{0, \phi_1 p, \phi_2 p, \ldots, \phi_{p-1} p\}$, where $\phi_i p = \phi_i \cdot p$ in the ring $(\mathbb{Z}_{p^2}, +, \cdot)$. For any $a \in \mathscr{U}(\mathbb{Z}_{p^2}, \cdot, 1)$, we have $N * a = \{0, \phi_1 a, \phi_2 a, \ldots, \phi_{p-1} a\}$, again, where $\phi_i a = \phi_i \cdot a$ in the ring $(\mathbb{Z}_{p^2}, +, \cdot)$. From the set of blocks $\mathscr{B} = \{N * c + d \mid c, d \in N,\ c \neq 0\}$, we will construct a family $(N, \mathscr{L}_i, \in)$ of affine planes.

Theorem 23.2. *Let $(\mathbb{Z}_{p^2}, +, *)$ be any IP-nearring constructed from the Ferrero pair $(\mathbb{Z}_{p^2}, \Phi)$ of Theorem 22.5. Then $(\mathbb{Z}_{p^2}, \mathscr{B}, \in)$ is an affine configuration with one pencil.*

Proof. Let $N = \mathbb{Z}_{p^2}$. Now $N * p = \{0, \phi_1 p, \phi_2 p, \ldots, \phi_{p-1} p\}$ is the only "basic block" in $\mathscr{B}$ that is a subgroup of $(N, +)$. For $a \in \mathscr{U}(\mathbb{Z}_{p^2}, \cdot, 1) = \mathbb{Z}_{p^2} \setminus N * p$, we have $N * a = \{0, \phi_1 a, \phi_2 a, \ldots, \phi_{p-1} a\} \subseteq \mathscr{U}(\mathbb{Z}_{p^2}, \cdot, 1) \cup \{0\}$. Certainly, $N * p$ and $\mathscr{U}(\mathbb{Z}_{p^2}, \cdot, 1)$ are disjoint. Define

$$\mathscr{L}_0 = N/N * p \cup \{N * a + \phi p \mid N * a \in \mathscr{B},\ N * a \neq N * p,\ \phi p \in N * p\}.$$

We will show that $(N, \mathscr{L}_0, \in)$ is an affine plane. The proof will be done in steps.

1. *If $x \in N$, then $N * p + x$ is the unique coset of $N * p$ in N which contains x.*

Proof of 1. This is the nature of cosets of subgroups. □

2. *If $x \in N$, $N * a \neq N * p$, and $a \neq 0$, then there is a unique $\phi p \in N * p$ such that $x \in N * a + \phi p$.*

Proof of **2**. For $N*a \neq N*p$, the blocks $N*a+y \in \mathscr{B} \cap \mathscr{L}_0$ are exactly

$$N*a+0, \quad N*a+\phi_1 p, \quad N*a+\phi_2 p, \ldots, N*a+\phi_{p-1} p.$$

We will now construct a matrix $A=(a_{ij})$ from these p distinct blocks. Let

$$a_{ij} = \begin{cases} 0, & \text{if } i=j=1; \\ \phi_{j-1}a, & \text{if } i=1 \text{ and } 2 \leq j \leq p; \\ \phi_{i-1}p, & \text{if } 2 \leq i \leq p \text{ and } j=1; \\ \phi_{j-1}a+\phi_{i-1}p, & \text{if } 2 \leq i \leq p \text{ and } 2 \leq j \leq p. \end{cases}$$

That is, the first row of A is $(0, \phi_1 a, \phi_2 a, \ldots, \phi_{p-1}a)$, whose entries make up the block $N*a$. For $2 \leq i \leq p$, the ith row is

$$(\phi_{i-1}p, \phi_1 a+\phi_{i-1}p, \phi_2 a+\phi_{i-1}p, \ldots, \phi_{p-1}a+\phi_{i-1}p),$$

whose entries make up the block $N*a+\phi_{i-1}p$. So,

$$A = \begin{pmatrix} 0+0 & \phi_1 a+0 & \phi_2 a+0 & \cdots & \phi_{p-1}a+0 \\ 0+\phi_1 p & \phi_1 a+\phi_1 p & \phi_2 a+\phi_1 p & \cdots & \phi_{p-1}a+\phi_1 p \\ 0+\phi_2 p & \phi_1 a+\phi_2 p & \phi_2 a+\phi_2 p & \cdots & \phi_{p-1}a+\phi_2 p \\ \vdots & \vdots & \vdots & \ddots & \vdots \\ 0+\phi_{p-1}p & \phi_1 a+\phi_{p-1}p & \phi_2 a+\phi_{p-1}p & \cdots & \phi_{p-1}a+\phi_{p-1}p \end{pmatrix}.$$

Note that the first column of A contains the elements of the coset and subgroup $0+N*p=N*p+0$; the second column of A contains the elements of the coset $\phi_1 a+N*p=N*p+\phi_1 a$; $\ldots$; and the pth column of A contains the elements of the coset $\phi_{p-1}a+N*p=N*p+\phi_{p-1}a$. We now want to show that these p cosets

$$N*p+0, \quad N*p+\phi_1 a, \ldots, N*p+\phi_{p-1}a \tag{23:5}$$

are all distinct.

If $N*p+0=N*p+\phi_i a$, then $\phi_i a \in N*p$. In the first paragraph of the proof of Theorem 23.2, we demonstrated that $\phi_i a \notin N*p$. If $N*p+\phi_i a=N*p+\phi_j a$, with $i \neq j$, then $\phi_i a - \phi_j a \in N * p$. Hence, $a - \phi_i^{-1}\phi_j a = (1 - \phi_i^{-1}\phi_j)a \in N * p$. Since $\phi_i \neq \phi_j$, we have $\phi_i^{-1}\phi_j = \phi_k$ for some $\phi_k \in \Phi$, and so $(1 - \phi_k)a \in N * p$. But $1 - \phi_k, a \in \mathscr{U}(\mathbb{Z}_{p^2}, \cdot, 1)$, and therefore $(1 - \phi_k)a \in \mathscr{U}(\mathbb{Z}_{p^2}, \cdot, 1)$. Since $\mathscr{U}(\mathbb{Z}_{p^2}, \cdot, 1)$ and $N * p$ are disjoint, we cannot have $N * P + \phi_i a = N * p + \phi_j a$ if $i \neq j$. So, the cosets in (23:5) are all distinct.

With these p cosets of $N * p$ being distinct, there are p^2 distinct entries in the matrix A. So, each element of $N = \mathbb{Z}_{p^2}$ is an entry in A exactly once.

With $x \in N$, there is a unique row of A which contains x. The elements of this row constitute the elements of some $N * a + \phi p$ or of $N * a$. This completes the proof of **2.** □

As a result of **1** and **2**, we see that the "lines" of $\mathscr{L}_0$ fall naturally into equivalence classes, which we can think of a families of parallel lines. For $N * p$, we let

$$[N * p]_{\|} = \{N * p, N * p + 1, \ldots, N * p + (p - 1)\},$$

the cosets of the subgroup $N * p$. For $a \in \mathscr{U}(\mathbb{Z}_{p^2}, \cdot, 1)$, we let

$$[N * a]_{\|} = \{N * a, N * a + \phi_1 p, \ldots, N * a + \phi_{p-1} p\}.$$

For any $[N * c]_{\|}$, if $L, M \in [N * c]_{\|}$, either $L = M$ or $L \cap M = \varnothing$.

We will need to show that two distinct points $x, y \in N$ are in exactly one $L \in \mathscr{L}_0$. Toward this goal, we first prove this special case.

3. *If $x, y \in N$, $x \neq y$, and $x, y \in N * c$ for some $N * c \in \mathscr{L}_0$, then there is no other $L \in \mathscr{L}_0$ with $x, y \in L$.*

Proof of **3**. First, suppose $N * c = N * p$. Certainly, because of **1**, we cannot have $x, y \in N * p + t \neq N * p$ for any other coset $N * p + t$ of $N * p$. Likewise, since $N * c_1 \neq N * c_2$ implies $N * c_1 \cap N * c_2 = \{0\}$, we cannot have $x, y \in N * a$ for any $a \in \mathscr{U}(\mathbb{Z}_{p^2}, \cdot, 1)$. So, if $x, y \in L$, $L \in \mathscr{L}_0$, and $L \neq N * p$, then $L = N * a + \phi_i p$ for some

$\phi_i \in \Phi$, and $0 \notin \{x,y\}$. With $x,y \in N*p \cap (N*a+\phi_i p)$, then $x = \phi p = \phi' a + \phi_i p$ and $y = \mu p = \mu' a + \phi_i p$ for some $\phi, \phi', \mu, \mu' \in \Phi_0 = \Phi \cup \{0\}$. So $(\mu - \phi_i)p = \mu' a$ and $(\phi - \phi_i)p = \phi' a$ are two distinct elements in $N*p \cap N*a$. They are distinct because $\mu' a = \phi' a$ implies $x = y$.

Next, suppose $N*c = N*a$ for some $a \in \mathscr{U}(\mathbb{Z}_{p^2}, \cdot, 1)$. From **2**, we know that $x,y \notin N*a+\phi_i p$ for any $\phi_i \in \Phi$. Suppose $x,y \in N*a \cap (N*p+t)$ for some coset $N*p+t \neq N*p$. Then $0 \notin \{x,y\}$, $x = \phi a = \phi' p + t$, and $y = \mu a = \mu' p + t$ for some $\phi, \phi', \mu, \mu' \in \Phi_0$. Consequently, we have $\phi a - \phi' p = t = \mu a - \mu' p$, and so $(\phi - \mu)a = (\phi' - \mu')p$. If $\phi = \mu$, or $\phi' = \mu'$, then $x = y$. So $(\phi' - \mu')p \in N*p$ and $(\phi - \mu)a = \phi(1 - \phi^{-1}\mu)a \in \mathscr{U}(\mathbb{Z}_{p^2}, \cdot, 1)$. Since $\mathscr{U}(\mathbb{Z}_{p^2}, \cdot, 1) \cap N*p = \varnothing$, we have a contradiction.

Finally, suppose $x,y \in N*a \cap (Nb+\phi_i p)$ for some $b \in \mathscr{U}(\mathbb{Z}_{p^2}, \cdot, 1)$ with $N*a \neq N*b$. Then there are $\phi, \phi', \mu, \mu' \in \Phi_0$ such that $x = \phi a = \phi' b + \phi_i p$ and $y = \mu a = \mu' b + \phi_i p$. With $x,y \in N*a$, there is a $\lambda \in \Phi$ such that $\lambda x = y$. (Again, we cannot have $0 \in \{x,y\}$ since $N*b+\phi_i p \neq N*b$.) From $\lambda x = y$ we get $\lambda\phi a = \lambda\phi' b + \lambda\phi_i p = \mu' b + \phi_i p = \mu a = y$. From **2** we get $\lambda\phi_i p = \phi_i p$, and this cannot be, since $\lambda\phi_i \neq \phi_i$. This completes the proof of step **3**. □

4. *If $x,y \in N$ and $x \neq y$, then there is at most one $L \in \mathscr{L}$ with $x,y \in L$.*

Proof of **4**. We have that $x,y \in (N*c+s) \cap (N*d+t)$ if and only if $x-s,\ y-s \in N*c \cap (N*d+(t-s))$. If $c,d \in \mathscr{U}(\mathbb{Z}_{p^2}, \cdot, 1)$, then we have $s,t \in N*p$, and so $t-s \in N*p$. From **3**, we cannot have $x-s,\ y-s \in N*c \cap (N*d+(t-s))$. If $N*c = N*d = N*p$, then we cannot have $x-s,\ y-s \in N*c \cap (N*d+(t-s))$ by the strength of **1**.

Finally, let $N*c+s = N*c+\phi_i p$ and $N*d+t = N*p+t$ with $c \in \mathscr{U}(\mathbb{Z}_{p^2}, \cdot, 1)$. With $x,y \in (N*c+\phi_i p) \cap (N*p+t)$, we have

$$x-\phi_i p,\ y-\phi_i p \in N*c \cap (N*p+t-\phi_i p),$$

which cannot be by the force of **3**. This completes the proof of **4**. □

5. *If $x, y \in N$ and $x \neq y$, then there is a unique $L \in \mathscr{L}_0$ with $x, y \in L$.*

Proof of **5**. Because of **4**, we have that there is at most one. We will count the pairs $\{a, b\}$ in the "lines" of $\mathscr{L}_0$ and compare the answer to the number of pairs $\{c, d\}$ in the "plain" N. There are $\binom{p^2}{2} = p^2(p^2-1)/2$ pairs $\{c, d\} \subset N$, $c \neq d$. For each of the $p(p+1)$ lines $L \in \mathscr{L}_0$, there are $\binom{p}{2} = p(p-1)/2$ pairs $\{a, b\} \subset L$, $a \neq b$. By **4**, each such pair $\{a, b\}$ belongs to at most one $L \in \mathscr{L}_0$. So, among the $p(p+1) \cdot p(p-1)/2 = p^2(p^2-1)/2$ pairs $\{a, b\}$ in some $L \in \mathscr{L}_0$, there are no duplications. Consequently, each such pair $\{x, y\} \subset N$, $x \neq y$, belongs to exactly one $L \in \mathscr{L}_0$. □

Of course, **5** provides an essential requirement for $(N, \mathscr{L}_0, \in)$ to be an affine plane. Take any $L \in \mathscr{L}_0$. Then L belongs to exactly one $[N * c]_{\|}$. If $x \in N$ and $x \notin L$, then there is a unique $M \in [N * c]_{\|}$ with $x \in M$, by **1** and **2**. If $t \in L \cap M$, then $L = M$, again by **1** and **2**. Since $x \notin L$ and $x \in M$,we have $L \cap M = \varnothing$. We now have (a) and (b) of Definitions 3.1. Certainly $L \in \mathscr{L}_0$ implies $|L| \geq 2$, and $|N \setminus L| = p^2 - p > 0$. Consequently, (c) and (d) of Exercise 3.2 are also satisfied. Hence, $(N, \mathscr{L}_0, \in)$ is an affine plane.

We shall now create $p-1$ additional affine planes $(N, \mathscr{L}_i, \in)$, $1 \leq i \leq p-1$. Let

$$\mathscr{L}_i = N/N * p \cup \{N * a + \phi p + i \mid N * a \in \mathscr{B},\ N * a \neq N * p,\ \phi p + i \in N * p + i\}.$$

It is easy to see that $\mathscr{L}_i = \{L + i \mid L \in \mathscr{L}_0\}$, and that the elements of the coset $N * p + i$ are used to select, for each $N * a \neq N * p$, the blocks $N * a + \phi p + i$ to be used as the lines in $\mathscr{L}_i$ parallel to $N * a + i$.

The mapping $i^+ : \mathbb{Z}_{p^2} \to \mathbb{Z}_{p^2}$ defined by $i^+(x) = x + i$ is a bijection such that is $L \in \mathscr{L}_0$, then $i^+(L) = L + i \in \mathscr{L}_i$. Hence, each $(N, \mathscr{L}_i, \in)$ is an affine plane. Certainly, $(N, N/N * p, \in)$ is a net, and so $(N, \mathscr{B}, \in)$ is an affine configuration with core $N/N * p$ and has one pencil. With this, the proof of Theorem 23.2 is complete. □

Appendix

Here, we wish to develop ideas from set theory and algebra or number theory which are used throughout the main body of the text. These are ideas one usually picks up by taking a one semester course in algebra or number theory.

A1 Elements of set theory

Hopefully, the idea of a set is not new to the reader. Roughly speaking, a set is just a collection of objects. One also uses the terms "family," "class," "collection," and "aggregate."

We often define a particular finite set by listing its elements inside braces; e.g.,

$$A = \{1, 2, a, 7, \sqrt{3}\}.$$

Infinite sets can sometimes be described in a similar way; e.g.,

$$\mathbb{N} = \{1, 2, 3, \ldots\},$$

and

$$C = \{1, 1/2, 1/3, \ldots, 1/n, \ldots\}.$$

Sets with an unspecified number of elements can be defined similarly; e.g.,

$$D = \{1, 2, \ldots, n\},$$

and

$$E = \{1, 1/2, 1/3, \ldots, 1/n\}.$$

Note that $\mathbb{N}$ and C denote infinite sets whereas D and E denote finite sets.

One often uses a peculiar notation to define a set. This notation has structure similar to

$$S = \{x \mid p(x)\} \tag{1:6}$$

or

$$U = \{f(r), g(r) \mid p(r)\}. \tag{1:7}$$

The notation (1:6) is read, "the set of all x such that $p(x)$ is true," and (1:7) is read, "the set of all $f(r)$ and $g(r)$ such that $p(r)$ is true. It may help the reader to translate the left brace { as "the set of all," and the vertical segment | as "such that." $p(x)$ is a proposition about x. For any particular x, the statement $p(x)$ is either true or false. The set S consists of all those objects x for which $p(x)$ is true. Mathematicians, including the author, are quite fussy about substitutions. For example, $(x \mid p(x))$ and $\langle x \mid p(x)\rangle$ are not acceptable substitutes for $\{x \mid p(x)\}$.

Now, let us look at some specific examples using this method of defining sets. Consider

$$S = \{x \mid x \text{ is an even integer }\}.$$

Here, $p(x)$ is the proposition "x is an even integer." So

$$S = \{\ldots, -2, 0, 2, 4, \ldots\}.$$

If

$$P = \{p \mid p > 0 \text{ and } p \text{ is a prime}\},$$

then $P = \{2, 3, 5, 7, 11, 13, \ldots\}$. A particularly important set is

$$\varnothing = \{x \mid x \neq x\}.$$

This is the *null set* or *empty set*, the set without any elements. It has also been defined by

$$\varnothing = \{x \mid x \text{ is an honest mathematician }\},$$

but we shall not tolerate that definition here.

There are some other sets which will have their own special notation in this work. They include $\mathbb{R}$ for the set of all real numbers, $\mathbb{C}$ for the set of complex numbers, $\mathbb{Q}$ for the set of rational numbers, i.e., the fractions, $\mathbb{Z}$ for the integers $\{\ldots,-2,-1,0,1,2,3,\ldots\}$, and $\mathbb{N}$ for the natural numbers $\{1,2,3,4,\ldots\}$.

We write $A \subseteq B$ if A is a *subset* of B. Now A is a subset of B if and only if every element of A is also an element of B. If a is an element of a set A, we write $a \in A$, and if x is not an element of A, we write $x \notin A$. We also say for $x \in A$ that "x is in A," or "A contains x" or "x is a member of A," or "x belongs to A," or something similar. Hence, $A \subseteq B$ if and only if $x \in A$ implies $x \in B$. Note that $\varnothing \subseteq A$ for every set A.

Two sets A and B are equal if they contain the same elements. So, $A = B$ if and only if $A \subseteq B$ and $B \subseteq A$. If $A = \{1,2,1\}$ and $B = \{2,1\}$, then $A = B$. Notice that A really only has two elements, namely 1 and 2, the same as B.

It may be that $A \subseteq B$ and $A \neq B$. We express this by writing $A \subset B$ and saying that A is a *proper* subset of B.

Given two sets A and B, there are ways to make new sets. First, there is the *union* of A with B, written $A \cup B$. Now

$$A \cup B = \{x \mid x \in A \text{ or } x \in B\}.$$

Notice that $A \cup B = B \cup A$. Next, there is the *intersection* of A and B, written $A \cap B$. So

$$A \cap B = \{x \mid x \in A \text{ and } x \in B\}.$$

Notice that $A \cap B = B \cap A$. The *set difference*, written $A \setminus B$, is

$$A \setminus B = \{x \mid x \in A \text{ and } x \notin B\}.$$

Notice that $A \setminus B \neq B \setminus A$ unless $A = B$. The *cartesian product* of A with B, written $A \times B$, is

$$A \times B = \{(a,b) \mid a \in A \text{ and } b \in B\}.$$

Notice that $A \times B \neq B \times A$, if $A \neq B$. But what is (a,b) in this definition? It is the *ordered pair* where the first element or first coordinate is a, and the second element or second coordinate is b. In general, $(a_1, a_2, \ldots, a_n)$ has a_i in the ith position; a_i is the ith element or ith coordinate of the *n-tuple* $(a_1, a_2, \ldots, a_n)$. One should not conclude, or even suggest, that $\{a,b\} = (a,b)$. The parentheses (and) suggest order, whereas the braces $\{$ and $\}$ do not suggest order. So, $(1,2) \neq (2,1)$, but $\{1,2\} = \{2,1\}$. Also, $(1,1) \neq \{1\}$.

Suppose for a set A we have $A = X \cup Y$ where $X \cap Y = \varnothing$. Then A is the *disjoint union* of X and Y, and we write $A = X \sqcup Y$. If $B \subseteq A$, then $A = (A \setminus B) \sqcup B$.

If A is a set, the *power set* of A is the family (set) of all subsets of A, and is denoted by $\mathscr{P}(A)$, and sometimes even by 2^A for reasons which may become clearer later. If $A = \{1,2\}$, then $\mathscr{P}(A) = \{\varnothing, \{1\}, \{2\}, A\}$, so $\varnothing, A \in \mathscr{P}(A)$ and $\mathscr{P}(A)$ has four elements. (This is the beginning of your understanding why the notation 2^A is sometimes used for $\mathscr{P}(A)$.) Notice that $\varnothing, B \in \mathscr{P}(B)$ for any set B, and that if $S, T \in \mathscr{P}(B)$, we also have $S \cup T$, $S \cap T$ and $S \setminus T$ are in $\mathscr{P}(B)$, but we usually do not have $S \times T \in \mathscr{P}(B)$.

With $A = B$, we define $A \times A = \{(a_1, a_2) \mid a_1, a_2 \in A\}$. This set is also denoted by A^2. Similarly, $A^3 = A \times A \times A = \{(a_1, a_2, a_3) \mid a_1, a_2, a_3 \in A\}$, and in general, for any positive integer n, $A^n = \{(a_1, \ldots, a_n) \mid a_1, \ldots, a_n \in A\}$. We then have $A^1 = \{(a) \mid a \in A\}$, which we identify with A by thinking $a \equiv (a)$. But what is A^0? Well, there are reasons to say that $A^0 = \{\varnothing\}$. Notice that $\varnothing \neq \{\varnothing\}$ since $\varnothing$ has zero elements and $\{\varnothing\}$ has exactly one element, namely the element $\varnothing$ itself. Finally, we define $A^\omega = \{(a_1, a_2, \ldots, a_n, \ldots,) \mid \text{ each } a_i \in A,\ i \in \mathbb{N}\}$. Sometimes we denote this set of all infinite sequences of A by $A^{\mathbb{N}}$, for reasons similar to denoting $\mathscr{P}(A)$ by 2^A.

A1.1 Relations

Let us begin with any two sets A and B, and construct the cartesian product $A \times B$. A *relation R from A to B* is any subset of $A \times B$. So R is a set of ordered pairs (a,b) with $a \in A$ and $b \in B$. We say that *a is related to b*, and write aRb for the fact that $(a,b) \in R$. The *domain* of a relation R is

$$\mathrm{dom}(R) = \{a \mid (a,b) \in R \text{ for some } b \in B\},$$

and the *range* of R is

$$\mathrm{rng}(R) = \{b \mid (a,b) \in R \text{ for some } a \in A\}.$$

If $A = B$ and $R \subseteq A \times A$, then we say that R is a *relation on A*. Such relations can have some important properties. If aRa for each $a \in A$, then R is *reflexive*. If aRb implies bRa, then R is *symmetric*. If aRb and bRc imply aRc, then R is *transitive*. If R is reflexive, symmetric, and transitive, then R is an *equivalence relation*. For an equivalence relation R on a set A, we define for each $a \in A$ its *equivalence class* $[a]_R$. Specifically,

$$[a]_R = \{x \in A \mid xRa\}.$$

So $[a]_R$ consists of all elements in A which are "equivalent to" a. For an equivalence relation R on a set A, we define

$$A/R = \{[a]_R \mid a \in A\}$$

and note that A/R consists of all the equivalence classes of R. Note that for an equivalence relation R on A, we have for $a,b \in A$, either $[a]_R = [b]_R$ or $[a]_R \cap [b]_R = \varnothing$.

The equivalence classes of an equivalence relation R on A make a *partition of A*. That is, a *partition of A* is a family of nonempty subsets $\mathscr{A} \subseteq \mathscr{P}(A)$ with the property that for each $a \in A$, there is exactly one member $T_a \in \mathscr{A}$ such that $a \in T_a$. Hence, if $S,T \in \mathscr{A}$, then either $S = T$ or $S \cap T = \varnothing$.

Just as an equivalence relation gives us a partition, we can make an equivalence relation from a partition. If $\mathscr{A}$ is a partition of A, we define R on A as follows. For $a \in A$, let $T_a \in \mathscr{A}$ be the unique set in $\mathscr{A}$ with $a \in T_a$. Define aRb if and only if $b \in T_a$ also. It is rather easy now to verify that R is an equivalence relation on A.

If $\mathscr{A}$ is a partition on A, then $\mathscr{A}$ is a family of *pairwise disjoint* sets, which is to say that for $S, T \in \mathscr{A}$, either $S = T$ or $S \cap T = \varnothing$. Since each $a \in A$ belongs to exactly one set $T_a \in \mathscr{A}$, we have $A = \cup\mathscr{A}$ if we define

$$\cup\mathscr{A} = \{x \mid x \in T \text{ for some } T \in \mathscr{A}\}.$$

This is the *union* of the family $\mathscr{A}$. Similarly, we define

$$\cap\mathscr{A} = \{x \mid x \in T \text{ for each } T \in \mathscr{A}\}.$$

This is the *intersection* of the family $\mathscr{A}$. This notation is used for any family of subsets $\mathscr{A}$, pairwise disjoint or not. Similarly, if $\mathscr{A} = \{A_i \mid i \in I\}$, an *indexed* family of sets with *index set* I, we define

$$\cup_{i \in I} A_i = \cup\mathscr{A}$$

and

$$\cap_{i \in I} A_i = \cap\mathscr{A}.$$

There is a family of equivalence relations of fundamental importance in this work, one for every integer $n \geq 2$. For such an integer, define $\equiv_n$ on $\mathbb{Z}$ by requiring that $a \equiv_n b$ if and only if n divides $a - b$, that is to say, if and only if $a - b = nq$ for some integer q. Since $a - a = n0$, we have that $a \equiv_n a$ for each and every $a \in \mathbb{Z}$. So $\equiv_n$ is reflexive. If $a \equiv_n b$, then $a - b = nq$ for some $q \in \mathbb{Z}$, and so $b - a = n(-q)$, which assures us that $b \equiv_n a$. This makes $\equiv_n$ symmetric. Next, suppose $a \equiv_n b$ and $b \equiv_n c$, and suppose $a - b = nq$ and $b - c = nq'$, with $q, q' \in \mathbb{Z}$. Then $(a - b) + (b - c) = nq + nq' = n(q + q')$. Since $q + q' \in \mathbb{Z}$, we can conclude that $a \equiv_n c$ and that $\equiv_n$ is transitive. Now we can be assured that such an $\equiv_n$ is an equivalence relation on $\mathbb{Z}$.

But what are the equivalence classes for an arbitrary $\equiv_n$, $n \geq 2$? Now

$$[0]_{\equiv_n} = \{\ldots, -2n, -n, 0, n, 2n, 3n, \ldots\},$$

which consists of all multiples of n. Also,

$$[1]_{\equiv_n} = \{\ldots, -2n+1, -n+1, 1, n+1, 2n+1, 3n+1, \ldots\},$$

and

$$[n-1]_{\equiv_n} = \{\ldots, -2n+(n-1), -n+(n-1),\\ n-1, n+(n-1), 2n+(n-1), 3n+(n-1), \ldots\}.$$

In fact, given any integer s, there are unique integers q and r, where $0 \leq r < n$, and $s = nq + r$. So $s - r = nq$, and we have $s \equiv_n r$ where

$$[r]_{\equiv_n} = \{\ldots, -2n+r, -n+r, r, n+r, 2n+r, 3n+r, \ldots\}.$$

Notice that $[r]_{\equiv_n}$ is related to $[0]_{\equiv_n}$ by the equations

$$[r]_{\equiv_n} = [0]_{\equiv_n} + r = r + [0]_{\equiv_n}, \tag{1:8}$$

where $[0]_{\equiv_n} + r = \{kn + r \mid k \in \mathbb{Z}\}$ and $r + [0]_{\equiv_n} = \{r + kn \mid k \in \mathbb{Z}\}$. This means that

$$\mathbb{Z}/\equiv_n = \{[0]_{\equiv_n}, [1]_{\equiv_n}, \ldots, [n-1]_{\equiv_n}\} = \{[0]_{\equiv_n}, 1 + [0]_{\equiv_n}, \ldots, (n-1) + [0]_{\equiv_n}\},$$

a set with n distinct equivalence classes. We should summarize this important result.

Theorem A1.1. *For each integer $n \geq 2$, the relation $\equiv_n$ is an equivalence relation on $\mathbb{Z}$, and there are exactly n distinct equivalence classes of $\equiv_n$, namely,*

$$\mathbb{Z}/\equiv_n = \{[0]_{\equiv_n}, [1]_{\equiv_n}, \ldots, [n-1]_{\equiv_n}\},$$

where for $0 \le r < n$, *and* $r \in \mathbb{Z}$,

$$[r]_{\equiv_n} = \{\ldots, -2n+r, -n+r, r, n+r, 2n+r, 3n+r, \ldots\}.$$

(Note: In place of $a \equiv_n b$, it is traditional to write $a \equiv b \pmod{n}$, and to say, "a is congruent to b modulo n.")

A1.2 Functions

Relations are very important in mathematics, but there are certain relations which are of paramount importance. These are *functions.*

Definition A1.2. Let A and B denote sets. A *function* f from A to B is a relation f from A to B satisfying the following properties.

(a) The domain of f is all of A.

(b) If $a \in A$, then there is exactly one $b \in B$ such that $(a,b) \in f$.

We often write $f : A \to B$ to denote that f is a function from A to B. The set A is the *domain* of f and the set B is the *codomain* of f.

From time to time we will get tired of using the word "function." As a substitute, we will use the term "map" or "mapping" or "transformation." These terms each carry the same meaning as "function."

For a function $f : A \to B$, we have a "rule" which assigns to each $a \in A$ exactly one $b \in B$. The "rule" is: assign to an $a \in A$ the unique $b \in B$ such that $(a,b) \in f$. Because of this rule, we write $b = f(a)$, usually, but we also write $b = fa$, $b = af$, and even $b = a^f$. All four of these notations are in use in mathematics, but $b = f(a)$ is certainly the one used most often, and is the one the reader is probably most familiar.

A function $f : A \to B$ is a relation, so the domain and the range of f already have meaning. Note that the range of f need not be the same as the codomain of f, but it is, of course, a subset of the codomain. If the range of f is equal to the codomain of f, that is, if $\text{rng}(f) = B$, then we say that the function f is *surjective*, and some say that the function f is *onto* B, and some simply say that f is "onto." We don't particularly care for the use of the word "onto" being used as an adjective, so we hope we will avoid its use in that way.

A function $f : A \to B$ is *injective*, or *one-to-one*, if $x, y \in A$ and $x \neq y$ imply that $f(x) \neq f(y)$. So different elements in A are associated to different elements in B. The function $g : \mathbb{R} \to [0, +\infty)$ defined by $g(x) = x^2$ is a surjective function with domain $\mathbb{R}$ and range equal to the set of nonnegative real numbers. The function $l : (0, +\infty) \to \mathbb{R}$ defined by $l(x) = \ln(x)$ is both surjective and injective. If a function $f : A \to B$ is both surjective and injective, then it is *bijective*.

For a set A, there is the important function $1_A : A \to A$ which doesn't do anything to anybody. To be exact, $1_A(x) = x$ for each $x \in A$. Now a bijective function $f : A \to B$ has an inverse $f^{-1} : B \to A$. That is, if we define $f' : B \to A$ by $f'(b) = a$ where a is the unique element of A for which $f(a) = b$, then $f'(f(a)) = a$ and $f(f'(b)) = b$ for each $a \in A$ and each $b \in B$. Hence, $f \circ f' = 1_B$ and $f' \circ f = 1_A$, where $\circ$ is defined in the next paragraph. Since this function f' is unique to f, we usually use the notation f^{-1} in place of f'. Try to realize that $(a, b) \in f$ if and only if $(b, a) \in f'$. Now let's try to understand the symbol $\circ$. It is really very easy.

If $f : B \to C$ and $g : A \to B$, then define $t : A \to C$ by $t(a) = f\big(g(a)\big)$ for each $a \in A$, and write $t = f \circ g$. The function t is the *composition of f with g*, or we also say that t *is f composed with g*. In the above paragraph, we have $f \circ f' = 1_B$ and $f' \circ f = 1_A$, or with the preferred notation of f^{-1} for f', we have $f \circ f^{-1} = 1_B$ and $f^{-1} \circ f = 1_A$. The function f^{-1} for the bijective function f is the *inverse* of f, the function 1_A is the *identity function* for the set A, and the function 1_B is the *identity function* for the set B. For $l : (0, +\infty) \to \mathbb{R}$ defined by $l(x) = \ln(x)$, we have $l^{-1} : \mathbb{R} \to (0, +\infty)$ defined by $l^{-1}(x) = e^x$. Notice that for any $h : X \to A$, we have $(f \circ g) \circ h = f \circ (g \circ h)$.

However, if we should choose to write xf in place of $f(x)$, then $\circ$ yields a slightly different appearance. With $f : B \to C$ and $g : A \to B$, then $a(g \circ f) = (ag)f = at$, and so $t = g \circ f$. This would be the same function as $f \circ g$ when we wrote $f(x)$.

When are two functions the same? There really is no agreement for the answer to this question, but all answers are close to one another. For $f : A \to B$ and $g : C \to D$ to be the same, or equal, we shall require (a) $A = C$, (b) $B = D$, and (c) $f(a) = g(a)$ for each $a \in A$. With this definition, $f : \mathbb{R} \to [0, +\infty)$ defined by $f(x) = x^2$ and $g : \mathbb{R} \to \mathbb{R}$ defined by $g(x) = x^2$ are different functions. That is because $\mathbb{R} \neq [0, +\infty)$. There are those, especially analysts, who would consider $f = g$ in this case. They have their reasons, but we shall consider $f \neq g$, nevertheless.

If $f : A \to B$ and $X \subset A$, then $f(x)$ has meaning for each $x \in X$. Hence, $g : X \to B$ defined by $g(x) = f(x)$ for each $x \in X$ is a function distinct from f, but agrees with f on X. We say that g is the *restriction of f to X* and write $g = f|X$. In particular, $1_A|X : X \to A$ defines a function which is called the *inclusion* function for X, or the *insertion* function for X. It is a rather simple function, but it is handy to have around when one needs it.

We abuse the notation $f(a)$ for $f : A \to B$ and $a \in A$. Actually, f defines two more functions $f_* : \mathscr{P}(A) \to \mathscr{P}(B)$ and $f^* : \mathscr{P}(B) \to \mathscr{P}(A)$. For $T \in \mathscr{P}(A)$, define $f_*(T) = \{f(t) \mid t \in T\}$, and for $U \in \mathscr{P}(B)$, $f^*(U) = \{s \in A \mid f(s) \in U\}$. Here, we are being honest and fairly careful, but we usually are not so honest and careful. For a subset $T \subseteq A$, we simply write $f(T)$ for $f_*(T)$, and for $U \subseteq B$, we simply write $f^{-1}(U)$ for $f^*(U)$. This is a deeply held tradition among mathematicians, and so it is best you adopt it, or at least learn how to deal with it. The set $f(T) \subseteq B$ is called the *image* of T under f, and the set $f^{-1}(U) \subseteq A$ is called the *inverse image* or *pre-image* of U. Sometimes $\mathrm{Im} f$ is used to denote $f(A)$, the range of f.

For two nonempty sets A and B, there could, of course, be lots of functions

$f : A \to B$. Let B^A denote all such functions. If $A = B$, then A^A denotes all functions $f : A \to A$. So, for $f, g, h \in A^A$, we have $f \circ g \in A^A$, and $f \circ (g \circ h) = (f \circ g) \circ h$. Also, note that $1_A \circ f = f \circ 1_A = f$ for every $f \in A^A$. These facts are worth highlighting.

Theorem A1.3. *Let A^A denote all the functions from a nonempty set A into A. If $f, g \in A^A$, then so is $f \circ g$, and if $h \in A^A$, also, then $f \circ (g \circ h) = (f \circ g) \circ h$. In addition, the function $1_A \in A^A$ defined by $1_A(x) = x$ for all $x \in A$ has the property that $1_A \circ f = f \circ 1_A = f$ for every $f \in A^A$.*

Every function $f : A \to B$ defines an equivalence relation on its domain A, which we shall here denote by $=_f$. For $a, a' \in A$, we say that $a =_f a'$ if and only if $f(a) = f(a')$. Certainly, for each $a \in A$, we have $a =_f a$ since $f(a) = f(a)$. This makes $=_f$ reflexive. If $a =_f b$, then $f(a) = f(b)$, and so $f(b) = f(a)$, which allows us to truthfully write $b =_f a$, which means that $=_f$ is symmetric. Now suppose $a =_f b$ and $b =_f c$. Then $f(a) = f(b)$ and $f(b) = f(c)$. Hence, $f(a) = f(c)$, from which we can conclude that $a =_f c$ and that $=_f$ is transitive. So, $=_f$ is an equivalence relation. The equivalence classes are $[a]_{=_f} = f^{-1}(\{f(a)\})$, which consists of exactly those $x \in A$ such that $f(x) = f(a)$.

There are times when it is somewhat troublesome to define a function. We need to take extra care when the domain of a function is a family of equivalence classes. It is common to define the function in terms of an arbitrary member of the equivalence class, so one must be assured of getting the same result independent of which member of the equivalence class is chosen. We will now illustrate this problem.

Suppose we want to define a function $f : \mathbb{Q} \to \mathbb{Z}$. Take $m/n \in \mathbb{Q}$. What should $f(m/n)$ be? Suppose we say that $f(m/n) = m$. Then $f(1/2) = 1$ and $f(2/4) = 2$. But $1/2$ is equivalent to $2/4$, and so this is not a well defined rule, and we do not have a function. (We write $m/n = s/t$ if and only if $mt = ns$.) Now, fractions really are equivalence classes, and $1/2$ and $2/4$ are from the same equivalence class. So, to define a function $f : \mathbb{Q} \to \mathbb{Z}$ one must make sure that the function f does the same thing to m/n as it does to s/t if m/n is equivalent to s/t, that is, if $mt = ns$. For example, we could define $f(m/n) = (m/n)^2$. Now you should convince yourself that $(m/n)^2 = (s/t)^2$ if $mt = ns$. Then you will have the confidence to say, "f is well-defined."

Every surjective function gives rise to this very problem. Every surjective function $f : A \to B$ defines the equivalence relation $=_f$ on A. There is a theorem which relates B and $A/{=_f}$ which is of fundamental importance.

Theorem A1.4. *Let $f : A \to B$ be a surjective function. Define $p_f : A \to A/{=_f}$ and $F : A/{=_f} \to B$ by $p_f(x) = [x]_{=_f}$, and $F([x]_{=_f}) = f(x)$. Then F is a bijection, and is the unique function $F : A/{=_f} \to B$ such that $F \circ p_f = f$.*

Proof. If F well defined? Suppose $[x]_{=_f} = [y]_{=_f}$. Then $x =_f y$, and so $f(x) = f(y)$. Hence, $F([x]_{=_f}) = F([y]_{=_f})$, and so F is indeed well defined.

If $b \in B$, then there is an $a \in A$ such that $f(a) = b$. Hence, $F([a]_{=_f}) = f(a) = b$. This makes F surjective. If $F([a]_{=_f}) = F([a']_{=_f})$, then $f(a) = f(a')$, and so $a =_f a'$. This means that $[a]_{=_f} = [a']_{=_f}$, and that F is injective. So, F is bijective.

Certainly, $F \circ p_f(a) = F([a]_{=_f}) = f(a)$, and so $F \circ p_f = f$. If $F' \circ p_f = f$, then we have $F([a]_{=_f}) = f(a) = (F' \circ p_f)(a) = F'([a]_{=_f})$. Consequently, $F = F'$, and F is indeed unique. □

The function p_f is called the *natural map*.

Definition A1.5. The *cardinality* of a set S is the number of elements in S. It is denoted by $|S|$. So $|\varnothing| = 0$, and if S is an infinite set, we write $|S| = \infty$.

If there is a bijection $f : X \to Y$ then certainly $|X| = |Y|$. Hence, by Theorem A1.4, if $f : X \to B$ is a surjection, then $|A/{=_f}| = |B|$.

A2 Elements of field and ring theory

By now, you have probably had experience working in various algebraic structures. You have added and multiplied natural numbers since your childhood. Then you did the same for all integers, as well as for the rational numbers. Later, you started dealing with real numbers which were not rational numbers, like π, $\sqrt{2}$, and e. You learned that you could add and multiply these as well: $\pi+\sqrt{2}$, 2π, and $e\pi$ probably do not trouble you in the least. To solve equations like $x^2+1=0$ and $x^2+x+1=0$, you expanded your horizon to include the complex numbers. In calculus, you started adding and multiplying continuous and differentiable functions. Perhaps you have even added and multiplied matrices. In §A1, we made new sets from old sets, like $A\cap B$ and $A\cup B$. This may be a short summary of your experiences in algebraic structures, but we want to focus even more exactly your experiences, and at the same time, enlarge your horizons to even wider algebraic experiences. To do this, we will express many of your experiences in terms of functions, functions having very special properties.

For any set A with several elements, we can find numerous functions $\mu : A\times A\to A$.

A2.1 Binary operations and examples

Definitions A2.1. Let A be a nonempty set. A function $\mu : A\times A\to A$ is called a *binary operation on A* and sometimes a *multiplication on A*. Instead of writing $c=\mu(a,b)$, we usually write $c=a\mu b$, or simply $c=ab$ and say that c is *a times b*, or c is the *product of a with b*.

Examples A2.2. (a) For $A=\mathbb{N}$, define $\mu:\mathbb{N}\times\mathbb{N}\to\mathbb{N}$ by $\mu(m,n)=mn$.

(b) For $A=\mathbb{N}$, define $\mu:\mathbb{N}\times\mathbb{N}\to\mathbb{N}$ by $\mu(m,n)=m+n$.

(c) For $A=\{-1,1\}$, define $\mu:A\times A\to A$ by $\mu(a,b)=ab$.

(d) For $A = \{1, i, -1, -i\}$, the four solutions to the equation $x^4 - 1 = 0$ in the complex numbers, define $\mu : A \times A \to A$ by $\mu(a,b) = ab$.

(e) For any nonempty set A, define $\mu : \mathscr{P}(A) \times \mathscr{P}(A) \to \mathscr{P}(A)$ by setting $\mu(S,T) = S \cup T$.

(f) For $A = \mathbb{Z}$, define $\mu : \mathbb{Z} \times \mathbb{Z} \to \mathbb{Z}$ by $\mu(m,n) = m - n$.

(g) Let $\mathscr{C}(\mathbb{R}) = \{f : \mathbb{R} \to \mathbb{R} \mid f \text{ is continuous }\}$. Set $\mu(f,g) = f + g$ for all $f, g \in \mathscr{C}(\mathbb{R})$. So, if $f(x) = x^2$ and $g(x) = \cos(x)$, then $\mu(f,g) = h$ where $h(x) = x^2 + \cos(x)$.

(h) Define $\mu : \mathscr{C}(\mathbb{R}) \times \mathscr{C}(\mathbb{R}) \to \mathscr{C}(\mathbb{R})$ by $\mu(f,g) = f \circ g$.

(i) For any nonempty set A, put $\mu(a,b) = b$ for any $a, b \in A$.

(j) For any nonempty set A, let $S \subset A$ and $z \in A \setminus S$. Define

$$\mu(a,b) = \begin{cases} z, & \text{if } a \notin S; \\ b, & \text{if } a \in S. \end{cases}$$

(k) For any nonempty set A, define $\mu : A^A \times A^A \to A^A$ by $\mu(f,g) = f \circ g$.

We will start right now to be more conventional, which means that we will write $a \cdot b$, or simply ab for a generic binary operation $\cdot : A \times A \to A$, but sometimes we may write $a + b$ for $+(a,b)$ where $+ : A \times A \to A$.

Binary operations can have various properties, and to describe some, let $\cdot$ be a binary operation on a nonempty set A. Then $\cdot$ is *commutative* if $a \cdot b = b \cdot a$ for every $a, b \in A$. (We would often just write $ab = ba$, eliminating the $\cdot$.) If $a(bc) = (ab)c$ for all $a, b, c \in A$, then $\cdot$ is *associative*. (If our binary operation is $+$, then we would write $a + (b + c) = (a + b) + c$. We would not even think of writing ab for $a + b$ when we denote our binary operation with any symbol suggesting $+$.) If there is a special element $e \in A$ for a binary operation $*$, one which has the property that $e * b = b * e = b$ for each and every $b \in A$, then e is the *identity* of A with respect to $*$. (Note that A cannot have more than one identity for the same $*$, for if e and e' are both identities of A with respect to $*$, then $e' = e' * e = e$.) If A has an identity e for a binary operation $\cdot$, then an $a \in A$ may have an *inverse*. If $a \cdot a' = a' \cdot a = e$, then a' is an *inverse of* a with respect to $\cdot$. If $\cdot$ is associative, then $a \in A$ can have at most one inverse. For if a' and a'' are inverses for a, then $aa' = a'a = e$ and $aa'' = a''a = e$. Hence, $a' = ea' = (a''a)a' = a''(aa') = a''e = a''$. In this case, let $a^{-1} = a'$. But if our binary operation is denoted by any symbol suggesting $+$, then $-a$ is used to denote the inverse of a. We also write $a - b$ for $a + (-b)$.

Often, we will consider A together with two of its binary operations, say $\bullet$ and $*$. Then $\bullet$ is *left distributive over* $*$ if $a \bullet (b * c) = (a \bullet b) * (a \bullet c)$ for all $a, b, c \in A$. Also, $\bullet$ is *right distributive over* $*$ if $(a * b) \bullet c = (a \bullet c) * (b \bullet c)$ for all $a, b, c \in A$.

Let us turn our attention to three very important algebraic structures, each having two binary operations. They are $(\mathbb{Q}, +, \cdot)$, $(\mathbb{R}, +, \cdot)$, and $(\mathbb{C}, +, \cdot)$ where $+$ and $\cdot$ are the usual binary operations of addition and multiplication on $\mathbb{Q}$, on $\mathbb{R}$, and on $\mathbb{C}$, respectively. In each of these three structures, we have the following properties.

(A1) $+$ is commutative.

(A2) $+$ is associative.

(A3) 0 is the identity with respect to $+$.

(A4) Each a has an inverse $-a$ with respect to $+$.

(M1) $\cdot$ is commutative.

(M2) $\cdot$ is associative.

(M3) 1 is the identity with respect to $\cdot$.

(M4) Each $a \neq 0$ has an inverse a^{-1} with respect to $\cdot$.

(DL) $\cdot$ is left distributive over $+$.

(DR) $\cdot$ is right distributive over $+$.

These properties lead to the following very important concept.

Definitions A2.3. Let F be a set having at least two distinct elements, denoted by 0 and 1. Also suppose that $+$ and $\cdot$ are two distinct binary operations on F. Then the algebraic structure $(F,+,\cdot)$ is a *field* if the binary operations have properties (A1)–(A4), (M1)–(M4), (DL) and (DR) as defined above.

Certainly, $(\mathbb{Q},+,\cdot)$, $(\mathbb{R},+,\cdot)$, and $(\mathbb{C},+,\cdot)$ are three distinct fields. There are finite fields, also. Let $\mathbb{Z}_2 = \{0,1\}$ and define commutative binary operations on $\mathbb{Z}_2$ by $0+0=0$, $0+1=1$, and $1+1=0$ for $+$, and $0\cdot 0=0$, $0\cdot 1=0$, and $1\cdot 1=1$ for $\cdot$. Then $(\mathbb{Z}_2,+,\cdot)$ is a field.

There are lots of fields, both infinite and finite. For example, consider $\mathbb{Q}(\sqrt{2}) = \{a+b\sqrt{2} \mid a,b \in \mathbb{Q}\}$. Then $(\mathbb{Q}(\sqrt{2}),+,\cdot)$ is a field and $\mathbb{Q} \subset \mathbb{Q}(\sqrt{2}) \subset \mathbb{R}$. Similarly, $(\mathbb{Q}(i),+,\cdot)$ is a field where $\mathbb{Q}(i) = \{a+bi \mid a,b \in \mathbb{Q}\}$ and $i^2 = -1$. These two infinite fields are models of things which we will also do with finite fields.

Our next goal will be to construct some more finite fields, but on the way, we shall also develop the important concept of a *ring*.

Definitions A2.4. Let R be a nonempty set. Suppose $+$ and $\cdot$ are two binary operations on R. Then $(R,+,\cdot)$ is a *ring* if conditions (A1)–(A4), (M2), (DL), and (DR) above are satisfied. If (M1) is also satisfied, then $(R,+,\cdot)$ is a *commutative ring*. If there is a special element $1 \in R$, $1 \neq 0$, for which (M3) is satisfied, then we say that the ring $(R,+,\cdot)$ has an *identity* 1.

Examples A2.5. (a) Each field $(F,+,\cdot)$ is a ring.

(b) The integers $(\mathbb{Z},+,\cdot)$ is a commutative ring with identity.

(c) The continuous functions $\mathscr{C}(\mathbb{R}) = \{f : \mathbb{R} \to \mathbb{R} \mid f \text{ is continuous }\}$, together with the well-known binary operations $\oplus$ and $\bullet$, defined by

$$(f \oplus g)(x) = f(x) + g(x) \quad \text{and} \quad (f \bullet g)(x) = f(x)\cdot g(x) \quad \text{for all } x \in \mathbb{R},$$

make $(\mathscr{C}(\mathbb{R}),\oplus,\bullet)$ a commutative ring with identity.

Our next step is to make each $\mathbb{Z}/{\equiv_n}$ of Theorem A1.1 into a finite commutative ring with identity. In preparation, let us make our notation simpler. We shall write $[r]_n$ for $[r]_{\equiv_n}$ since the $\equiv$ is cumbersome. Also, we shall write $\mathbb{Z}/(n)$ for $\mathbb{Z}/{\equiv_n}$ in order to be more traditional. So

$$\mathbb{Z}/(n) = \{[0]_n, [1]_n, \ldots, [n-1]_n\}.$$

Define $+_n : \mathbb{Z}/(n) \times \mathbb{Z}/(n) \to \mathbb{Z}/(n)$ by $+_n([a]_n, [b]_n) = [a+b]_n$, or simply

$$[a]_n + [b]_n = [a+b]_n. \tag{2:9}$$

Similarly, define $\cdot_n : \mathbb{Z}/(n) \times \mathbb{Z}/(n) \to \mathbb{Z}/(n)$ by

$$\cdot_n([a]_n, [b]_n) = [ab]_n,$$

or simply

$$[a]_n \cdot [b]_n = [a]_n[b]_n = [ab]_n. \tag{2:10}$$

We want to show that for $n \geq 2$, the algebraic structure $(\mathbb{Z}/(n), +_n, \cdot_n)$ is a commutative ring with identity. Such a ring will have exactly n elements, and so it will be finite.

Theorem A2.6. *For $n \geq 2$, $(\mathbb{Z}/(n), +_n, \cdot_n)$ is a commutative ring with identity $[1]_n$.*

Proof. Our first task will be to prove that $+_n$ and $\cdot_n$ are well defined. (Recall our discussion at the end of §A1.) The mappings $+_n$ and $\cdot_n$ are defined in terms of representatives of the equivalence classes. Suppose $[a]_n = [x]_n$ and $[b]_n = [y]_n$. Can we be assured that

$$[a]_n + [b]_n = [x]_n + [y]_n$$

and that

$$[a]_n \cdot [b]_n = [x]_n \cdot [y]_n?$$

Since $[a]_n = [x]_n$ and $[b]_n = [y]_n$, then $a - x = qn$ and $b - y = q'n$ for some $q, q' \in \mathbb{Z}$. So $(a+b) = (q+q')n + (x+y)$, and so $(a+b) - (x+y) = (q+q')n$, which assures us that $[a+b]_n = [x+y]_n$. Hence, $[a]_n + [b]_n = [x]_n + [y]_n$.

Similarly, $ab = (qn+x)(q'n+y) = (qq'n+qy+q'x)n+xy$, and so $[ab]_n = [xy]_n$, which in turn assures us that $[a]_n[b]_n = [x]_n[y]_n$. This means that the binary operations $+_n$ and $\cdot_n$ are well defined.

The properties of a ring will be inherited from the corresponding properties of the ring $(\mathbb{Z}, +, \cdot)$. Take arbitrary $[a]_n, [b]_n \in \mathbb{Z}/(n)$. Then

$$[a]_n + [b]_n = [a+b]_n = [b+a]_n = [b]_n + [a]_n.$$

Similarly,

$$[a]_n[b]_n = [ab]_n = [ba]_n = [b]_n[a]_n.$$

Hence, properties (A1) and (M1) are valid. (Notice how commutativity of $+_n$ and $\cdot_n$ are inherited from the properties of the ring of integers $(\mathbb{Z}, +, \cdot)$.) Next, for any $[a]_n \in \mathbb{Z}/(n)$, we have

$$[0]_n + [a]_n = [a]_n = [a]_n + [0]_n,$$

and

$$[1]_n[a]_n = [a]_n = [a]_n[1]_n.$$

This gives us (A3) and (M3).

Take any $[a]_n, [b]_n, [c]_n \in \mathbb{Z}/(n)$. Then

$$[a]_n + \{[b]_n + [c]_n\} = [a+b+c]_n = \{[a]_n + [b]_n\} + [c]_n.$$

Similarly,

$$[a]_n\{[b]_n[c]_n\} = [abc]_n = \{[a]_n[b]_n\}[c]_n.$$

Now we also have (A2) and (M2).

Since

$$[a]_n + [-a]_n = [0]_n = [-a]_n + [a]_n$$

for any $[a]_n \in \mathbb{Z}/(n)$, we have $-[a]_n = [-a]_n$. Hence, (A4).

To see that (DL) is valid, take any $[a]_n, [b]_n, [c]_n \in \mathbb{Z}/(n)$. Then

$$[a]_n\{[b]_n + [c]_n\} = [a(b+c)]_n = [ab+ac]_n = \{[a]_n[b]_n\} + \{[a]_n[c]_n\}.$$

This gives us (DL) as desired. To see that (DR) is also valid, we will apply the next proposition.

Proposition A2.7. *Let* $(R, +, \cdot)$ *be an algebraic structure satisfying* (A1)–(A4), (M1)–(M2), *and* (DL). *Then* (DR) *is also satisfied.*

Proof of Proposition A2.7. Take $a, b, c \in R$. Then $(a+b) \cdot c = c \cdot (a+b) = (c \cdot a) + (c \cdot b) = (a \cdot c) + (b \cdot c)$. Hence, (DR) is satisfied. □

Now, to complete the proof of Theorem A2.6, we only need to notice that Proposition A2.7 is applicable to each $(\mathbb{Z}/(n), +_n, \cdot_n)$, $n \geq 2$. We have now finished the proof of Theorem A2.6, and we have infinitely many finite rings. □

Using the notation of (1:8), we have that $[a]_n + [b]_n = [a+b]_n$ becomes

$$([0]_n + a) + ([0]_n + b) = [0]_n + (a+b), \tag{2:11}$$

and that $[a]_n \cdot [b]_n = [ab]_n$ becomes

$$[0]_n + a) \cdot ([0]_n + b) = [0]_n + (ab). \tag{2:12}$$

This notation will be extended later.

The notation we have used above is still too cumbersome. Let us define $\mathbb{Z}_n = \{0, 1, \ldots, n-1\}$. If we identify each $r \in \mathbb{Z}_n$ with $[r]_n = [r]_{\equiv_n}$ and if we identify $\mathbb{Z}_n$ with $\mathbb{Z}/(n)$, and in place of $+_n$ and $\cdot_n$, we simply write $+$ and $\cdot$, then for $a, b \in \mathbb{Z}_n$, we write $a+b$ to represent $[a]_n + [b]_n$, and $a \cdot b$ or even ab to represent $[a]_n[b]_n$. But we will be aware that $ab = r$ where $[ab]_n = [r]_n$, and $0 \leq r < n$. Similarly, $a+b = r'$ where $[a+b]_n = [r']_n$. We speak of $(\mathbb{Z}_n, +, \cdot)$ as the *ring of integers modulo n*.

A2.2 Ideals and quotient rings

One very important skill to develop is the ability to look at an interesting example, or class of examples, and extract from them the properties that made them interesting. We have introduced the rings $(\mathbb{Z}/(n), +_n, \cdot_n) = (\mathbb{Z}_n, +, \cdot)$, constructed from the ring $(\mathbb{Z}, +, \cdot)$, not only because we will use the rings $(\mathbb{Z}_n, +, \cdot)$ specifically, but we will use them also for a model for further development. We shall explore now one such use.

Notice that $[0]_n = \{kn \mid k \in \mathbb{Z}\}$ gives us a ring also, namely $([0]_n, +, \cdot)$ with $+$ and $\cdot$ as for the usual way of adding and multiplying integers. So, $([0]_n, +, \cdot)$ is a *subring*, and even an *ideal* of the ring $(Z, +, \cdot)$.

Definitions A2.8. For a ring $(R, +, \cdot)$, suppose S is a nonempty subset of R with the properties that $a+b$, $ab \in S$ whenever $a, b \in S$. (We then say that S is *closed with respect to* $+$ *and* $\cdot$.) If $(S, +, \cdot)$ is a ring, then $(S, +, \cdot)$ is a *subring* of $(R, +, \cdot)$. If $(I, +, \cdot)$ is a subring of $(R, +, \cdot)$ with the additional property that ra, $ar \in I$ for each $a \in I$ and each $r \in R$, then I is an *ideal* of the ring R. For an ideal I of a ring R, and an $r \in R$, define

$$I + r = \{a + r \mid a \in I\}$$

and

$$r + I = \{r + a \mid a \in I\},$$

and note that $I + r = r + I$, and call this the *coset of I with representative r*. Let

$$R/I = \{r + I \mid r \in R\},$$

the family of all cosets of I in R.

Note that $\mathbb{Z}/(n) = \{[0]_n + r = r + [0]_n \mid r \in \mathbb{Z}\}$ is just the set of all the cosets of the ideal $[0]_n$ of the ring $\mathbb{Z}$.

In $\mathbb{Z}/(n)$, we say that for two cosets $r + [0]_n$, $r' + [0]_n \in \mathbb{Z}/(n)$, we have $r + [0]_n = r' + [0]_n$ exactly when $r - r' \in [0]_n$. (The reader should now perform the mental exercises required to be satisfied that what we actually said is indeed what we have just said we said.) This brings us to our next theorem.

Theorem A2.9. *Let I be an ideal of a ring $(R,+,\cdot)$. Define $\equiv_I$ on R by $x \equiv_I y$ if and only if $x-y \in I$, where $x-y = x+(-y)$. Then $\equiv_I$ is an equivalence relation on R and the equivalence classes of $\equiv_I$ are just the cosets of I. That is, if $x \in R$, then $[x]_{\equiv_I} = x+I$.*

Proof. Take any $x \in R$. Since I is a subring of R, we know $0 \in I$. Hence $x-x \in I$, and so $x \equiv_I x$. This makes $\equiv_I$ reflexive. If $x \equiv_I y$, then $x-y = a \in I$ for some $a \in I$. Hence, $-a = -(x-y) = y-x \in I$ since I is a subring. We can then conclude that $y \equiv_I x$, which makes $\equiv_I$ symmetric. Finally, if $x \equiv_I y$ and $y \equiv_I z$, then there are $a, b \in I$ such that $x-y = a$ and $y-z = b$. Hence, $x-z = (x-y)+(y-z) = a+b \in I$ since I is a ring. Now we can conclude that $x \equiv_I z$, making $\equiv_I$ transitive, and therefore $\equiv_I$ really is an equivalence relation on R.

To show that $[x]_{\equiv_I} = x+I$, we first take a $y \in [x]_{\equiv_I}$. Then $y \equiv_I x$, and so $y-x = a \in I$ for some a. Hence, $y = a+x = x+a$, which is an element of $x+I$. We now have that $[x]_{\equiv_I} \subseteq x+I$. Next, take $y = x+a \in x+I$ with $a \in I$. Then $y = a+x$ and so $y-x = a \in I$, from which we conclude that $y \equiv_I x$. Hence $x+I \subseteq [x]_{\equiv_I}$. This is now enough to conclude that $[x]_{\equiv_I} = x+I$. (Note that we have used only the fact that I is a subring. We did not use the property that $ra, ar \in I$ whenever $r \in R$ and $a \in I$. Could you guess why we did not make a stronger statement in the statement of the theorem?) So, now we arrive at the conclusion of our proof of Theorem A2.9. □

Theorem A2.10. *Let I be an ideal of a ring $(R,+,\cdot)$. Define $+$ and $\cdot$ on R/I by*

$$(x+I)+(y+I) = (x+y)+I \tag{2:13}$$

and

$$(x+I)\cdot(y+I) = xy+I. \tag{2:14}$$

Then + and · are binary operations on R/I and $(R/I,+,\cdot)$ is a ring, called the quotient ring of I in R.

Proof. The first thing to do is to demonstrate that $+$ and $\cdot$ are well-defined binary operations. Suppose $r+I=x+I$ and $s+I=y+I$. Our task is to show that $(r+s)+I=(x+y)+I$ and that $rs+I=xy+I$. With $r=x+a$ and $s=y+b$ for suitable $a,b\in I$, we have

$$r+s=(x+a)+(y+b)=(x+y)+(a+b).$$

Since $a+b\in I$, we have $(r+s)+I=(x+y)+I$. Also,

$$rs=(x+a)(y+b)=xy+(xb+ay+ab),$$

and since I is an ideal, we have $xb,ay,ab\in I$, and so $xb+ay+ab\in I$. Hence, $rs+I=xy+I$, and now we have that (2:13) and (2:14) give us well-defined binary operations.

As with $\mathbb{Z}/(n)$, we will get that R/I obtains the required properties of a ring from the corresponding properties of R. The proof is almost identical.

Take $x+I$, $y+I$, $z+I\in R/I$. Then

$$\begin{aligned}(x+I)+[(y+I)+(z+I)]&=(x+I)+(y+z+I)\\&=x+(y+z)+I=(x+y)+z+I=[(x+I)+(y+I)]+(z+I).\end{aligned}$$

Similarly,

$$(x+I)[(y+I)(z+I)]=xyz+I=[(x+I)(y+I)](z+I).$$

So now we can truthfully say that $+$ and $\cdot$ are associative. Easily,

$$(x+I)+(y+I)=(x+y)+I=(y+x)+I=(y+I)+(x+I),$$

making $+$ commutative. For $\cdot$ to be commutative, we need

$$xy+I=(x+I)(y+I)=(y+I)(x+I)=yx+I,$$

and this will be so if and only if $xy - yx \in I$ for every $x, y \in R$. But we really do not need $\cdot$ to be commutative. (Observations such as this made during the proof of a theorem are signals for corollaries.) Trivially, $0 + I = I$ is the identity with respect to $+$, since

$$(0+I)+(a+I) = (a+I)+(0+I) = a+I$$

for any $a + I \in R/I$. Also,

$$(-a+I)+(a+I) = I = (a+I)+(-a+I),$$

yielding the fact that an arbitrary $a + I \in R/I$ has its inverse with respect to $+$ given by $-(a+I) = -a+I$.

With $a+I$, $b+I$, $c+I \in R/I$, we have

$$\begin{aligned}(a+I)[(b+I)+(c+I)] &= a(b+c)+I \\ &= (ab+ac)+I \\ &= (a+I)(b+I)+(a+I)(c+I),\end{aligned}$$

and similarly,

$$[(a+I)+(b+I)](c+I) = (a+I)(c+I)+(b+I)(c+I).$$

This established (DL) and (DR), and completes the proof that $(R/I, +, \cdot)$ is a ring. □

If the ring R has an identity 1 and $1 \notin I$, then

$$(1+I)(x+I) = x+I = (x+I)(1+I)$$

for any $x + I \in R/I$, and so $1 + I$ is the identity for the ring $(R/I, +, \cdot)$. This, together with the observation made in the proof of the last theorem, give us the following corollary.

Corollary A2.11. *Let I be an ideal of a ring $(R,+,\cdot)$. Then $(R/I,+,\cdot)$ is a commutative ring if and only if $xy-yx \in I$ for every $x,y \in R$. If the ring R has identity 1, and if $1 \notin I$, then the ring R/I has an identity, namely $1+I$.*

In Corollary A2.11, we required $1 \notin I$ to be assured that $1+I$ was the identity of R/I. In fact, if I is an ideal of a ring R with identity 1, and if $I \neq R$, then we can be assured that $1 \notin I$. For if $1 \in I$ and $r \in R$, then $1 \cdot r = r \in I$. Hence, $R \subseteq I \subseteq R$, forcing the conclusion that $I = R$. One can say more. If $I \subset R$, then I cannot contain any element with a multiplicative inverse. For if $a \in I$ and a^{-1} exists in R, then $aa^{-1} = 1 \in I$. We have just seen that $1 \notin I$ if $I \subset R$ and I is an ideal.

This seems like a nice place to collect together several important technical properties of a ring. (Actually, we may have used some of these slightly in the proofs of the above theorems. So, logically, the best place for these results is immediately after the definition of a ring.)

Theorem A2.12. *Let $(R,+,\cdot)$ be a ring. Then the following properties are valid.*

(a) *If $a \in R$ and $a+a=a$, then $a=0$.*
(b) *If $a \in R$, then $0 \cdot a = a \cdot 0 = 0$.*
(c) *If $a,b \in R$, then $a(-b) = (-a)b = -(ab)$.*
(d) *If $a,b \in R$, then $-(a+b) = (-a)+(-b) = -a-b$.*
(e) *If $I \subset R$ and I is an ideal, then I cannot contain an element a which has a multiplicative inverse $a^{-1} \in R$. In particular, $1 \notin I$.*
(f) *If $(R,+,\cdot)$ is a field, $a,b \in R$, and $ab=0$, then $a=0$ or $b=0$.*

Proof. If $a+a=a$, then $(a+a)+(-a) = a+(-a) = 0$. But

$$(a+a)+(-a) = a+(a+(-a)) = a+0 = a.$$

Hence $a=0$. This gives us (a). Next, if $a \in R$, then

$$a \cdot 0 = a \cdot (0+0) = (a \cdot 0) + (a \cdot 0).$$

By (a), we conclude that $a \cdot 0 = 0$. Similarly,

$$0 \cdot a = (0+0) \cdot a = (0 \cdot a) + (0 \cdot a),$$

and so $0 \cdot a = 0$. Now we have (b). For $a, b \in R$, then

$$0 = a \cdot 0 = a \cdot (b + (-b)) = (ab) + (a(-b)).$$

So $-(ab) = a(-b)$. Similarly,

$$0 = 0 \cdot b = [a + (-a)] \cdot b = (ab) + [(-a)b].$$

Consequently, $-(ab) = (-a)b$. Now we have (c). Finally,

$$(a+b) + (-a-b) = (a+b) + (-b-a) = 0,$$

hence (d) is valid. The validity of (e) was demonstrated immediately after Corollary A2.11. For (f), suppose $a \neq 0$. Then $b = 1b = (a^{-1}a)b = a^{-1}0 = 0$. Suppose $b \neq 0$, then $ab = ba = 0$, and we conclude that $a = 0$. So, (f) is as claimed. □

We have already introduced the finite field $(\mathbb{Z}_2, +, \cdot)$ where $\mathbb{Z}_2 = \{0, 1\}$. This is just a special case of $(\mathbb{Z}_n, +, \cdot)$, $n \geq 2$. One can also quickly be assured that $(\mathbb{Z}_3, +, \cdot)$ and $(\mathbb{Z}_5, +, \cdot)$ are finite fields, but that $(\mathbb{Z}_4, +, \cdot)$ and $(\mathbb{Z}_6, +, \cdot)$ are not fields. Remember, we want to develop skills in making interesting observations and then try to understand our observations in order to extend these ideas to the limit, or at least to a pleasing limit. So, it is natural to ask at this point the following questions. For which values of n is $(\mathbb{Z}_n, +, \cdot)$ a field? For which ideals I of a ring R is $(R/I, +, \cdot)$ a field? This next proposition will be helpful in answering these questions.

Proposition A2.13. *Let $(R, +, \cdot)$ be a commutative ring with identity 1, and take an arbitrary $a \in R$. Let $(a) = \{ar \mid r \in R\}$. Then (a) is an ideal of R.*

Proof. Take $ar \in (a)$ and $x \in R$. Then $(ar)x = a(rx) \in (a)$ and $x(ar) = (ar)x$, and so $xb = bx \in (a)$ for every $x \in R$ and every $b \in (a)$. Next for any $ar, ar' \in (a)$, it holds that $ar + ar' = a(r + r') \in (a)$, and $(ar)(ar') = a(rar') \in (a)$. Furthermore, $0 = a \cdot 0 \in (a)$, and $a(-r) = -(ar) \in (a)$. Certainly the associative and distributive laws are valid, so (a) is a subring of R, and also an ideal of R. □

Definitions A2.14. The ideal (a) of Proposition A2.13 is called the *principal ideal with generator a.* If I is an ideal of a ring R, and if $\{0\} \subset I \subset R$, then I is a *proper* ideal of R. (This is because $\{0\}$ and R are *trivial* ideals of R.) Now, if I is an ideal of R, $I \subset R$, and $I \subset J \subseteq R$, when J is an ideal, forces $J = R$, then I is a *maximal* ideal of R. So, there are no ideals strictly between a maximal ideal I and the entire ring R. Similarly, if I is an ideal, $\{0\} \subset I$, and $\{0\} \subseteq J \subset I$, when J is an ideal, forces $J = \{0\}$, then I is a *minimal* ideal. So, there are no ideals strictly between a minimal ideal I and $\{0\}$.

Proposition A2.15. *Let $(R, +, \cdot)$ be a commutative ring with identity* 1. *For an ideal I of R, $I \subset R$, and any $m \in R \setminus I$, let*

$$(I, m) = \{a + rm \mid a \in I,\ r \in R\}.$$

Then (I, m) is an ideal and $I \subset (I, m)$ properly.

Proof. Take any $a \in I$. Then $a = a + 0m \in (I, m)$, and so $I \subset (I, m)$. Take $r, x \in R$. Then $a + rm \in (I, m)$. Since $ax, xa \in I$, we have $(a + rm)x = ax + (rx)m \in (I, m)$ and $x(a + rm) = xa + (xr)m \in (I, m)$. There remains to show that (I, m) is a subring of R.

Take $a, a' \in I$ and $r, r' \in R$. Then

$$(a + rm) + (a' + r'm) = (a + a') + (r + r')m \in (I, m).$$

Also,

$$-(a + rm) = (-a) + (-r)m \in (I, m) \quad \text{and} \quad 0 + 0m = 0 \in (I, m).$$

Also,

$$(a+rm)(a'+r'm) = [aa' + a(r'm) + (rm)a'] + (rr'm)m \in (I,m),$$

since I is an ideal. The associative and distributive properties are valid since they are valid in R. Hence, (I,m) is a subring and an ideal of R. □

In the process of demonstrating that a certain subset of a ring is an ideal, we have to show that it is a subring. This has required us to check numerous little details, and perhaps you have felt that there must be a better way. Such a better way is provided by this next proposition.

Proposition A2.16. *Let S be a nonempty subset of a ring R. Then S is a subring of R if $a,b \in S$ imply that $a-b \in S$ and that $ab \in S$.*

Proof. With $a \in S$, we have $a,a \in S$, and so $0 = a - a \in S$. Thus, $0,a \in S$, and so $-a = 0 - a \in S$ as well. If $a,b \in S$, then $a,-b \in S$, and so $a+b = a-(-b) \in S$. Hence, S is closed with respect to $+$ and $\cdot$, $0 \in S$, and $-a \in S$ for every $a \in S$. Associativity of $+$ and $\cdot$ for S comes from associativity in R. The same is true for the left and right distributive properties. Hence, $(S,+,\cdot)$ is a subring of $(R,+,\cdot)$. □

Corollary A2.17. *Let I be a nonempty subset of a ring R. Then I is an ideal of R if $a,b \in I$ and $r \in R$ imply $a-b$, ra, $ar \in I$.*

Proof. By Proposition A2.16, I is a subring. The extra requirements to be an ideal are part of the hypotheses. □

Now, let us return to our task of making fields $(R/I,+,\cdot)$ from ideals I of a ring $(R,+,\cdot)$.

Theorem A2.18. *Let M be a maximal ideal of a commutative ring $(R,+,\cdot)$ with identity 1. Then $(R/M,+,\cdot)$ is a field. Conversely, if M is an ideal of a commutative ring R with identity 1, and if $(R/M,+,\cdot)$ is a field, then M is a maximal ideal.*

Proof. Let us begin with a maximal ideal M of a commutative ring $(R,+,\cdot)$ with identity 1. Now $a \in R \setminus M$ if and only if $a+M \neq M$. For an $a+M \neq M$, we need to show the existence of a multiplicative inverse $(a+M)^{-1}$, that is, an $a'+M \in R/M$ such that $(a'+M)(a+M) = a'a+M = 1+M$. By Proposition A2.15, (M,a) is an ideal, and since M is maximal and since $M \subset (M,a)$, we must conclude that $(M,a) = R$. Hence, $1 = m+a'a$ for suitable $m \in M$ and $a' \in R$. Thus, $(a'+M)(a+M) = a'a+M = a'a+m+M = m+aa'+M = 1+M$. This shows that any $a+M \neq M$ has a multiplicative inverse, namely $(a+M)^{-1} = a'+M$. (Remember, $0+M = M$ is the additive identity in the ring $(R/M,+,\cdot)$, i.e. M is the "zero" of the ring R/M.) Hence, $(R/M,+,\cdot)$ is a field.

Conversely, suppose that $(R/M,+,\cdot)$ is a field. If we take the position that there is an ideal I, where $M \subset I \subset R$, that is, the position, that M is not a maximal ideal, then certainly M is also an ideal of the ring I, and if $a+M \neq M$ and $a+M \in I/M$, then there is an $a'+M = (a+M)^{-1} \in R/M$ such that $1+M = (a+M)(a'+M) = aa'+M$. Since $a \in I$, we have $aa' \in I$, and so $1+M \in I/M$, which tells us that $1 \in I$. With $1 \in I$, then $I = R$, as shown just before Corollary A2.11. Hence, M is a maximal ideal of R. □

Theorem A2.19. *For any integer $n \geq 2$, $(\mathbb{Z}_n,+,\cdot)$ is a field if and only if n is a prime.*

Proof. If n is not a prime, then $n = st$ where $s,t \in \mathbb{Z}$, $1 < s$, and $1 < t$. So, as ideals, $(n) \subset (s)$, and (n) is not a maximal ideal, therefore $\mathbb{Z}_n = \mathbb{Z}/(n)$ is not a field. With that, we can honestly say that if $(\mathbb{Z}_n,+,\cdot)$ is a field, then n is a prime.

Next, consider the converse. Let p be a prime integer, $p \geq 2$. If there is an ideal I such that $(p) \subset I \subset \mathbb{Z}$, then there is a smallest positive integer in I, which we will denote by m. Divide p by m to get a quotient q and a remainder r, so $p = mq+r$, where $0 \leq r < m$, and r is an integer. Since $p, mq \in I$, so is $r = p-mq \in I$. But m is the smallest positive integer in I, so $r = 0$, and then $p = mq$. But p is a prime, hence $m = 1$ or $m = p$. If $m = 1$, then $I = \mathbb{Z}$. So

$m = p$. We have seen that the smallest positive integer in I is p.

Let t be any positive integer in I. Then divide t by p to get quotient q' and remainder r', so $t = pq' + r'$, where $0 \le r' < p$. Hence, $0 \le r' = t - pq' \in I$, and so $r' = 0$ by the conclusion of the preceding paragraph. Hence $t = pq' \in (p)$, a maximal ideal, and so $\mathbb{Z}_p = \mathbb{Z}/(p)$ is a field. □

A2.3 Morphisms

Sometimes two rings can be closely related. Their relationship is often described by a *morphism*.

Definitions A2.20. Let $(R, +, \cdot)$ and $(S, +, \cdot)$ be two rings. A function $\phi : R \to S$ is a *ring homomorphism* if (a) $\phi(a+b) = \phi(a) + \phi(b)$ for all $a, b \in R$, and (b) $\phi(ab) = \phi(a)\phi(b)$ for all $a, b \in R$. If $\mathrm{Im}\phi = S$, that is, if ϕ is surjective, then ϕ is a *ring epimorphism*. If ϕ is injective, then ϕ is a *ring monomorphism*. If ϕ is both injective and surjective, then ϕ is a *ring isomorphism*, and we say that R and S are *isomorphic*. (One can show that ϕ^{-1} exists if ϕ is an isomorphism, and that $\phi^{-1} : S \to R$ is also an isomorphism.) If $(R, +, \cdot) = (S, +, \cdot)$, then ϕ is a *ring endomorphism*. If ϕ is a ring endomorphism and also a ring isomorphism, then ϕ is a *ring automorphism*.

Examples A2.21. (a) Consider the ring $(\mathscr{C}(\mathbb{R}), \oplus, \bullet)$ of Examples A2.5 and the field $(\mathbb{R}, +, \cdot)$ of real numbers. Fix an arbitrary $a \in \mathbb{R}$ and define $\phi_a : \mathscr{C}(\mathbb{R}) \to \mathbb{R}$ by $\phi_a(f) = f(a)$ for any $f \in \mathscr{C}(\mathbb{R})$. Suppose $f, g \subset \mathscr{C}(\mathbb{R})$. Then

$$\phi_a(f \oplus g) = (f \oplus g)(a) = f(a) + g(a) = \phi_a(f) + \phi_a(g),$$

and

$$\phi_a(f \bullet g) = (f \bullet g)(a) = f(a) \cdot g(a) = \phi_a(f) \cdot \phi_a(g).$$

Hence, each ϕ_a is a ring homomorphism. Given $r \in \mathbb{R}$, there is the constant function $\mathbf{r} \in \mathscr{C}(\mathbb{R})$ defined by $\mathbf{r}(x) = r$ for every $x \in \mathbb{R}$. So $\phi_a(\mathbf{r}) = \mathbf{r}(a) = r$. Hence, ϕ_a is surjective, and ϕ_a is an epimorphism.

(b) Define for any positive integer $n \geq 2$, the map $\phi_n : \mathbb{Z} \to \mathbb{Z}_n$ by $\phi_n(k) = [k]_n$. The proof of Theorem A2.6 shows that each such ϕ_n is a ring epimorphism.

Definition A2.22. Let $\phi : R \to S$ be a ring homomorphism from the ring $(R, +, \cdot)$ onto the ring $(S, +, \cdot)$. The *kernel of* ϕ denoted by $\ker \phi$, is defined by $\ker \phi = \{r \in R \mid \phi(r) = 0\}$.

Theorem A2.23. *Let $\mu : R \to S$ be a ring homomorphism. Then $(\mu(R), +, \cdot)$ is a subring of S, and so we could think of $\mu : R \to \mu(R)$ as a ring epimorphism.*

Let $\phi : R \to S$ be a ring epimorphism from the ring $(R, +, \cdot)$ onto the ring $(S, +, \cdot)$. Then (a) $\ker \phi$ *is an ideal of R, and* (b) *there is an isomorphism $\Phi : R/\ker \phi \to S$ defined by $\Phi(a + \ker \phi) = \phi(a)$.*

Conversely, if I is an ideal of the ring R, then $p_I : R \to R/I$ defined by $p_I(x) = x + I$ is a ring epimorphism and $\ker p_I = I$.

Hence, ideals are kernels of ring epimorphisms, and conversely.

Proof. Take $\mu(a), \mu(b) \in \mu(R)$. Then

$$\mu(a) - \mu(b) = \mu(a - b) \text{ and } \mu(a)\mu(b) = \mu(ab).$$

So $\mu(R)$ is a subring of S.

We will first demonstrate that $\ker \phi$ is an ideal. Now

$$\phi(0) = \phi(0 + 0) = \phi(0) + \phi(0),$$

so $\phi(0) = 0$ by (a) of Theorem A2.12. Hence,

$$0 = \phi(0) = \phi(a + (-a)) = \phi(a) + \phi(-a).$$

This forces $\phi(-a) = -\phi(a)$. So, if $a, b \in \ker \phi$, then

$$\phi(a - b) = \phi[a + (-b)] = \phi(a) + \phi(-b) = 0 + [-\phi(b)] = -0 = 0.$$

This puts $a - b \in \ker \phi$. If $a \in \ker \phi$ and $r \in R$, then

$$\phi(ar) = \phi(a)\phi(r) = 0\phi(r) = 0$$

by (b) of Theorem A2.12. Hence, $ar \in \ker\phi$. Similarly,

$$\phi(ra) = \phi(r)\phi(a) = \phi(r)\cdot 0 = 0,$$

and so $ra \in \ker\phi$. Hence, $\ker\phi$ is an ideal by Corollary A2.17.

Next, consider $\Phi : R/\ker\phi \to S$ defined by $\Phi(a+\ker\phi) = \phi(a)$. First of all, is this a well-defined mapping? Let $K = \ker\phi$. If $a+K = b+K$, can we be assured that $\phi(a) = \Phi(a+K) = \Phi(b+K) = \phi(b)$? Since $a+K = b+K$, we have $a-b \in K$. Hence,

$$0 = \phi(a-b) = \phi[a+(-b)] = \phi(a)+\phi(-b) = \phi(a)-\phi(b),$$

from which we can infer that $\phi(a) = \phi(b)$. Hence, Φ is well-defined.

Since

$$\begin{aligned}\Phi[(a+K)+(b+K)] &= \Phi[(a+b)+K] = \phi(a+b)\\ &= \phi(a)+\phi(b) = \Phi(a+K)+\Phi(b+K),\end{aligned}$$

and

$$\begin{aligned}\Phi[(a+K)(b+K)] &= \Phi(ab+K) = \phi(ab)\\ &= \phi(a)\phi(b) = \Phi(a+K)\Phi(b+K),\end{aligned}$$

we can rightfully conclude that Φ is a ring homomorphism. If $s \in S$, then there is a $b \in R$ such that $\phi(b) = s$. Hence, $\Phi(b+K) = \phi(b) = s$, and we get that Φ is a ring epimorphism. If $\Phi(a+K) = \Phi(b+K)$, then $\phi(a) = \phi(b)$, which means that $a-b \in \ker\phi = K$. Hence, $a+K = b+K$, and this means that Φ is injective, and in turn that Φ is an isomorphism.

Now, take any ideal I of R, and consider the map $p_I : R \to R/I$ defined by $p_I(x) = x+I$. With

$$p_I(x+y) = (x+y)+I = (x+I)+(y+I) = p_I(x)+p_I(y)$$

and

$$p_I(xy) = xy + I = (x+I)(y+I) = p_I(x)p_I(y)$$

we conclude that p_I is a ring homomorphism, and with $x+I \in R/I$, we have $p_I(x) = x+I$, and so p_I is an epimorphism.

Remember that the "zero" of R/I is the coset $0+I = I$. So, if $a \in I$, then $p_I(a) = a+I = I$, which tells us that $I \subseteq \ker p_I$. If $I = p_I(u) = u+I$, then $u \in I$, and so $\ker p_I \subseteq I$. Now we have that $\ker p_I = I$. This completes the proof of Theorem A2.23. □

The examples (a) of Examples A2.21 provides ring epimorphisms ϕ_a from $\mathscr{C}(\mathbb{R})$ to $\mathbb{R}$, and so $\Phi_a : \mathscr{C}(\mathbb{R})/\ker\phi_a \to \mathbb{R}$ defined by

$$\Phi_a(f + \ker\phi_a) = \phi_a(f) = f(a)$$

is a ring isomorphism. Since $\mathbb{R}$ is a field, so is $\mathscr{C}(\mathbb{R})/\ker\phi_a$, and so $\ker\phi_a$ is a maximal ideal. In particular, $\ker\phi_a = \{f \in \mathscr{C}(\mathbb{R}) \mid f(a) = 0\}$. This means that the ring $\mathscr{C}(\mathbb{R})$ has infinitely many maximal ideals, one for each $a \in \mathbb{R}$.

A2.4 Polynomials

From any commutative ring R with identity, we can construct other rings with identity in numerous ways. One way, in particular, is to construct the ring of all polynomials

$$a = a_0 + a_1x + a_2x^2 + \cdots + a_nx^n \tag{2:15}$$

where each $a_i \in R$ and n is a nonnegative integer. Notice that the a above yields the infinite sequence

$$(a_0, a_1, a_2, \ldots, a_n, 0, 0, 0, \ldots), \tag{2:16}$$

and that such an infinite sequence yields a, so we identify one with the other.

At this point, we wish to emphasize that our polynomials are not functions. To understand this, consider the ring $(\mathbb{Z}_2, +, \cdot)$. There are only four distinct functions $\phi : \mathbb{Z}_2 \to \mathbb{Z}_2$, whereas there are infinitely many polynomials a as in (2:15), or equivalently, there are infinitely many infinite sequences as in (2:16). Each polynomial a as in (2:15), with each $a_i \in R$, defines a function $a^* : R \to R$ defined by $a^*(r) = a_0 + a_1 r + a_2 r^2 + \cdots + a_n r^n$ for arbitrary $r \in R$. (a^* is the polynomial function defined by the polynomial a.) For example, $p = x + x^2$ and $q = x + x^3$ are clearly different polynomials for $R = \mathbb{Z}_2$, but they define the same function, since $p^*(0) = 0$, $p^*(1) = 0$, $q^*(0) = 0$, and $q^*(1) = 0$.

To be exact, one defines the set $R[x]$ of all *polynomials with coefficients in R and indeterminate x* as all sequences

$$(a_0, a_1, a_2, \ldots, a_n, \ldots) \tag{2:17}$$

where each $a_i \in R$ and where $a_i = 0$ for all but finitely many indices i. (Without this last finite condition on the a_i, the sequence in (2:17) would be called a *formal power series.*) Such a sequence in (2:17) is a *polynomial*. The a_i in (2:17) are the *coefficients*, and a_i is the coefficient of x^i. And just what is the indeterminate x? We define

$$x = (0, 1, 0, 0, \ldots). \tag{2:18}$$

This may understandably seem very strange, but with the identification of the a of (2:15) and the infinite sequence of (2:16), one can see the connection. The problem lies in giving precise meaning to expressions as in (2:15), but there really is no harm in thinking of $R[x]$ as all expressions as those in (2:15) where n can be any nonnegative integer.

And just how are polynomials added and multiplied? If $a = \sum_{i=0}^{m} a_i x^i$ and $b = \sum_{i=0}^{m} b_i x^i$, then $a + b = \sum_{i=0}^{m} (a_i + b_i) x^i$ and $ab = \sum_{i=0}^{2m} c_i x^i$ where $c_i = \sum_{j+k=i} a_j b_k$. So, part of the problem begins with the definition of c_i. For example, in c_{2m} (namely $\sum_{j+k=2m} a_j b_k$) exactly what are the coefficients a_{m+1},

$a_{m+2}, \ldots, a_{2m}, b_{m+1}, b_{m+2}, \ldots, b_{2m}$? If we have

$$a = (a_0, a_1, a_2, \ldots)$$

and

$$b = (b_0, b_1, b_2, \ldots),$$

then

$$a + b = (a_0 + b_0, a_1 + b_1, a_2 + b_2, \ldots) \tag{2:19}$$

and

$$ab = (c_0, c_1, c_2, \ldots) \tag{2:20}$$

where each

$$c_i = \sum_{j=0}^{i} a_j b_{i-j} = a_0 b_i + a_1 b_{i-1} + \cdots + a_{i-1} b_1 + a_i b_0 = \sum_{j+k=i} a_j b_k.$$

Now, everything makes sense, even without the restrictions that $a_i = 0$ for all but finitely many i and that $b_j = 0$ for all but finitely many j. But with the knowledge that there is an integer N such that $a_i = b_i = 0$ for all $i \geq N$, we are assured that $c_k = 0$ for all but finitely many k. This assures us that $+$ and $\cdot$ are well defined binary operations on $R[x]$.

Again, if the reader would rather think of a polynomial

$$a = (a_0, a_1, a_2, \ldots, a_n, 0, 0, 0, \ldots)$$

as

$$a = a_0 + a_1 x + a_2 x^2 + \cdots + a_n x^n,$$

then there really is no harm in doing so. In fact, when we take x as defined in (2:18), then it is easy to see that

$$x^n = (\delta_{0,n}, \delta_{1,n}, \delta_{2,n}, \ldots, \delta_{k,n}, \ldots)$$

where $\delta_{i,j}$ is the Kronecker delta function defined by

$$\delta_{i,j} = \begin{cases} 0, & \text{if } i \neq j; \\ 1, & \text{if } i = j. \end{cases}$$

The main point to remember is that distinct polynomials may define the same function. (Of course, this is not true if $R = \mathbb{R}$, $R = \mathbb{C}$, or $R = \mathbb{Q}$.)

Our first milestone is to show that $(R[x], +, \cdot)$ is a commutative ring with identity.

Theorem A2.24. *Let $(R, +, \cdot)$ be a commutative ring with 1. Then $(R[x], +, \cdot)$ is also a commutative ring with identity $\mathbf{1}$, where $+$ and $\cdot$ are defined by (2:19) and (2:20), respectively.*

Proof. We need arbitrary elements $a, b, c \in R[x]$, say

$$a = (a_0, a_1, a_2, \ldots),\ b = (b_0, b_1, b_2, \ldots) \text{ and } c = (c_0, c_1, c_2, \ldots).$$

Also, let $\mathbf{0} = (0, 0, 0, \ldots)$ and $\mathbf{1} = (1, 0, 0, \ldots)$. Since $a_i + b_i = b_i + a_i$ for each i, we certainly have $a + b = b + a$. Also, $a_i + (b_i + c_i) = (a_i + b_i) + c_i$ for each i, so $a + (b + c) = (a + b) + c$. Easily, $\mathbf{0} + a = a + \mathbf{0} = a$, and if $-a = (-a_0, -a_1, -a_2, \ldots)$, then $a + (-a) = (-a) + a = \mathbf{0}$.

Multiplicative properties are more difficult. First, $ab = (d_0, d_1, d_2, \ldots)$ where $d_i = \sum_{s+t=i} a_s b_t = \sum_{t+s=i} b_t a_s = e_i$ where $ba = (e_0, e_1, e_2, \ldots)$. Hence, $\cdot$ is commutative.

Associativity of $\cdot$ is really quite easy, but requires some thought. With $ab = (d_0, d_1, d_2, d_3, \ldots)$ as above, then $(ab)c = (f_0, f_1, f_2, f_3, \ldots)$ where

$$f_i = \sum_{s+t=i} d_s c_t = \sum_{s+t=i} \left(\sum_{u+v=s} a_u b_v \right) c_t = \sum_{u+v+t=i} a_u b_v c_t.$$

Similarly, if $bc = (g_0, g_1, g_2, \ldots)$ with $g_k = \sum_{s+t=k} b_s c_t$, then $a(bc) = (h_0, h_1, h_2, \ldots)$ where

$$h_i = \sum_{j+k=i} a_j g_k = \sum_{j+k=i} a_j \left(\sum_{s+t=k} b_s c_t \right) = \sum_{j+s+t=i} a_i b_s c_t = f_i.$$

This assures us that $(ab)c = a(bc)$.

Perhaps this argument that $\cdot$ is associative is not so convincing, and perhaps the following remarks will help. What are a_i, b_j, and c_k of a, b, and c, respectively? Well, a_i stands for the term $a_i x^i$, b_j stands for the term $b_j x^j$, and c_k stands for the term $c_k x^k$. Hence, $a_i b_j c_k$ stands for $a_i x^i b_j x^j c_k x^k = a_i b_j c_u x^{i+j+k}$. Consequently, $\sum_{u+v+t=i} a_u b_v c_t$ represents the sum of all the products $a_u x^u b_v x^v c_t x^t = a_u b_v c_t x^i$, which is exactly what we want for the coefficient of x^i.

Because of Proposition A2.7, we only need to show that (DL) is valid, and then (DR) will be valid also. So, $a(b+c) = (m_0, m_1, m_2, \ldots)$ where $m_k = \sum_{i+j=k} a_i(b_j + c_j) = \sum_{i+j=k} a_i b_j + \sum_{i+j=k} a_i c_j$, and so $a(b+c) = (ab) + (ac)$. This gives us (DL), and consequently, (DR).

Finally, $\mathbf{1} \cdot a = (1, 0, 0, \ldots)(a_0, a_1, a_2, \ldots) = (n_0, n_1, n_2, \ldots)$ where

$$n_i = 1 \cdot a_i + \sum_{j=1}^{i} 0 a_{i-j} = a_i.$$

Hence, $\mathbf{1} \cdot a = a$. This makes $\mathbf{1}$ the identity for $\cdot$ in $R[x]$. □

Corollary A2.25. *For a commutative ring $(R, +, \cdot)$ with identity* 1*, the function $\varepsilon_0 : R \to R[x]$ defined by $\varepsilon_0(r) = (r, 0, 0, \ldots)$ is a ring monomorphism. (We say that R is embedded in $R[x]$, or the R is a subring of $R[x]$.)*

Proof. For $r, r' \in R$,

$$\varepsilon_0(r + r') = (r + r', 0, 0, \ldots) = (r, 0, 0, \ldots) + (r', 0, 0, \ldots) = \varepsilon_0(r) + \varepsilon_0(r'),$$

and

$$\varepsilon_0(rr') = (rr', 0, 0, \ldots) = (r, 0, 0, \ldots)(r', 0, 0, \ldots) = \varepsilon_0(r)\varepsilon_0(r').$$

If $\varepsilon_0(r) = \varepsilon_0(r')$, then $(r, 0, 0, \ldots) = (r', 0, 0, \ldots)$, and so $r = r'$. Hence, ε_0 is injective, making ε_0 a monomorphism. □

We will focus on polynomial rings $F[x]$, where the coefficients are in a field F. Such rings have properties very close to the properties of the ring $(\mathbb{Z},+,\cdot)$. To understand some of these, we introduce a little more terminology.

For a commutative ring R with identity, suppose $a=(a_0,a_1,a_2,\ldots)\in R[x]$. If $a\neq \mathbf{0}$, we wish to assign a nonnegative integer to a, called the *degree of a*, and denoted by $\deg a$. To be exact, if

$$a=(a_0,\ldots,a_n,0,0,\ldots)$$

and $a_n\neq 0$, then $\deg a=n$. Hence $\deg \mathbf{1}=\deg(1,0,0,\ldots)=0$. We notice three different ways of handling $\mathbf{0}$. These are (a) $\deg \mathbf{0}=-\infty$, (b) $\deg \mathbf{0}=-1$, and (c) $\deg \mathbf{0}$ is left undefined. We do not like (b) and can live with (a). In practice, we seem to follow (c), and so we will not talk about $\deg \mathbf{0}$. The largest integer n for which $a_n\neq 0$ in $a=(a_0,a_1,a_2,\ldots)\in R[x]$ gives us the $\deg a$ if $a\neq \mathbf{0}$. The coefficient a_n is the *leading coefficient of a*. The *leading coefficient of $\mathbf{0}$* is 0.

Proposition A2.26. *For a field F and $a,b\in F[x]$, $\mathbf{0}\notin\{a,b\}$, we have:*

(a) $\deg a\cdot b=\deg a+\deg b$.

(b) *If $a\neq\mathbf{0}$ and $b\neq\mathbf{0}$, then $ab\neq\mathbf{0}$.*

(c) $\deg(-a)=\deg a$.

(d) *If* $\deg a<\deg b$, *then* $\deg(a+b)=\deg b$.

(e) *If* $\deg a=\deg b=m$, *and if $a_m=-b_m$, then* $\deg(a+b)<\deg a$.

(f) *If* $\deg a=\deg b=m$, *and if $a_m\neq -b_m$, then* $\deg(a+b)=\deg a$.

Proof. If $a=(a_0,a_1,a_2,\ldots)$, $b=(b_0,b_1,b_2,\ldots)$, $\deg a=m$, and $\deg b=n$, then $ab=(c_0,c_1,c_2,\ldots)$ where each $c_i=\sum_{j+k=i}a_jb_k$. For $t>m+n$, we have

$$\begin{aligned}c_t&=a_0b_t+a_1b_{t-1}+\cdots+a_mb_{t-m}+a_{m+1}b_{t-m-1}+\cdots+a_tb_0\\&=a_0\cdot 0+a_1\cdot 0+\cdots+a_m\cdot 0+0\cdot b_{t-m-1}+\cdots 0\cdot b_0\\&=0.\end{aligned}$$

For $t = m+n$, we have

$$\begin{aligned} c_t &= a_0 b_t + a_1 b_{t-1} + \cdots + a_m b_{t-m} + a_{m+1} b_{t-m-1} + \cdots + a_t b_0 \\ &= a_0 \cdot 0 + a_1 \cdot 0 + \cdots + a_{m-1} \cdot 0 + a_m b_n + 0 \cdot b_{t-m-1} + \cdots 0 \cdot b_0 \\ &= a_m b_n \neq 0. \end{aligned}$$

Hence, $\deg a \cdot b = m+n = \deg a + \deg b$, and so $ab \neq \mathbf{0}$. We have (a) and (b).

For (c), suppose $a = (a_0, a_1, \ldots, a_m, 0, 0, \ldots)$ and $a_m \neq 0$. Then $\deg a = m$ and $-a = (-a_0, -a_1, \ldots, -a_m, 0, 0, \ldots)$ and $-a_m \neq 0$. Hence, $\deg -a = m$. So we now have (c), also. If $b = (b_0, b_1, \ldots, b_n, 0, 0, \ldots)$ with $b_n \neq 0$, then $\deg b = n$. From this it follows, since $\deg a < \deg b$, that

$$a + b = (a_0 + b_0, \ldots, a_m + b_m, b_{m+1}, \ldots, b_n, 0, 0, \ldots)$$

and so $\deg (a+b) = \deg b$. So (d) is obtained. Now let $m = n$ for the a and b above. Then $a + b = (a_0 + b_0, \ldots, a_m + b_m, 0, 0, \ldots)$. When $a_m = -b_m$, we have $a_m + b_m = 0$, and so $\deg (a+b) < \deg a$. When $a_m \neq -b_m$, we have $a_m + b_m \neq 0$, and so $\deg (a+b) = \deg a$. This gives us (e) and (f). □

We are in a position now to show some analogies between the ring $(\mathbb{Z}, +, \cdot)$ and any $(F[x], +, \cdot)$ for a field F. There are two important properties of $\mathbb{Z}$ which we now state. The first is *the well-ordering principle.*

A2.27. For any $k \in \mathbb{Z}$, let $I_k = \{k, k+1, k+2, \ldots\}$. Suppose $\varnothing \neq S \subseteq I_k$. Then S has a smallest member $m(S)$. That is, there is an $m(S) \in S$ with the property that $m(S) \leq t$ for every $t \in S$.

This rather innocent looking property makes lots of proofs much easier. The second property is *the principle of mathematical induction for integers.*

A2.28. Let S be a set of integers such that (a) $k \in S$, and (b) if the integer n is in S, then the integer $n+1$ is also in S. Then $I_k = \{k, k+1, k+2, \ldots\} \subseteq S$.

If the reader is willing to accept these two principles, we can proceed.

Theorem A2.29 (The Division Algorithm for $\mathbb{Z}$). *Let $d, m \in \mathbb{Z}$, with $d \neq 0$. Then there are unique integers q and r such that*

$$m = dq + r$$

where $0 \leq r < |d|$.

Proof. We will first show the existence of such a pair q and r. Then we will show their uniqueness.

Let $S = \{m - dx \mid m - dx \geq 0,\ x \in \mathbb{Z}\}$. Since $d \neq 0$, either $d \geq 1$ or $d \leq -1$. If $d \geq 1$, then $d|m| \geq |m|$, and

$$m - d(-|m|) = m + d|m| \geq m + |m| \geq 0.$$

With $x = -|m|$, we have $m - dx \in S$. So, when $d \geq 1$, we are assured that $S \neq \varnothing$. If $d \leq -1$, then $-d|m| \geq |m|$, and

$$m - d|m| \geq m + |m| \geq 0.$$

With $x = |m|$, we have $m - dx \in S$. So, when $d \leq -1$, we are also assured that $S \neq \varnothing$. There is no choice now but to conclude the $S \neq \varnothing$.

From **A2.27**, S has a smallest member $r = m(S)$. Let q be the x such that $m - dx = r$, so $m - dq = r$, with $r \geq 0$. This gives us the equation $m = dq + r$, as desired, but we must still show that $r < |d|$. We will consider two cases: (1) $d \geq 1$, and (2) $d \leq -1$.

With $d \geq 1$, $m - d(q+1) = m - dq - d < m - dq = r$. Since $r = m(S)$, we must conclude that $m - d(q+1) \notin S$, and therefore, $m - d(q+1) < 0$. Hence, $r = m - dq = m - d(q+1) + d < d = |d|$, since $m - d(q+1) < 0$. This gives us the desired inequality $0 \leq r < |d|$ when $d \geq 1$.

Next, with $d \leq -1$, we have $m - d(q-1) = m - dq + d < m - dq = r$. Again, with $r = m(S)$, we must conclude that $m - d(q-1) \notin S$, and so $m - d(q-1) < 0$. Thus, $r = m - dq = m - d(q-1) - d < -d = |d|$ since $m - d(q-1) < 0$. Again, we arrive at the desired inequality $0 \leq r < |d|$.

We have existence; we desire uniqueness. Suppose $m = dq' + r'$ where $q', r' \in \mathbb{Z}$ and $0 \le r' < |d|$. So $dq + r = dq' + r'$ from which we obtain the equation $d(q - q') = r' - r$, and then $|d||q - q'| = |r' - r|$. From the facts that $-|d| < -r \le 0$ and $0 \le r' < |d|$, we get $-|d| < r' - r < |d|$, and $-|d| < r - r' < |d|$. Consequently, $|r' - r| < |d|$. So we are in the position of having $|d||q - q'| = |r' - r| < |d|$. This can only be true if $|q - q'| = 0$, or equivalently, $q = q'$. But $q = q'$ implies $r = r'$ because $|d||q - q'| = |r' - r|$. We now have uniqueness and an end to our proof of Theorem A2.29. □

Before we provide the analogous result for $F[x]$, we will present a useful lemma.

Lemma A2.30. *Let F be a field and let $a, b \in F[x]$ where $a \neq \mathbf{0}$ and $b \neq \mathbf{0}$. Suppose* $\deg a \le \deg b$. *Then there is a $d \in F[x]$ such that $b = da$ or else* $\deg(b - da) < \deg b$.

Proof. Let $a = a_0 + a_1 x + \cdots + a_n x^n$, $b = b_0 + b_1 x + \cdots + b_m x^m$ where $a_n \neq 0$, $b_m \neq 0$, and $n \le m$. Let $d = b_m a_n^{-1} x^{m-n}$. Then

$$\begin{aligned} b - da &= (b_0 + b_1 x + \cdots + b_m x^m) - (b_m a_n^{-1} a_0 x^{m-n} + b_m a_n^{-1} a_1 x^{m-n+1} + \cdots \\ &\quad + b_m a_n^{-1} a_{n-1} x^{m-1} + b_m a_n^{-1} a_n x^m) \\ &= b_0 + b_1 x + \cdots + b_{m-n-1} x^{m-n-1} + (b_{m-n} - b_m a_n^{-1} a_0) x^{m-n} \\ &\quad + (b_{m-n+1} - b_m a_n^{-1} a_1) x^{m-n+1} + \cdots + (b_{m-1} - b_m a_n^{-1} a_{n-1}) x^{m-1}. \end{aligned}$$

If $b - da = \mathbf{0}$, then $b = da$. Otherwise, $\deg b - da \le m - 1 < m = \deg b$, as was to be shown. □

Now we are ready to prove

Theorem A2.31 (The Division Algorithm of $F[x]$)**.** *Let $d, m \in F[x]$ where F is a field. Suppose that $d \neq \mathbf{0}$. Then there are unique $q, r \in F[x]$ such that*

$$m = dq + r$$

where either $r = \mathbf{0}$ or $\deg r < \deg d$.

Proof. Our proof will be analogous to the proof of Theorem A2.29. We will first prove the existence of such q and r in $F[x]$. If there is a $q \in F[x]$ such that $m = dq$, then let $r = \mathbf{0}$, and we have the required existence. Otherwise, $m \neq dg$ for every $g \in F[x]$, and so $m - dg \neq \mathbf{0}$ for every $g \in F[x]$. Hence, deg $(m - dg)$ is defined for every $g \in F[x]$. This gives us a nonempty set of nonnegative integers. Let

$$S(m,d) = \{\deg\,(m - dg) \mid g \in F[x]\}.$$

By **A2.27**, the set $S(m,d)$ has a smallest member which we shall denote by k. This means that there is a $q \in F[x]$ such that deg $(m - dq) = k$, and k is less than or equal to deg $(m - dg)$ for every $g \in F[x]$.

Apply Lemma A2.30 to $m - dq$ if $k \geq \deg d$. So, either

$$m - dq = fd$$

or else

$$\deg\,(m - dq - fd) < \deg\,(m - dq).$$

If $m - dq = fd$, then $m = dq + df = d(q + f)$, which is contrary to our assumption that $m \neq dg$ for every $g \in F[x]$. This leaves us with deg $(m - d(q + f)) <$ deg $(m - dq) = k$, which we cannot accept, either. This forces us to conclude that $k < \deg d$. Let $r = m - dq$, an element of $F[x]$, and so $m = dq + r$ with $\deg r < \deg d$. This provides our desired existence of q and r.

For the uniqueness, suppose $m = dq + r = dq' + r'$ where we also have $q', r' \in F[x]$, and either $r' = \mathbf{0}$ or $\deg r' < \deg d$. Then $d(q - q') = r' - r$. Suppose $r' - r \neq \mathbf{0}$. Then deg $d(q - q') = \deg d + \deg\,(q - q') = \deg\,(r' - r)$, and since deg $(r' - r) < \deg d$, we have an impossibility. Hence, $r' - r = \mathbf{0}$ and $r' = r$. Since $d \neq \mathbf{0}$, the only way for $d(q - q') = \mathbf{0}$ is $q - q' = \mathbf{0}$, because of (b) of Proposition A2.26. Hence, $q = q'$. Now we have uniqueness of q and r, and an end to the proof of Theorem A2.31. □

The q and r of Theorem A2.29 or Theorem A2.31 are called, respectively, the *quotient* and *remainder* on dividing m by d.

A2.5 Fields from irreducible polynomials

For any field F, $(F[x],+,\cdot)$ is a commutative ring with identity. By Proposition A2.13, if $a \in F[x]$, then (a) is an ideal of $F[x]$. We are keenly interested in knowing which $a \in F[x]$ make $F[x]/(a)$ into a field. Fix any $a \neq \mathbf{0}$, $a \in F[x]$. If $b \in F[x]$, then there are quotient q and remainder r where $b = aq + r$, and where $r = \mathbf{0}$ or $\deg r < \deg a$. Hence, $b + (a) = r + aq + (a) = r + (a)$. We get that $F[x]/(a) = \{r + (a) \mid r = \mathbf{0} \text{ or } \deg r < \deg a\}$. Suppose $a = bc$ with $b, c \in F[x]$ and with $\deg b \geq 1$ and $\deg c \geq 1$. Then $b + (a) \neq (a)$ and $c + (a) \neq (a)$, but $[b + (a)][c + (a)] = bc + (a) = (a)$. By Theorem A2.12, we can conclude that $F[x]/(a)$ is not a field. So, if $F[x]/(a)$ is to be a field, any factorization of a, say $a = bc$, with $b, c \in F[x]$, must have factors b and c where $\deg b = 0$ or $\deg c = 0$. Hence, b or c is a *constant polynomial*, that is, a polynomial of the form $(u, 0, 0, \ldots)$. If this is so, we say that a is *irreducible*. (Being irreducible in $F[x]$ is analogous to being a prime in $\mathbb{Z}$, and being *reducible* is not being irreducible.) We can then say that if $F[x]/(a)$ is a field, then a is irreducible. What we want to see next is that the converse is also true.

A remark about an $r + (a) \in F[x]/(a)$ is in order. This representation is unique. For suppose $r + (a) = r' + (a) \in F[x]/(a)$ where $r = \mathbf{0}$ or $\deg r < \deg a$ and $r' = \mathbf{0}$ or $\deg r' < \deg a$. Then $r - r' \in (a)$, and so $r - r' = ab$ for some $b \in F[x]$. But $r - r' = \mathbf{0}$ or $\deg\,(r - r') < \deg a$. If $b \neq \mathbf{0}$, then $\deg a \leq \deg ab$. Hence, $r - r' = \mathbf{0}$, and so $r = r'$.

So, we start with an irreducible $a \in F[x]$. So, if $a = bc$ with $b, c \in F[x]$, then b or c is a constant polynomial, $b \neq \mathbf{0}$, and $c \neq \mathbf{0}$. (If $a = \mathbf{0}$, then $F[x]/(a)$ cannot be a field. For example, $1 + x^2 \in F[x]$ has no multiplicative inverse by Proposition A2.26.) To show that $F[x]/(a)$ is a field for any irreducible $a \in F[x]$, we need only show that (a) is a maximal ideal. Then we will apply Theorem A2.18.

For an irreducible $a \in F[x]$, suppose $(a) \subset I \subset F[x]$ and that I is an ideal of $F[x]$. If any $b \in I$ has $\deg b = 0$, then $b \neq \mathbf{0}$ and b is a constant polynomial. So b^{-1} exists, and then $I = F[x]$ by Theorem A2.12. Hence, $\mathbf{0} \neq b \in I$ implies $\deg b \geq 1$. By **A2.27**, we may choose a $b \in I$, $b \neq \mathbf{0}$, with smallest degree. Dividing a by b we get $q, r \in F[x]$ so that $a = bq + r$ and so $r = a - bq \in I$. If $r = \mathbf{0}$, then $a = bq$. If $r \neq \mathbf{0}$, then $\deg r < \deg b$, contrary to our choice of b, since $r \in I$. Hence $r = \mathbf{0}$ and $a = bq$. But remember that a is irreducible. Hence q is a constant polynomial and a polynomial b in I of smallest degree is actually $q^{-1}a$, a constant times a. This means that $a \in I$ and a has smallest degree of all nonzero elements of I. (Note the analogy to the proof of Theorem A2.19.)

Next, let $t \in I$, $t \neq \mathbf{0}$. Divide t by a to get quotient q' and remainder r', so $t = aq' + r'$, where $r' = \mathbf{0}$ or $\deg r' < \deg a$. Now $r' = t - aq' \in I$, and so, since $a \in I$ has smallest degree, we must conclude that $r' = \mathbf{0}$, and this means that $t = aq'$. We have just seen that $t \in (a)$, and so $I \subseteq (a)$. This makes (a) a maximal ideal, and $F[x]/(a)$ a field. In summary, we get

Theorem A2.32. *For a field F and $\mathbf{0} \neq a \in F[x]$, the quotient ring*

$$F[x]/(a) = \{r + (a) \mid r = \mathbf{0} \text{ or } \deg r < \deg a\}.$$

This representation of an $r + (a) \in F[x]/(a)$, where $r = \mathbf{0}$ or $\deg r < \deg a$, is unique. Also, the quotient ring $F[x]/(a)$ is a field if and only if a is irreducible in $F[x]$.

One value of Theorem A2.32 is that it provides a way of making fields. For example, $1 + x + x^2 \in \mathbb{Z}_2[x]$ is irreducible. For if $1 + x + x^2 = (a + bx)(c + dx)$, then $x = -ab^{-1}$ and $-cd^{-1}$ would be solutions to the equation $1 + x + x^2 = 0$. But $1 + 0 + 0^2 = 1 \neq 0$ and $1 + 1 + 1^2 = 1 \neq 0$. Hence, $1 + x + x^2 = 0$ has no solution in $\mathbb{Z}_2$, and so $\mathbb{Z}_2[x]/(1 + x + x^2)$ is a field. Also, by Theorem A2.32, the field $\mathbb{Z}_2[x]/(1 + x + x^2)$ has exactly four elements.

The polynomial $2 + x^2$ is irreducible in $\mathbb{Q}[x]$ for similar reasons, that is $2 + x^2 = 0$ has no solutions in $\mathbb{Q}$. Hence, $\mathbb{Q}[x]/(2 + x^2)$ is a field. This field is isomorphic to $\mathbb{Q}(\sqrt{2})$ as described before Definitions A2.4. Similarly, $\mathbb{Q}(i)$ is isomorphic to $\mathbb{Q}[x]/(1 + x^2)$.

How about a field with exactly nine elements? We want $\mathbb{Z}_3[x]/(a)$ for some irreducible $a \in \mathbb{Z}_3[x]$ with deg $a = 2$. It is easy to see that $1+x^2$ is irreducible in $\mathbb{Z}_3[x]$, and so $\mathbb{Z}_3[x]/(1+x^2)$ is a field with exactly nine elements. Also, $2+x^2$ is irreducible in $\mathbb{Z}_5[x]$. So $\mathbb{Z}_5[x]/(2+x^2)$ is a field with exactly 25 elements. Now, does the reader suspect that there is a field with exactly p^2 elements for every prime p?

Another value of Theorem A2.32 is finding a field K for which a polynomial $a = a_0 + a_1x + \ldots + a_nx^n \in F[x]$ has a *root*, that is, a solution to the equation

$$a = a_0 + a_1x + \ldots + a_nx^n = 0$$

for x.

The following proposition will prove to be useful.

Proposition A2.33. *Let $f : R \to S$ be a ring homomorphism. Then f is a monomorphism if and only if* $\ker f = \{0\}$.

Proof. Suppose that $\ker f = \{0\}$. To see that f is injective, we let $f(a) = f(b)$. Then $0 = f(a) - f(b) = f(a-b)$, and so $a - b \in \ker f = \{0\}$. Thus, $a - b = 0$ and so $a = b$. Hence, f is injective, and so f is a monomorphism. Conversely, if $x \in \ker f$, then $f(x) = 0$. But $f(0) = 0$. Since f is a monomorphism, we have $x = 0$. Hence, $\ker f = \{0\}$. □

For any $a \in F[x]$ with deg $a \geq 1$, $F[x]/(a)$ contains an $r + (a) \neq (a)$. Let $u \in F$. Then u defines the coset $u + (a)$ where $u \equiv (u, 0, 0, \ldots)$. This defines the mapping $\phi : F \to F[x]/(a)$ where $\phi(u) = u + (a)$. Certainly

$$\phi(u+v) = (u+v) + (a) = (u + (a)) + (v + (a)) = \phi(u) + \phi(v)$$

and

$$\phi(uv) = uv + (a) = (u + (a))(v + (a)) = \phi(u)\phi(v).$$

So ϕ is a ring homomorphism. Since F is a field, its only ideals are $\{0\}$ and F. Hence, $\ker f = \{0\}$. This means that ϕ is a monomorphism. Since ϕ is a monomorphism, we have that F is embedded in $F[x]/(a)$. So, each such ring $F[x]/(a)$ has a subring isomorphic to F.

We can think of the field F as being a subring of the ring $F[x]/(a)$ because of the monomorphism $\phi(u) = u + (a)$. In particular, if $a \in F[x]$ is irreducible, then we think of F as a subfield of the field $K = F[x]/(a)$. and that K is an *extension field* of the field F. We even write $F \subset K$. If $b = b_0 + b_1x + \cdots + b_nx^n \in F[x]$, then we think of $b = b_0 + b_1x + \cdots + b_nx^n \in K[x]$. Now $y = x + (a) \in K = F[x]/(a)$, and so $b_0 + b_1y + \cdots + b_ny^n = a_0 + a_1x + \cdots + a_nx^n + (a) = a + (a) = (a)$. Hence, the equation $b_0 + b_1y + \cdots + b_ny^n = 0$ has solution $x + (a) \in K$.

Since $a \in F[x]$ is irreducible in $F[x]$, the equation $a = a_0 + a_1x + \cdots + a_nx^n = 0$ has no solution in F, if $\deg a \geq 2$. For if u is a solution to $a = 0$, that is, if $a_0 + a_1u + \cdots + a_nu^n = 0$, then upon dividing a by $x - u$, we get $a = q(x-u) + r$ where $r = \mathbf{0}$ or $\deg r = 0$. Since u is a solution to $a = 0$, we have $0 = q \cdot 0 + r$, and so $r = 0$. Hence, $a = q \cdot (x-u)$. Since $\deg a \geq 2$, we have $\deg q \geq 1$, $\deg\,(x-u) = 1$, and so a is not irreducible. Hence, if we insist upon the irreducibility of a and if $\deg a \geq 2$, then $a = 0$ has no solution in F. But it does have a solution in K. So, we have

Theorem A2.34. *For a field F, if $a \in F[x]$ is irreducible and* $\deg a \geq 2$, *then the equation*

$$a = a_0 + a_1x + \cdots + a_nx^n = 0$$

has a solution in $K = F[x]/(a)$.

Corollary A2.35. *For a field F, if $a \in F[x]$ has* $\deg a \geq 2$, *and if $u \in F$ is a solution to the equation*

$$a = a_0 + a_1x + \cdots + a_nx^n = 0,$$

then $a = q \cdot (x-u)$ for some $q \in F[x]$ with $\deg q \geq 1$.

Proof. Apply Theorem A2.31 to get $a = q \cdot (x-u) + r$, where $q, r \in F[x]$ and $r = \mathbf{0}$ or $\deg r < \deg\,(x-u) = 1$. So, $r = \mathbf{0}$ or r is a constant polynomial of degree 0.

This means

$$a_0 + a_1x + \cdots + a_nx^n = (b_0 + b_1x + \cdots + b_{n-1}x^{n-1})(x - u) + r,$$

for $q = b_0 + b_1x + \cdots + b_{n-1}x^{n-1}$. Since

$$0 = a_0 + a_1u + \cdots a_nu^n = (b_0 + b_1u + \cdots + b_{n-1}u^{n-1})(u - u) + r,$$

we must conclude that $r = \mathbf{0}$. □

Notice that $\deg q = n - 1$ for the q in Corollary A2.35. This leads us naturally to

Theorem A2.36. *For a field F, if $a \in F[x]$ has degree $\deg a = n \geq 1$, then there are at most n distinct solutions to the equation*

$$a = a_0 + a_1x + \cdots + a_nx^n = 0$$

in the field F

Proof. Suppose $u_1, u_2, \ldots, u_n, u_{n+1}$ are $n+1$ distinct solutions to

$$a = a_0 + a_1x + \cdots + a_nx^n = 0.$$

By Corollary A2.35,

$$a_0 + a_1x + \cdots + a_nx^n = (b_0 + b_1x + \cdots + b_{n-1}x^{n-1})(x - u_1)$$

for some $q_1 = b_0 + b_1x + b_2x^2 + \cdots + b_{n-1}x^{n-1} \in F[x]$. So $u_2, u_3, \ldots, u_n, u_{n+1}$ are solutions to

$$q_1 = b_0 + b_1x + \cdots + b_{n-1}x^{n-1} = 0.$$

By an inductive argument on n, we could finish our proof here and now, but let's go on a step or so.

Using Corollary A2.35 again, we have

$$b_0 + b_1x + \cdots + b_{n-1}x^{n-1} = (c_0 + c_1x + \cdots + c_{n-2}x^{n-2})(x - u_2)$$

for some $q_2 = c_0 + c_1x + \cdots + c_{n-2}x^{n-2} \in F[x]$, and that $u_3, \ldots, u_n, u_{n+1}$ are solutions to

$$q_2 = c_0 + c_1x + \cdots + c_{n-2}x^{n-2} = 0.$$

Continuing in this way we would have that u_n and u_{n+1} are distinct solutions to some

$$q_{n-1} = d_0 + d_1x = 0,$$

but $d_0 + d_1x = 0$ has only one solution. The assumption that the solutions u_1, $u_2, \ldots, u_n, u_{n+1}$ are distinct got us into this unacceptable situation, so we have no choice but to conclude that the equation

$$a = a_0 + a_1x + \cdots + x_nx^n = 0$$

has at most n distinct solutions in F. □

We will be particularly interested in constructing fields $(F[x]/(a), +, \cdot)$ from a field F and an irreducible $a \in F[x]$ with $\deg a = 2$. The reason being is that they are relatively easy to construct. We really need, then, to find irreducible $a \in F[x]$ with $\deg a = 2$. The following results will be useful in proving that there are plenty of irreducible $a \in F[x]$ with $\deg a = 2$.

Lemma A2.37. *For a field F, a polynomial $ax^2 + bx + c \in F[x]$, with $a \neq 0$, is reducible if and only if $x^2 + (b/a)x + (c/a)$ is reducible. Consequently, a polynomial $ax^2 + bx + c \in F[x]$ is irreducible if and only if $x^2 + (b/a)x + (c/a)$ is irreducible.*

Proof. If $x^2 + (b/a)x + (c/a) = (dx + e)(fx + g)$, then

$$ax^2 + bx + c = (adx + ae)(fx + g).$$

Conversely, if $ax^2 + bx + c = (dx + e)(fx + g)$, then

$$x^2 + (b/a)x + (c/a) = [(d/a)x + (e/a)][fx + g].$$

□

Corollary A2.38. *For a field F, if $x^2+bx+c \in F[x]$ is irreducible, and if $d \in F \setminus \{0\}$, then $dx^2+dbx+dc$ is irreducible, and conversely.*

Lemma A2.39. *For a field F, if $x^2+ax+b \in F[x]$ is reducible, then it has factorization*

$$x^2+bx+c=(x+d)(x+e)$$

for some $d,e \in F$.

Proof. If

$$x^2+bx+c=(\alpha x+\beta)(\gamma x+\delta)=(\alpha\gamma)x^2+(\alpha\delta+\beta\gamma)x+(\beta\gamma),$$

then $\alpha\gamma=1$. So

$$\begin{aligned} x^2+bx+c &= (\alpha x+\beta)(\gamma x+\delta)=\alpha(x+\alpha^{-1}\beta)\gamma(x+\gamma^{-1}\delta) \\ &= \alpha\gamma(x+\alpha^{-1}\beta)(x+\gamma^{-1}\delta)=(x+d)(x+e), \end{aligned}$$

where $d=\alpha^{-1}\beta$ and $e=\gamma^{-1}\delta$. □

A2.6 Fields isomorphic to quotient fields

For an irreducible polynomial $p=x^2+mx+n \in F[x]$, F a field, we take two arbitrary elements $a+bx+(p)$ and $c+dx+(p)$ in the quotient field $F[x]/(p)$. (See Theorem A2.32.) Addition is easy:

$$[a+bx+(p)]+[c+dx+(p)]=(a+c)+(b+d)x+(p).$$

Multiplication is almost as easy:

$$\begin{aligned} [a+bx+(p)]\cdot[c+dx+(p)] &= (ac)+(ad+bc)x+bdx^2+(p) \\ &= (ac)+(ad+bc)x+bdx^2-bdp+(p) \\ &= (ac-bdn)+(ad+bc-bdm)x+(p). \end{aligned}$$

This suggests taking a field F and defining two binary operations on $F \times F$ as

$$(a,b)+(c,d) = (a+c,\ b+d) \tag{2:21}$$

and

$$(a,b)\cdot(c,d) = (ac-bdn,\ ad+bc-bdm). \tag{2:22}$$

Theorem A2.40. *For a field F, if $m,n \in F$, and if $+$ and $\cdot$ are defined by (2:21) and (2:22), respectively, then $(F \times F,+,\cdot)$ is a ring. If $p = x^2+mx+n \in F$ is irreducible, then $(F \times F,+,\cdot)$ is a field. In any event, $(F \times F,+,\cdot)$ is isomorphic to $(F[x]/(p),+,\cdot)$.*

Proof. The isomorphism $\phi : F \times F \to F[x]/(p)$ defined by $\phi(a,b) = a+bx+(p)$. The details are tedious, but easy. They are left to the reader who needs to develop confidence with working with these ideas. □

An irreducible polynomial x^2+mx+n can often have $m=0$. In that case, (2:22) becomes

$$(a,b)\cdot(c,d) = (ac-bdn, ad+bc).$$

How do you multiply the complex numbers $a+bi$ and $c+di$? Well,

$$(a+bi)\cdot(c+di) = (ac-bd)+(ad+bc)i.$$

Do you see the connection? With $m=0$ and $n=1$, then $(\mathbb{R} \times \mathbb{R},+,\cdot)$ is isomorphic to $(\mathbb{C},+,\cdot)$. In particular, $(\mathbb{R}/(x^2+1),+,\cdot)$ is isomorphic to $(\mathbb{C},+,\cdot)$.

It should be much easier to work with fields as in Theorem A2.40 rather than those as in Theorem A2.32. At least, the notation is not as clumsy. Let $F' = \{(a,0) \mid a \in F\}$. Then for $(a,0)$, $(b,0) \in F'$, we have $(a,0)+(b,0) = (a+b,0)$ and $(a,0)(b,0) = (ab,0)$. This shows that the map $e : F \to F'$, where $e(a) = (a,0)$, is a field isomorphism. Hence, F' is a subfield of $F \times F$ isomorphic to F. This gives us

Theorem A2.41. *For a field F and an irreducible $p = x^2 + mx + n \in F[x]$, the fields $F[x]/(p)$ and $F \times F$ have a subfield F' isomorphic to F. We say that F is a subfield, or that these fields $F[x]/(p)$ and $F \times F$ are extensions of F.*

Definition A2.42. For F and p as in Theorem A2.41, we say that $F[x]/(p)$ or $F \times F$ is a *quadratic extension of F*. So, a field K is a *quadratic extension* of F if K is isomorphic to $F[x]/(p)$ for some irreducible $p = x^2 + mx + n \in F[x]$.

A2.7 Matrix rings

From a commutative ring R with identity, we constructed the commutative ring of polynomials $(R[x], +, \cdot)$ with identity. There are many ways to construct a new ring from a given ring R, and another way is to construct matrix rings. We consider here only two by two matrices over a ring R with identity.

Definition A2.43. For a ring R with identity, we define

$$M_2(R) = \left\{ \begin{pmatrix} a & b \\ c & d \end{pmatrix} \middle| a, b, c, d \in R \right\}$$

and call any such $\begin{pmatrix} a & b \\ c & d \end{pmatrix}$ a *two by two matrix over R*.

Our main objectives are (1) to show that each $A \in M_2(R)$ defines a mapping $A : R^2 \to R^2$, and (2) to show that $(M_2(R), +, \circ)$ is a ring when $+$ and $\circ$ are suitably defined.

An element $(x, y) \in R^2$ will also be written as $\begin{pmatrix} x \\ y \end{pmatrix}$. If $(x, y), (x', y') \in R^2$, define binary operations $+$ and $\cdot$ on R^2 by

$$(x, y) + (x', y') = (x + x', y + y'),$$

and

$$(x, y) \cdot (x', y') = (xx', yy').$$

Exercise A2.44. Show, for any ring $(R,+,\cdot)$, that $(R^2,+,\cdot)$ is a ring.

Now take $A = \begin{pmatrix} a & b \\ c & d \end{pmatrix} \in M_2(R)$. Define a mapping $A : R^2 \to R^2$ by

$$A\begin{pmatrix} x \\ y \end{pmatrix} = \begin{pmatrix} a & b \\ c & d \end{pmatrix}\begin{pmatrix} x \\ y \end{pmatrix} = \begin{pmatrix} ax+by \\ cx+dy \end{pmatrix}.$$

Suppose $A = \begin{pmatrix} a & b \\ c & d \end{pmatrix}$, $A' = \begin{pmatrix} a' & b' \\ c' & d' \end{pmatrix}$, and $A \neq A'$. Then A and A' define different mappings. For if $a \neq a'$ or $c \neq c'$, then $A\begin{pmatrix} 1 \\ 0 \end{pmatrix} = \begin{pmatrix} a \\ c \end{pmatrix}$ and $A'\begin{pmatrix} 1 \\ 0 \end{pmatrix} = \begin{pmatrix} a' \\ c' \end{pmatrix}$. Hence, A and A' define different mappings. Similarly, if $b \neq b'$ or $d \neq d'$, then we have $A\begin{pmatrix} 0 \\ 1 \end{pmatrix} = \begin{pmatrix} b \\ d \end{pmatrix}$ and $A'\begin{pmatrix} 0 \\ 1 \end{pmatrix} = \begin{pmatrix} b' \\ d' \end{pmatrix}$. If we had not required that R have an identity, then we would not have this property that A and A' define different functions. For example, if we had $(\mathbb{Z},+,*)$, where $a*b = 0$ for any $a,b \in \mathbb{Z}$ and where $+$ is the usual addition of integers, then $(\mathbb{Z},+,*)$ is a ring, and every $A \in M_2(\mathbb{Z})$ defines the zero mapping. Since we are using a ring R with identity, we identify each $A \in M_2(R)$ with its own function $A : R^2 \to R^2$. Hence, our first objective is obtained.

Exercise A2.45. For $\begin{pmatrix} x \\ y \end{pmatrix}, \begin{pmatrix} x' \\ y' \end{pmatrix} \in R^2$ and $A \in M_2(R)$, prove that $A\left[\begin{pmatrix} x \\ y \end{pmatrix} + \begin{pmatrix} x' \\ y' \end{pmatrix}\right] = A\begin{pmatrix} x \\ y \end{pmatrix} + A\begin{pmatrix} x' \\ y' \end{pmatrix}$.

Lemma A2.46. *For* $A = \begin{pmatrix} a & b \\ c & d \end{pmatrix}$ *and* $B = \begin{pmatrix} a' & b' \\ c' & d' \end{pmatrix}$, *then* $A \circ B = \begin{pmatrix} aa'+bc' & ab'+bd' \\ ca'+dc' & cb'+dd' \end{pmatrix}$.

Proof. For any $\begin{pmatrix} x \\ y \end{pmatrix} \in R^2$, we have

$$\begin{aligned} A \circ B\begin{pmatrix} x \\ y \end{pmatrix} &= \begin{pmatrix} a & b \\ c & d \end{pmatrix} \circ \begin{pmatrix} a' & b' \\ c' & d' \end{pmatrix}\begin{pmatrix} x \\ y \end{pmatrix} = \begin{pmatrix} a & b \\ c & d \end{pmatrix}\begin{pmatrix} a'x+b'y \\ c'x+d'y \end{pmatrix} \\ &= \begin{pmatrix} aa'x+ab'y+bc'x+bd'y \\ ca'x+cb'y+dc'x+dd'y \end{pmatrix} = \begin{pmatrix} (aa'+bc')x+(ab'+bd')y \\ (ca'+dc')x+(cb'+dd')y \end{pmatrix} \\ &= \begin{pmatrix} aa'+bc' & ab'+bd' \\ ca'+dc' & cb'+dd' \end{pmatrix}\begin{pmatrix} x \\ y \end{pmatrix} \\ &= C\begin{pmatrix} x \\ y \end{pmatrix}. \end{aligned}$$

□

Lemma A2.47. *For* $A = \begin{pmatrix} a & b \\ c & d \end{pmatrix}$ *and* $B = \begin{pmatrix} a' & b' \\ c' & d' \end{pmatrix}$, *then* $A+B = \begin{pmatrix} a+a' & b+b' \\ c+c' & d+d' \end{pmatrix}$.

Proof. For any $\begin{pmatrix} x \\ y \end{pmatrix} \in R^2$, we have

$$\begin{aligned}[A+B]\begin{pmatrix} x \\ y \end{pmatrix} &= A\begin{pmatrix} x \\ y \end{pmatrix} + B\begin{pmatrix} x \\ y \end{pmatrix} = \begin{pmatrix} ax+by \\ cx+dy \end{pmatrix} + \begin{pmatrix} a'x+b'y \\ c'x+d'y \end{pmatrix} \\ &= \begin{pmatrix} (a+a')x+(b+b')y \\ (c+c')x+(d+d')y \end{pmatrix} = D\begin{pmatrix} x \\ y \end{pmatrix},\end{aligned}$$

where $D = \begin{pmatrix} a+a' & b+b' \\ c+c' & d+d' \end{pmatrix}$. □

Theorem A2.48. *For a ring R, let* $M(R)$ *denote the set of all functions* $f: R \to R$. *Define* $\oplus$ *on* $M(R)$ *by*

$$(f \oplus g)(x) = f(x) + g(x)$$

for arbitrary $x \in R$. *With* $\circ$ *denoting composition of functions, then* $(M(R), \oplus, \circ)$ *satisfies all the properties of a ring except the left distributive law. Hence,* $(M(R), \oplus, \circ)$ *is a nearring.*

Proof. Let $\mathbf{z} \in M(R)$ be defined by $\mathbf{z}(x) = 0$ for every $x \in R$. Then

$$f \oplus \mathbf{z} = \mathbf{z} \oplus f = f \text{ for every } f \in M(R).$$

If $f \in M(R)$, define $-f : R \to R$ by $(-f)(x) = -[f(x)]$. Then

$$f \oplus (-f) = (-f) \oplus f = \mathbf{z}.$$

For $f, g, h \in M(R)$, we have

$$\begin{aligned}[(f \oplus g) \oplus h](x) &= (f \oplus g)(x) + h(x) = (f(x) + g(x)) + h(x) \\ &= f(x) + (g(x) + h(x)) = [f \oplus (g \oplus h)](x).\end{aligned}$$

Hence, $(f \oplus g) \oplus h = f \oplus (g \oplus h)$. Similarly,

$$\begin{aligned}[f \circ (g \circ h)](x) &= f[(g \circ h)(x)] = f[g(h(x))] \\ &= (f \circ g)(h(x)) = [(f \circ g) \circ h](x),\end{aligned}$$

and so $f \circ (g \circ h) = (f \circ g) \circ h$. Finally,

$$\begin{aligned}[(f \oplus g) \circ h](x) &= (f \oplus g)(h(x)) = f(h(x)) + g(h(x)) \\ &= (f \circ h)(x) + (g \circ h)(x) = [(f \circ h) \oplus (g \circ h)](x).\end{aligned}$$

Consequently, $[f \oplus g] \circ h = (f \circ h) \oplus (g \circ h)$.

To see that the left distributive law does not hold in general, we take $(R, +, \cdot)$ to be the usual ring of integers, $(\mathbb{Z}, +, \cdot)$. Let $g(x) = h(x) = x$ for every $x \in \mathbb{Z}$, and let $f(x) = x^2$. Then $[f \circ (g \oplus h)](x) = 4x^2$ and $[(f \circ g) \oplus (f \circ h)](x) = 2x^2$. Consequently, $f \circ (g \oplus h) \neq (f \circ g) \oplus (f \circ h)$. □

Corollary A2.49. *For a ring $(R, +, \cdot)$, let*

$$E(R) = \{f \in M(R) \mid f(x+y) = f(x) + f(y) \text{ for all } x, y \in R\}.$$

Then $(E(R), \oplus, \circ)$ is a subring of the nearring $(M(R), \oplus, \circ)$.

Proof. If $f, g \in E(R)$, and $x, y \in R$, then

$$\begin{aligned}(f \oplus g)(x+y) &= f(x+y) + g(x+y) = f(x) + f(y) + g(x) + g(y) \\ &= f(x) + g(x) + f(y) + g(y) = (f \oplus g)(x) + (f \oplus g)(y),\end{aligned}$$

and so $f \oplus g \in E(R)$. Similarly,

$$\begin{aligned}(f \circ g)(x+y) &= f(g(x+y)) = f(g(x) + g(y)) \\ &= f(g(x)) + f(g(y)) = (f \circ g)(x) + (f \circ g)(y),\end{aligned}$$

and so $f \circ g \in E(R)$. It remains to show that $\circ$ is right distributive over $\oplus$. For $f, g, h \in E(R)$, then

$$\begin{aligned}[(f \oplus g) \circ h)](x) &= (f \oplus g)(h(x)) = f(h(x)) + g(h(x)) \\ &= (f \circ h)(x) + (g \circ h)(x) = [(f \circ h) \oplus (g \circ h)](x).\end{aligned}$$

Hence, $f \circ (g \oplus h) = (f \circ g) \oplus (f \circ h)$. It is easy to show that $\mathbf{z} \in E(R)$ and that $-f \in E(R)$ if $f \in E(R)$. All the remaining properties for a ring are satisfied because they are valid in the nearring $M(R)$. □

Corollary A2.50. *$M_2(R)$ is a subring of the ring $E(R^2)$.*

Proof. From Lemma A2.46 and Lemma A2.47, we know that $M_2(R)$ is closed with respect to $\circ$ and $+$. From Exercise A2.45, we see that $M_2(R) \subseteq E(R^2)$. Certainly, if $A = \left(\begin{smallmatrix} a & b \\ c & d \end{smallmatrix}\right) \in M_2(R)$, then so is $-A = \left(\begin{smallmatrix} -a & -b \\ -c & -d \end{smallmatrix}\right) \in M_2(R)$. Hence, $M_2(R)$ is a subring of $E(R^2)$. □

With Corollary A2.50, we have obtained our second objective.

We have been using $\circ$ to denote the multiplication in $M_2(R)$. It is more traditional to use nothing, or perhaps $\cdot$ if it is felt that something is needed. So, for $A, B \in M_2(R)$, we will also write AB, or perhaps $A \cdot B$ in place of $A \circ B$. The reason for our use of $A \circ B$ here was that the reader could identify distinct $A, B \in M_2(R)$ with distinct functions $A : R^2 \to R^2$ and $B : R^2 \to R^2$. This required R to have an identity. Even if R does not have an identity, we could see that $(M_2(R), +, \circ)$ is a ring, but our proof is not valid as it stands. The curious reader may find it rewarding to verify that $(M_2(R), +, \circ)$ is a ring even if R does not have an identity. In so doing, he may even discover why we have chosen to require our ring R to have an identity.

There are a few results about the ring $(\mathbb{Z}, +, \cdot)$ which everybody should know, and which frequently prove to be useful.

If $m, n, k \in \mathbb{Z}$ and $n = mk$, we say that m is a *divisor* of n, and that m *divides* n. Let m and n be positive integers. The *greatest common divisor of m and n* is a positive integer d with the following two properties. (1) There are positive integers m_1 and n_1 such that $m = m_1 d$ and $n = n_1 d$, i.e., d is a common divisor of both m and n. (2) If $n = n'd'$ and $m = m'd'$ for positive integers d', m' and n', then $d \geq d'$. i.e., d is larger than any other common divisor.

A2.51. Using Theorem A2.29, we can compute the greatest common divisor d of positive integers m and n. We write $d = \gcd\{m, n\}$. Let m and n be positive integers. If $m = n$, then $m = n = d = \gcd\{m, n\}$. For $q = 1$ and $r = 0$, we have

$n = m \cdot 1 + 0$ of Theorem A2.29. If m divides n, that is, if there is an integer q such that $n = mq$, then $d = m$. For $r = 0$, we again have $n = mq + r$ of Theorem A2.29.

We next consider the case that $m \neq n$. Since one of them must be smaller than the other, there is no loss of generality in assuming that $0 < m < n$ and that m does not divide n. By Theorem A2.29, there are unique integers q_1 and r_1 such that $0 \leq r_1 < r_0$, if we define $r_0 = m$, and where

$$n = mq_1 + r_1 = r_0 q_1 + r_1. \tag{2:23}$$

If $r_1 > 0$, then there are unique integers q_2 and r_2 such that $0 \leq r_2 < r_1 < r_0 = m$ and

$$m = r_0 = r_1 q_2 + r_2. \tag{2:24}$$

Continuing, if $r_2 > 0$, then there are unique integers q_3 and r_3 such that $0 \leq r_3 < r_2 < r_1 < r_0 = m$ and

$$r_1 = r_2 q_3 + r_3.$$

If we have

$$r_{i-2} = r_{i-1} q_i + r_i$$

for integers r_{i-2}, r_{i-1}, q_i, and r_i, with $0 < r_i < r_{i-1} < r_{i-2} < \cdots < r_1 < r_0 = m$, then there are unique integers q_{i+1} and r_{i+1} such that $0 \leq r_{i+1} < r_i < \cdots < r_0 = m$ and

$$r_{i-1} = r_i q_{i+1} + r_{i+1}. \tag{2:25}$$

Let s be the smallest positive integer for which $r_s = 0$. From

$$r_{s-2} = r_{s-1} q_s + r_s = r_{s-1} q_s,$$

we have, that r_{s-1} is a divisor of r_{s-2}. Let $d = r_{s-1}$.

From $r_{s-3} = r_{s-2} q_{s-1} + r_{s-1}$ and the fact that $d = r_{s-1}$ is a divisor of r_{s-2}, we conclude that d is a divisor of both r_{s-2} and r_{s-1}. From knowing that d is a divisor of both r_{i+1} and r_i, we infer from (2:25) that d is a divisor of both r_i

and r_{i-1}. By induction, we have that d divides each r_i, $0 \le i \le s-1$. From $r_0 = m$ and (2:23), we have that d divides both m and n. So, our d is a common divisor of both m and n.

Suppose now that d' divides both m and n and that $d' > 0$. From (2:23), we infer that d' divides both $m = r_0$ and r_1. From (2:24) and the fact that d' divides both r_0 and r_1, we infer that d' divides r_2. Assuming that d' divides both r_{i-1} and r_i, we infer from (2:25) that d' divides r_{i+1}. By induction, d' divides $r_{s-1} = d$. Consequently, $d = \gcd\{m,n\}$.

Theorem A2.52. *Let m and n be positive integers. Then there are integers a and b such that*

$$d = am + bn,$$

where $d = \gcd\{m,n\}$.

Proof. Suppose one divides the other, e.g., suppose m divides n. Then $d = m$, and $d = 1 \cdot m + 0 \cdot n$.

Without loss of generality, we assume that $0 < m < n$. From (2:23), we have $r_1 = (-q_1)m + 1 \cdot n$. From (2:24), we have $r_2 = (-q_2)r_1 + 1 \cdot r_0 = (-q_2)r_1 + m$. Hence, $r_2 = (-q_2)[-q_1 m + n] + m = (q_1 q_2 + 1)m + (-q_2)n$. Assume there are integers a_j and b_j, for $1 \le j \le i$, such that $r_j = a_j m + b_j n$. Then from (2:25), we have

$$\begin{aligned} r_{i+1} &= r_{i-1} - r_i q_{i+1} \\ &= (a_{i-1}m + b_{i-1}n) - (a_i m + b_i n)q_{i+1} \\ &= (a_{i-1} - a_i q_{i+1})m + (b_{i-1} - b_i q_{i+1})n \\ &= a_{i+1}m + b_{i+1}n, \end{aligned}$$

for some integers a_{i+1} and b_{i+1}. Hence, $d = r_{s-1} = a_{s-1}m + b_{s-1}n$. Let $a = a_{s-1}$ and $b = b_{s-1}$. So, we have $d = am + bn$, as desired. □

If $\gcd\{m,n\} = 1$, then we say that m and n are *relatively prime*.

Corollary A2.53. *Let m and n be positive integers. Then m and n are relatively prime if and only if there are integers a and b such that $am+bn=1$.*

Proof. If m and n are relatively prime, then $\gcd\{m,n\}=1=am+bn$, for some $a,b\in\mathbb{Z}$, by Theorem A2.52. Conversely, suppose $a,b\in\mathbb{Z}$ and $am+bn=1$. If $1<d=\gcd\{m,n\}$, then $m=dm_1$ and $n=dn_1$ for some integers m_1 and n_1. Hence, $1=adm_1+bdn_1=d(am_1+bn_1)$. This forces us into the unacceptable position of saying that $d>1$ and that d divides 1. So, we are forced to accept that the $\gcd\{m,n\}=d=1$. □

A3 Elements of semigroup and monoid theory

We have assumed that the reader is comfortable while working in the fields $(\mathbb{Q},+,\cdot)$ and $(\mathbb{R},+,\cdot)$, and the rings $(\mathbb{Z},+,\cdot)$, $(\mathbb{R}[x],+,\cdot)$, and $(\mathscr{C}(\mathbb{R}),\oplus,\bullet)$. Hopefully, the transition to the fields $(\mathbb{Z}_p,+,\cdot)$, p a prime, and the rings $(\mathbb{Z}_n,+,\cdot)$ were reasonably comfortable. Perhaps the rings $(F[x]/(a),+,\cdot)$, $a\in F[x]$, F a field, and even the fields $(F[x]/(a),+,\cdot)$, where $a\in F[x]$ is irreducible, are still rather strange. We now enter an area of algebra which may seem very strange and awkward at first. Those who show determination often find beauties available only to relatively few.

We will be primarily concerned with a nonempty set S and one associative binary operation. Being a rather primitive structure, many algebraists believe such a place is the natural starting place to commence ones study of abstract algebra. We have tried to make the transition more natural.

Are you ready? Let us begin.

Definition A3.1. Let S be a nonempty set and let $*$ be an associative binary operation on the set S. Then $(S,*)$ is a *semigroup*.

Semigroups are found nearly everywhere, if you look for them. That's one of the main problems about studying them. There are so many and they are so diverse in their properties, that it is very difficult to develop a coherent theory. Those who specialize in semigroups need the ability to handle diversity, and the ability to organize chaos.

Examples A3.2. (a) Let $S = \mathscr{C}(\mathbb{R})$ and let $* = \circ$, the composition of functions. That is $f \circ g = h$ where $h(x) = f(g(x))$. Certainly $f \circ (g \circ h) = (f \circ g) \circ h$. The function $1_{\mathbb{R}} : \mathbb{R} \to \mathbb{R}$ defined by $1_{\mathbb{R}}(x) = x$ for $x \in \mathbb{R}$ is the identity for $\circ$.

(b) Let X be any set, and let $S = \mathscr{P}(X)$. With $* = \cup$, we have $(A \cup B) \cup C = A \cup (B \cup C)$, and so $\cup$ is associative. Since $\varnothing \in \mathscr{P}(X)$ and since $\varnothing \cup A = A \cup \varnothing = A$ for arbitrary $A \in \mathscr{P}(X)$, we have the identity $\varnothing \in \mathscr{P}(X)$ with respect to $\cup$.

(c) Let $S = \{A \subseteq \mathbb{Z} \mid |A| < \infty\}$. That is, S is the family of all finite subsets of $\mathbb{Z}$. If $* = \cap$, then $A \cap (B \cap C) = (A \cap B) \cap C$ for all $A, B, C \in S$. If A and B are finite subsets of $\mathbb{Z}$, then so is $A \cap B$. Hence, $(S, \cap)$ is a semigroup without identity.

(d) Let $\mathbb{N} = \{1, 2, 3, \ldots\}$, the natural numbers, and let $* = +$, the ordinary addition of natural numbers. Since $a + (b + c) = (a + b) + c$, we certainly have that $(\mathbb{N}, +)$ is a semigroup without an identity. (If we had taken $\mathbb{N}_0 = \mathbb{N} \cup \{0\}$, then $(\mathbb{N}_0, +)$ would be a semigroup with identity. This trick can be extended to any semigroup $(S, \cdot)$, and we do so in below.)

(e) For any ring $(R, +, \cdot)$, both $(R, +)$ and $(R, \cdot)$ are semigroups, and if R is a field, then $(R^*, \cdot)$ is a semigroup where $R^* = R \setminus \{0\}$.

(f) If $2\mathbb{Z} = \{2n \mid n \in \mathbb{Z}\}$, the set of even integers, and if $* = \cdot$, the ordinary multiplication of integers, then $(2\mathbb{Z}, \cdot)$ is a semigroup without an identity.

(g) Take any nonempty set S. Define $*$ on S by $a * b = b$ for any $a, b \in S$. Then $a * (b * c) = b * c = c = (a * b) * c$, so $*$ is associative, and $(S, *)$ is a semigroup.

(h) Take any set S with $|S| \geq 2$. Take any $A \subset S$, $A \neq \varnothing$. Choose any $z \in S \setminus A$ and fix it. Define $*_A$ on S by

$$a *_A b = \begin{cases} z, & \text{if } a \notin A; \\ b, & \text{if } a \in A. \end{cases}$$

With a little thought and patience, you should be able to verify that $(S, *_A)$ is a semigroup.

(i) Let X be any nonempty set, and let $M(X)$ be the set of all functions $f : X \to X$. If $\circ$ is composition of functions, then $(M(X), \circ)$ is a semigroup with identity 1_X.

Many of the examples in Examples A3.2 have an identity. Such semigroups have a special name.

Definitions A3.3. Let $(S, *)$ be a semigroup. If there is an element $e \in S$ such that $e * x = x * e = x$ for every $x \in S$, then $(S, *, e)$ is a *monoid*. (Of course, such an $e \in S$ is the *identity* of the monoid $(S, *, e)$. Some say simply that $(S, *, e)$ is a *semigroup with identity* e.

If a semigroup $(S, *)$ does not have an identity, then we can add one to it.

Theorem A3.4. *Any semigroup $(S, *)$ can be embedded in a monoid. In particular, choose any element $e \notin S$ and let $T = S \cup \{e\}$. Define $\cdot$ on T by*

$$x \cdot y = \begin{cases} x * y, & \textit{if } x, y \in S; \\ y, & \textit{if } x = e; \\ x, & \textit{if } y = e. \end{cases}$$

Then $(T, \cdot, e)$ is a monoid.

Proof. Sometimes it is very tedious to prove associativity. To prove that $x \cdot (y \cdot z) = (x \cdot y) \cdot z$ for every $x, y, z \in T$, we must consider eight cases. For $x, y, z \in T$, one has $x \in S$ or $x = e$, and $y \in S$ or $y = e$, and finally $z \in S$ or $z = e$. For simplicity, we say in this proof only that (x_1, x_2, x_3) is $[a_1; a_2; a_3]$, where $a_i \in \{0, 1\}$, where $a_i = 0$ means $x_i = e$. If $a_i = 1$, then $x_i \in S$. So we now consider the eight cases.

1. (x, y, z) is $[1; 1; 1]$. Then $x \cdot (y \cdot z) = x * (y * z) = (x * y) * z = (x \cdot y) \cdot z$.
2. (x, y, z) is $[1; 1; 0]$. Then $x \cdot (y \cdot z) = x \cdot (y \cdot e) = (x \cdot y) = (x \cdot y) \cdot e = (x \cdot y) \cdot z$.

3. (x,y,z) is $[1;0;1]$. Then $x\cdot(y\cdot z)=x\cdot(e\cdot z)=x\cdot z=(x\cdot e)\cdot z=(x\cdot y)\cdot z$.
4. (x,y,z) is $[1;0;0]$. Then $x\cdot(y\cdot z)=x\cdot(e\cdot e)=x\cdot e=x=(x\cdot e)\cdot e=(x\cdot y)\cdot z$.
5. (x,y,z) is $[0;1;1]$. Then $x\cdot(y\cdot z)=e\cdot(y\cdot z)=y\cdot z=(e\cdot y)\cdot z=(x\cdot y)\cdot z$.
6. (x,y,z) is $[0;1;0]$. Then $x\cdot(y\cdot z)=e\cdot(y\cdot e)=y=(e\cdot y)\cdot e=(x\cdot y)\cdot z$.
7. (x,y,z) is $[0;0;1]$. Then $x\cdot(y\cdot z)=e\cdot(e\cdot z)=z=(e\cdot e)\cdot z=(x\cdot y)\cdot z$.
8. (x,y,z) is $[0;0;0]$. Then $x\cdot(y\cdot z)=e\cdot(e\cdot e)=e=(e\cdot e)\cdot e=(x\cdot y)\cdot z$.

In all cases, $x\cdot(y\cdot z)=(x\cdot y)\cdot z$, so $(T,\cdot,e)$ is a semigroup. Since $e\cdot x=x\cdot e=x$ for every $x\in T$, $(T,\cdot,e)$ is a monoid. □

Note. It may be that $(S,*)$ already had an identity ι. In $(T,\cdot)$, ι is no longer the identity, since $\iota\cdot e=\iota\neq e$. If ι were an identity in T, then $\iota\cdot e=e$.

If there is an element z in a semigroup S such that $az=za=z$ for all $a\in S$, then there is only one such elements. For if z' is another such element, then $z=zz'=z'$. Such an element is called the *zero* of the semigroup. If a semigroup $(S,\cdot)$ does not have a zero, then we can tack one onto it.

Theorem A3.5. *Let $(S,\cdot)$ be a semigroup, and suppose $z\notin S$. Then on $U=S\cup\{z\}$ define*

$$a*b=\begin{cases} z, & \text{if } z\in\{a,b\};\\ a\cdot b, & \text{otherwise.}\end{cases}$$

Then $(U,*)$ is a semigroup with zero z.

Proof. Certainly $z*a=a*z=z$ for all $a\in S$. So, we need only to show that $*$ is associative. Here, associativity is quite easy to see. We need to show that $a*(b*c)=(a*b)*c$ for all $a,b,c\in T$. If $a,b,c\in S$, then

$$a*(b*c)=a\cdot(b\cdot c)=(a\cdot b)\cdot c=(a*b)*c.$$

Now we will let $z\in\{a,b,c\}$. If $z=a$, then

$$a*(b*c)=z \text{ and } (a*b)*c=z*c=z.$$

If $z = b$, then

$$(a * b) * c = z * c = z \text{ and } a * (b * c) = a * z = z.$$

Finally, if $z = c$, then

$$(a * b) * c = z \text{ and } a * (b * c) = a * z = z.$$

In all cases we have $a * (b * c) = (a * b) * c$. Hence $(U, *)$ is a semigroup with zero z. □

Corollary A3.6. *Let $(S, *)$ be a semigroup. First apply (A3.4) to get the monoid $(T, \cdot, e)$. Then apply (A3.5) to get the semigroup $(U, \bullet, z)$ with zero z. Then $(U, \bullet, e, z)$ is a monoid with identity e and zero z.*

Proof. Certainly $(U, \bullet, z)$ is a semigroup with zero z, so we only need to show that e is the identity. Take $a \in S$. Then by A3.4, $e \bullet a = a \bullet e = a$. Certainly $e \bullet e = e$. Finally, $e \bullet z = z = z \bullet e$ by Theorem A3.5. Hence, e is the identity for $(U, \bullet)$. This completes the proof of Corollary A3.6. □

We can start with a semigroup $(S, \cdot)$. If we don't like its size, then we can make it bigger using A3.4 and/or Theorem A3.5. If the process is too slow, we can add as many copies of $(S, \cdot)$ to itself as we wish.

Theorem A3.7. *Let $(S, \cdot)$ be any semigroup. Let I be any nonempty index set so that if $a \in S$ and $i \in I$, then $(a, i) \notin S$. Let $M = S \times I$. Define $*$ on M by*

$$(a, i) * (b, j) = (a \cdot b, j).$$

Then $(M, *)$ is a semigroup. If $(S, \cdot, e)$ is a monoid, then for each $i \in I$, (e, i) is a left identity of M, that is, $(e, i) * (b, j) = (b, j)$ for any $(b, j) \in M$.

Proof. Take (a, i), (b, j), $(c, k) \in M$. Then

$$(a, i) * [(b, j) * (c, k)] = (a, i) * (bc, k) = (abc, k)$$

and

$$[(a,i)*(b,j)]*(c,k) = (ab,j)*(c,k) = (abc,k).$$

If $e \in S$ is an identity, and if $i \in I$, then $(e,i)*(b,j) = (eb,j) = (b,j)$ for any $(b,j) \in M$. □

If $(T,*)$ and $(S,\cdot)$ are semigroups, $T \subseteq S$, and $a*b = a\cdot b$ for all $a,b \in T$, then $(T,*)$ is a *subsemigroup* of $(S,\cdot)$. We also say that T is *closed with respect to* $\cdot$.

Corollary A3.8. *Consider the situation as described in Theorem A3.7. For any $i \in I$, let $S_i = \{(a,i) \mid a \in S\}$. Then $(S_i,*)$ is a subsemigroup of M.*

Definitions A3.9. Suppose $(S,\cdot)$ and $(T,*)$ are semigroups. We call a function $\phi : S \to T$ a *semigroup homomorphism* if $\phi(a\cdot b) = \phi(a)*\phi(b)$ for every $a,b \in S$. The word "homomorphism" can be replaced by epimorphism, monomorphism, endomorphism, or automorphism, if the appropriate conditions, as described in Definitions A2.20, are valid. For example, a semigroup homomorphism $\phi : S \to S$ is a *semigroup endomorphism*. When context is clear, we will not add the word "semigroup" or "ring," etc., before a morphism word.

Corollary A3.10. *Consider the situation as described in Corollary A3.8. Define $\phi_i : S \to M$ by $\phi_i(a) = (a,i)$. Then ϕ_i is a semigroup monomorphism. Define $\lambda_i : S \to S_i$ by $\lambda_i(a) = (a,i)$. Then λ_i is a semigroup isomorphism.*

Theorem A3.11. *Let $\phi : S \to T$ be a semigroup homomorphism. Then $\phi(S)$ is a subsemigroup of T.*

Proof. Take $\phi(a),\phi(b) \in \phi(S)$. Then $\phi(a)*\phi(b) = \phi(a\cdot b) \in \phi(S)$. Hence, $\phi(S)$ is closed with respect to $*$ for T. Associativity for $*$ in $\phi(S)$ follows from associativity of $*$ in T. □

Theorem A3.12. *For a semigroup $(S,\cdot)$, let*

$$E(S) = \{\phi : S \to S \mid \phi \text{ is a semigroup endomorphism }\}.$$

Then $(E(S),\circ)$ is a monoid.

Proof. For $f,g,h \in E(S)$, we have $f \circ (g \circ h) = (f \circ g) \circ h$ since f, g, and h are functions. The function 1_S, where $1_S(a) = a$ for all $a \in S$, has the property that $1_S \circ f = f \circ 1_S = f$ for every $f \in E(S)$. Now $1_S \in E(S)$, and $f \circ g \in E(G)$, for if $a,b \in S$, then

$$1_S(ab) = ab = 1_S(a)1_S(b),$$

and

$$(f \circ g)(ab) = f(g(ab)) = f(g(a)g(b)) = f(g(a))f(g(b)) = (f \circ g)(a)(f \circ g)(b).$$

Hence $(E(S), \circ, 1_S)$ is a monoid. □

Theorem A3.13. *Let $\phi : S \to S$ be a semigroup automorphism. Then so is $\phi^{-1} : S \to S$ a semigroup automorphism.*

Proof. Take $x, y \in S$. Then there are unique $a,b \in S$ such that $\phi(a) = x$ and $\phi(b) = y$. Hence, $\phi^{-1}(x) = a$ and $\phi^{-1}(y) = b$. Since $\phi(ab) = \phi(a)\phi(b) = xy$, we have $\phi^{-1}(xy) = ab = \phi^{-1}(x)\phi^{-1}(y)$. We can now be assured that ϕ^{-1} is an automorphism. □

Theorem A3.14. *Let $(M, \cdot, e)$ be a monoid. Define*

$$\mathscr{U}(M) = \{a \in M \mid a^{-1} \in M\}.$$

That is, $\mathscr{U}(M)$ consists of all elements $a \in M$ which has an element $a' \in M$ such that $aa' = a'a = e$. We denote a' by a^{-1}. Then $(\mathscr{U}(M), \cdot, e)$ is a monoid and a subsemigroups of $(M, \cdot)$. That is, $(\mathscr{U}(M), \cdot, e)$ is a submonoid of $(M, \cdot, e)$.

Proof. For $a \in M$, suppose $as = sa = e$ and $at = ta = e$ with $s,t \in M$. Then from $at = e$, we have $s = se = s(at) = (sa)t = et = t$. So, if there is an element $a' \in M$ such that $aa' = a'a = e$, then there is only one. That is why we can denote a' by a^{-1}.

Take $a, b \in \mathscr{U}(M)$. Hence, $a^{-1}, b^{-1} \in M$, and so

$$(ab)(b^{-1}a^{-1}) = a[b(b^{-1}a^{-1})] = a[(bb^{-1})a] = a(ea^{-1}) = aa^{-1} = e.$$

Also,

$$(b^{-1}a^{-1})(ab) = b^{-1}[a^{-1}(ab)] = b^{-1}[(a^{-1}a)b] = b^{-1}(eb) = e.$$

So, we conclude that $(ab)^{-1} = b^{-1}a^{-1} \in M$. This assures us that $ab \in \mathscr{U}(M)$. Hence, $\mathscr{U}(M)$ is closed with respect to $\cdot$, and since associativity comes from M, we have that $(\mathscr{U}(M), \cdot)$ is a subsemigroup of $(M, \cdot)$. Since $e \cdot e = e$, we have $e \in \mathscr{U}(M)$. Hence, $(\mathscr{U}(M), \cdot, e)$ is a monoid. □

Corollary A3.15. *Let $(M, \cdot, e)$ be a monoid.*

(1) *Then $e \in \mathscr{U}(M)$.*
(2) *Also, $e^{-1} = e$.*
(3) *If $a \in \mathscr{U}(M)$, then $a^{-1} \in \mathscr{U}(M)$.*
(4) *If $a \in \mathscr{U}(M)$, then $(a^{-1})^{-1} = a$.*
(5) *If $a, b \in \mathscr{U}(M)$, then $ab \in \mathscr{U}(M)$ and $(ab)^{-1} = b^{-1}a^{-1}$.*

Proof. Since $e \cdot e = e$, we have $e^{-1} = e$ and $e \in \mathscr{U}(M)$. This gives us (1) and (2). Does a^{-1} have an element $a^* \in M$ so that $a^{-1}a^* = a^*a^{-1} = e$. Why yes! Let $a^* = a$. Certainly $a^{-1}a = aa^{-1} = e$. Consequently $a^{-1} \in \mathscr{U}(M)$ and $(a^{-1})^{-1} = a$. Also $e \cdot e = e$ implies $e^{-1} \in \mathscr{U}(M)$. Hence (3) and (4). The proof of Theorem A3.14 gives us (5). □

Definition A3.16. For a monoid $(M, \cdot, e)$, define $\mathscr{U}(M) = \{a \in M \mid a^{-1} \in M\}$. Then $(\mathscr{U}(M), \cdot, e)$ is the *group of units* of $(M, \cdot, e)$.

Why the phrase "group of units?" Well, a *unit* is an element that has a multiplicative inverse, and monoids in which every element has an inverse is a *group*. The concept of a group is so important that we give it a number.

Definition A3.17. A *group* is a monoid $(G, \cdot, e)$ for which $\mathscr{U}(G) = G$. That is, if $a \in G$, then $a^{-1} \in G$.

Theorem A3.18. *For every monoid $(M, \cdot, e)$, $(\mathscr{U}(M), \cdot, e)$ is a group.*

Proof. This is immediate from Theorem A3.14 and Definition A3.17. □

Corollary A3.19. *For a semigroup $(S,\cdot)$, let*

$$A(S) = \{\phi \in E(S) \mid \phi \text{ is a semigroup automorphism}\}.$$

Then $(A(S),\circ)$ is a group.

Proof. This follows from Theorem A3.13 and Theorem A3.18. □

We would strongly recommend that one do the following two exercises very carefully.

Exercise A3.20. Let $X = \{a,b\}$. List all four functions $f : X \to X$ of the monoid $M(X)$ as described in (i) of Examples A3.2 Find the elements of the group of units $\mathscr{U}(M(X))$. Notice that $\mathscr{U}(M(X))$ and $M(X) \setminus \mathscr{U}(M(X))$ are each subsemigroups of $M(S)$.

Exercise A3.21. Here, do the same thing for $Y = \{a,b,c\}$ as you did for X in Exercise A3.20. That is, list all 27 functions $f : Y \to Y$ of the monoid $(M(Y),\circ,1_Y)$ as described in (i) of Examples A3.2. Find the six elements of the group of units $\mathscr{U}(M(Y))$. Notice that $\mathscr{U}(M(Y))$ and $M(Y) \setminus \mathscr{U}(M(Y))$ are each subsemigroups of $M(Y)$. (You may want to save your results for future reference.)

Exercise A3.22. Can you generalize Exercise A3.20 and Exercise A3.21? Consider $S = \{a_1,\ldots,a_n\}$ with $|S| = n$. How many functions $f : S \to S$ are there in $M(S)$? How many elements does the group of units $\mathscr{U}(M(S))$ have? Are $\mathscr{U}(M(S))$ and $M(S) \setminus \mathscr{U}(M(S))$ each subsemigroups of $M(S)$?

For a semigroup $(S,\cdot)$, define the binary operation on $\mathscr{P}(S)$ by

$$AB = \{ab \mid a \in A,\ b \in B\}$$

for $A,B \in \mathscr{P}(S)$.

Theorem A3.23. *If $(S,\cdot)$ is a semigroup, then so is $(\mathscr{P}(S),\cdot)$. In addition, $(\mathscr{P}(S),\cdot)$ has a zero, namely $\varnothing$. If $(M,\cdot,e)$ is a monoid, then so is $(\mathscr{P}(M),\cdot,\{e\})$.*

Proof. If $A, B, C \in \mathscr{P}(S)$, then

$$A(BC) = \{a(bc) \mid a \in A,\ b \in B\ c \in C\} = \{(ab)c \mid a \in A,\ b \in B\ c \in C\} = (AB)C.$$

Since $\varnothing A = \{xa \mid x \in \varnothing,\ a \in A\}$, and since there is no $x \in \varnothing$, we have $\varnothing A = \varnothing$. Similarly, $A\varnothing = \{ax \mid a \in A,\ x \in \varnothing\} = \varnothing$. This is exactly what is required for $\varnothing$ to be the zero of $(\mathscr{P}(S), \cdot)$. If e is the identity and $A \in \mathscr{P}(M)$, then $\{e\}A = \{ea \mid a \in A\} = A = \{ae \mid a \in A\} = A\{e\}$. Hence, $\{e\}$ is the identity. □

Exercise A3.24. For a semigroup $(S, \cdot)$, define $f : S \to \mathscr{P}(S)$ by $f(a) = \{a\}$. Show that f is a semigroup monomorphism.

There are various ways to extract from a monoid various subsemigroups. Here, we introduce a very important one.

Definition A3.25. Let $(M, \cdot, e)$ be a monoid. The *center* of M is defined by

$$Z(M) = \{c \in M \mid cx = xc \text{ for every } x \in M\}.$$

So, the center of M consists of all those elements of M which commute with each element $x \in M$. In particular, $e \in Z(M)$, so $Z(M) \neq \varnothing$.

Theorem A3.26. *Let $(M, \cdot, e)$ be a monoid. Then the center $Z(M)$ is a submonoid.*

Proof. If $c, c' \in Z(M)$ and $x \in M$, then

$$(cc')x = c(c'x) = c(xc') = (cx)c' = (xc)c' = x(cc').$$

Hence, $cc' \in Z(M)$. (The associativity of $\cdot$ carries over to $Z(M)$.) Certainly $e \in Z(M)$ since $ex = xe = x$ for every $x \in M$. □

Each element $g \in \mathscr{U}(M)$, for any monoid $(M, \cdot, e)$, defines a very important function. Let $(M, \cdot, e)$ be a monoid, and let $g \in \mathscr{U}(M)$. Define $\phi_g : M \to M$ by $\phi_g(x) = gxg^{-1}$. The following theorem is really quite easy to prove, and is a very important result.

Theorem A3.27. *For a monoid $(M,\cdot,e)$ and $g \in \mathscr{U}(M)$, the map ϕ_g is a monoid automorphism.*

Proof. Take $x,y \in M$. Then

$$\phi_g(xy) = g(xy)g^{-1} = g(xey)g^{-1} = g(xg^{-1}gy)g^{-1} = (gxg^{-1})(gyg^{-1}) = \phi_g(x)\phi_g(y).$$

Hence, ϕ_g is an endomorphism.

Suppose $\phi_g(x) = \phi_g(y)$. Then $gxg^{-1} = gyg^{-1}$. Multiplying by g^{-1} on the left of the last equation, and by g on the right gives us $x = y$. Hence, ϕ_g is a monomorphism.

Finally, let $t \in M$. Then $g^{-1}tg \in M$, and

$$\phi_g(g^{-1}tg) = g(g^{-1}tg)g^{-1} = (gg^{-1})t(gg^{-1}) = ete = t.$$

This makes ϕ_g an epimorphism, and altogether we get that ϕ_g is an automorphism. □

In Corollary A3.19, we learned that the semigroup of automorphisms of a semigroup $(S,\cdot)$ give us a group $(A(S),\circ)$. Then

$$\mathrm{Inn}M = \{\phi_g \mid g \in \mathscr{U}(M)\}$$

is the collection of the ϕ_g described above, if $(M,\cdot,e)$ is a monoid. Such a ϕ_g is an *inner automorphism*. These inner automorphisms form a *subgroup* of the group $A(M)$. That is, $\mathrm{Inn}M \subseteq A(M)$, and $(\mathrm{Inn}M,\circ)$ is a group.

Theorem A3.28. *Let $(M,\cdot,e)$ be a monoid. Then* Inn*M is a subgroup of the group $A(M)$. Furthermore, if $\phi_g,\phi_h \in \mathrm{Inn}M$, then $\phi_g \circ \phi_h = \phi_{gh}$, and $\phi_g^{-1} = \phi_{g^{-1}}$.*

Proof. We already have, from Theorem A3.26, that $\mathrm{Inn}M \subseteq A(M)$. Take $\phi_g,\phi_h \in \mathrm{Inn}M$. Then

$$\phi_g \circ \phi_h(x) = \phi_g(hxh^{-1}) = g(hxh^{-1})g^{-1} = (gh)x(h^{-1}g^{-1}) = (gh)x(gh)^{-1}$$

by Corollary A3.15. So $\phi_g \circ \phi_h = \phi_{gh} \in \text{Inn}M$ and $\text{Inn}M$ is closed with respect to $\circ$. If $h = g^{-1}$, then $\phi_g \circ \phi_h = \phi_e$, and $\phi_e(x) = exe^{-1} = x$. This means that $\phi_e = 1_M$, and so $\phi_g^{-1} = \phi_{g^{-1}} \in \text{Inn}M$. Hence, $(\text{Inn}M, \circ)$ is a subgroup of $(A(G), \circ)$. $\square$

This puts us in a position to describe an important homomorphism.

Theorem A3.29. *Let $(M, \cdot, e)$ be a monoid. If $T : \mathscr{U}(M) \to \text{Inn}(M)$ is given by $T(g) = \phi_g$ for all $g \in \mathscr{U}(M)$, then T is a group epimorphism.*

Proof. Take $g, h \in \mathscr{U}(M)$. Then $T(gh) = \phi_{gh} = \phi_g \circ \phi_h = T(g) \circ T(h)$. Here, we have used Theorem A3.28 to get $\phi_{gh} = \phi_g \circ \phi_h$. This is all we need to conclude that T is a group homomorphism. A $\phi_g \in \text{Inn}(M)$ comes from a $g \in \mathscr{U}(M)$, and with this g, we have $T(g) = \phi_g$. Hence, T is surjective, making T an epimorphism. $\square$

Could it be that $g \neq h$, but yet $\phi_g = \phi_h$?

Theorem A3.30. *Let $g, h \in \mathscr{U}(M)$ for a monoid $(M, \cdot, e)$. Then $\phi_g = \phi_h$ if and only if $g^{-1}h \in Z(M)$.*

Proof. Suppose $\phi_g = \phi_h$. Then, for any $x \in M$, we have $gxg^{-1} = hxh^{-1}$. Hence, $gxg^{-1}h = hx$, and so $x(g^{-1}h) = (g^{-1}h)x$ for every $x \in M$. This is exactly what is needed to put $g^{-1}h \in Z(M)$. Conversely, if $g^{-1}h \in Z(M)$, then for every $x \in M$, we have $(g^{-1}h)x = x(g^{-1}h)$. Working from the reverse of the above argument, we have $gxg^{-1} = hxh^{-1}$ for every $x \in M$. This is exactly what is meant by saying that $\phi_g = \phi_h$. $\square$

Exercise A3.31. Return to your work from Exercise A3.21. Hopefully, you found exactly six elements in $\mathscr{U}(M(Y))$ For each $g \in \mathscr{U}(M(Y))$, one gets $\phi_g \in A(M(Y))$. That is ϕ_g actually defines an automorphism of $M(Y)$. How many distinct automorphisms do you get from the various ϕ_g? What does Theorem A3.28 say about the relationship between $\mathscr{U}(M(Y))$ and $\text{Inn}M(Y)$?

We seem to be focusing more and more upon groups. That's because groups are at the foundation of nearly everything we do within these notes. Actually, group theory is an extremely rich theory. There are numerous applications of group theory throughout the other branches of mathematics, and they have significant applications outside of mathematics. Our knowledge about groups is expanding rapidly, but this only adds to their fascination. The requirements to be a group are really quite modest, but these modest requirements allow for some truly beautiful consequences. It is hoped that the reader will begin to appreciate this active branch of mathematics.

A4 Elements of group theory

We only plan to present here enough group theory for the reader to begin to see the potential of the theory, both for applications and for its own intrinsic beauty. So, we will present little more than what is needed to develop the main themes of these notes.

Groups are not as abundant as semigroups, or even monoids, but they seem to arise all around us. Certainly, every group is a semigroup, and even a monoid, and, as we have already seen, every monoid has a group of units. So, groups are monoids whose groups of units is the entire monoid. See Definition A3.16. We have also seen that there are plenty of monoids. Recall (i) of Examples A3.2. And then each semigroup $(S, \cdot)$ has its own group of automorphisms $A(S)$. See Corollary A3.19. A monoid $(M, \cdot, e)$ allows us to select from $A(M)$ the group of inner automorphisms Inn$M$. Recall Theorem A3.27. And then, from a monoid $(M, \cdot, e)$, we also get the group homomorphism $T : \mathscr{U}(M) \to \text{Inn}M$ of Theorem A3.28. We plan to examine all these ideas afresh within a group theory context, but we will feel free to use our knowledge about semigroups and monoids. Remember always that a group is a semigroup and even a monoid, so the material of §A3 is valid for groups.

For focus, we now state again the requirements for a group.

Definition A4.1. A *group* is an ordered pair $(G,\cdot)$, where G is a nonempty set and $\cdot$ is a binary operation defined on G that has the following properties.

(1) The binary operation $\cdot$ is associative.

(2) There is an identity e with respect to the binary operation $\cdot$.

(3) Each $a \in G$ has an inverse $a^{-1} \in G$.

Associativity of $\cdot$ means that $a\cdot(b\cdot c) = (a\cdot b)\cdot c$ for every $a,b,c \in G$. This can be very difficult to prove sometimes, but often, a binary operation will get its associativity from the associativity of another binary operation.

An identity e with respect to a binary operation $\cdot$ is an element that does not change any element. That is, $e\cdot a = a\cdot e = a$ for every $a \in G$. If we write $+$ for our binary operation, then instead of writing $e+a = a+e = a$ for every $a \in G$, we write $0+a = a+0 = a$ or perhaps $z+a = a+z = a$ for every $a \in G$. In this case, our identity 0 or z is often called "zero." Sometimes we might write 1 for e for a binary operation $\cdot$, and sometimes we may refer to $\cdot$ as a "multiplication," and to $+$ as an "addition." Also, we will often write ab for $a\cdot b$, just to make life easier.

An inverse takes an element back to the identity. For $a \in G$, the inverse $a^{-1} \in G$ must satisfy $a\cdot a^{-1} = a^{-1}\cdot a = e$. Of course, if we use $+$, then we use $-a$ for the inverse of $a \in G$. That is, $a+(-a) = (-a)+a = 0$. We will feel free to say "minus a" for $-a$ instead of "the inverse of a" or "the additive inverse of a." Also, we will free to write $a-b$ for $a+(-b)$.

If a set G has an associative binary operation $\cdot$, then $(G,\cdot)$ is a semigroup. If $\cdot$ also has an identity e, then $(G,\cdot,e)$ is a monoid. So, we need an inverse a^{-1} for each element $a \in G$ to get a group from a monoid.

Examples A4.2. (1) If $(F,+,\cdot)$ is a field, then $(F,+)$ and $(F^*,\cdot)$ are groups, where $F^* = F\setminus\{0\}$.

(2) If $(\mathbb{R},+,\cdot)$ is the field of real numbers, then $\mathbb{R}_+ = \{r \in \mathbb{R} \mid r > 0\}$ gives us a group $(\mathbb{R}_+,\cdot)$, where $\cdot$ is the usual multiplication of real numbers. One charming thing about group theory is that the groups $(\mathbb{R}_+,\cdot)$ and $(\mathbb{R},+)$ are essentially the same group. That is, they are isomorphic as groups.

(3) If $(F,+,\cdot)$ is a field, and if n is any positive integer, then all the solutions to the equation $x^n = 1$ in F form a group C_n with multiplication $\cdot$. That is,

$$C_n = \{a \in F \mid a^n = 1\}.$$

(4) For any field F, let

$$\mathrm{Aff}(F) = \{[b,m] : F \to F \mid (b,m) \in F \times F^*,\ [b,m](x) = b + mx\}.$$

Then $(\mathrm{Aff}(F),\circ)$ is a group.

(5) Let

$$GL(2,F) = \left\{ \begin{pmatrix} a & b \\ c & d \end{pmatrix} \middle| a,b,c,d \in F,\ ad - bc \neq 0 \right\},$$

where $(F,+,\cdot)$ is a field. Then $GL(2,F)$ is the set of all two by two matrices A over F such that the determinant of A is not 0. Now $(GL(2,F),\cdot)$ is a group which has numerous interesting subgroups.

(6) Let

$$T = \left\{ \begin{pmatrix} \cos\alpha & \sin\alpha \\ -\sin\alpha & \cos\alpha \end{pmatrix} \middle| \alpha \in \mathbb{R} \right\}.$$

Then T is a subgroup of $GL(2,\mathbb{R})$.

(7) Let

$$K = \left\{ \begin{pmatrix} a & b \\ c & d \end{pmatrix} \middle| ad - bc = 1 \right\}.$$

Then K is a subgroup of $GL(2,F)$.

(8) Let

$$M = T \cup \left\{ \begin{pmatrix} \cos\alpha & \sin\alpha \\ \sin\alpha & -\cos\alpha \end{pmatrix} \middle| \alpha \in \mathbb{R} \right\}.$$

Then M is a subgroup of $GL(2,\mathbb{R})$.

In Examples A4.2, we used the term "subgroup," and we have referred to matrices, their multiplication, and their determinant. We have also used the notation $[b,m] : F \to F$. Hopefully, these do not cause the reader any problems, but we will provide a little explanation now, hopefully for a reminder of things learned before.

First consider the term "subgroup." In Definitions A2.8, we talked about a subring of a ring. Just before Corollary A3.8, we introduced the idea of a subsemigroup. In Theorem A3.14, we used the term "submonoid." Then just before Theorem A3.27, we used the term "subgroup." This was all done with a purpose. Mathematicians are usually rather careful to define their specialized terminology, but there are some families of words which are related. Sometimes we do not define each word in this family, hoping that others will be able to determine exactly what we mean. But to avoid taking too big of a risk, we will define the term "subgroup."

Definition A4.3. Let $(G,\cdot)$ be a group. A nonempty subset $H \subseteq G$ is a *subgroup* of G provided $(H,*)$ is a group, where $* : H \times H \to H$ is the restriction of $\cdot : G \times G \to G$ to $H \times H$. This means that $a*b = a\cdot b = ab \in H$ whenever $a,b \in H$, and that $a^{-1} \in H$ whenever $a \in H$. (Associativity of $*$ comes from the fact that $H \subseteq G$ and that $\cdot$ is associative.) If H is a subgroup of G, we often write $H < G$.

So, to be a subgroup of $(G,\cdot)$, H must satisfy

(1) $H \subseteq G$,

(2) $H \neq \varnothing$,

(3) $a \cdot b \in H$ whenever $a,b \in H$, and

(4) $a^{-1} \in H$ whenever $a \in H$.

If $a \cdot b \in H$ whenever $a,b \in H$, we say that H is *closed* with respect to the multiplication $\cdot$.

Next, let's review matrix multiplication. We won't have need to go beyond two by two matrices. So, let's take $\left(\begin{smallmatrix} a & b \\ c & d \end{smallmatrix}\right)$ and $\left(\begin{smallmatrix} x & y \\ z & u \end{smallmatrix}\right)$. Then

$$\begin{pmatrix} a & b \\ c & d \end{pmatrix} \cdot \begin{pmatrix} x & y \\ z & u \end{pmatrix} = \begin{pmatrix} ax+bz & ay+bu \\ cx+dz & cy+du \end{pmatrix}.$$

The determinant of a matrix $A = \left(\begin{smallmatrix} a & b \\ c & d \end{smallmatrix}\right)$ is $\det A = ad - bc$.

Finally, there is the notation $[b,m] : F \to F$. Certainly, we have used notation like $f : X \to Y$ or $\phi : A \to B$. Well, this is the same thing. We are using $[b,m]$ in place of f or ϕ. So $[b,m]$ is a function with domain F of a field $(F,+,\cdot)$ and codomain F. It sends an $x \in F$ to $b + mx \in F$.

Each group $(G,\cdot)$ has its *trivial* subgroups, namely $\{e\}$ and G. Any other subgroup is a *proper* subgroup. Since a group $(G,\cdot)$ is a monoid, then Definition A3.25 applies. That is, for a group $(G,\cdot)$,

$$Z(G) = \{c \in G \mid cx = xc \text{ for every } x \in G\}$$

is the *center* of the group G. From Theorem A3.26, we know that $Z(G)$ is a submonoid. It is actually a subgroup of G.

Theorem A4.4. *For any group $(G,\cdot)$, $Z(G)$ is a subgroup.*

Proof. From Theorem A3.26, $Z(G)$ is a monoid. If $c \in Z(G)$, then $cx = xc$ for each $x \in G$. Multiplying on the left by c^{-1} we get $x = c^{-1}xc$ for every $x \in G$. Multiplying this last equation on the right by c^{-1} we get $xc^{-1} = c^{-1}x$ for every $x \in G$. Hence, $c^{-1} \in Z(G)$. This makes $Z(G)$ a subgroup. □

Definitions A4.5. A group $(G,\cdot)$ for which $Z(G) = G$ is *abelian*, or *commutative*. This is because $ab = ba$ for every $a,b \in G$, since $Z(G) = G$. If a group is not abelian, then it is *nonabelian*, or *noncommutative*.

We will be dealing mostly with abelian groups. In (4) of Examples A4.2, we have that $\mathrm{Aff}(F)$ is a nonabelian group if $|F| > 2$. Also, if $|X| > 2$, then $\mathscr{U}(M(X))$ is nonabelian.

It might be helpful to go through very carefully the details required to show that something is a group. It is often nontrivial. The details required to show that $(\text{Aff}(F), \circ)$ is a group are only moderately involved.

Theorem A4.6. *Let $(F, +, \cdot)$ be a field. (For example, let $(F, +, \cdot) = (\mathbb{R}, +, \cdot)$, the field of real numbers.) Then $(\text{Aff}(F), \circ)$ is a group, and it is nonabelian if $|F| > 2$.*

Proof. Since each $[b,m]$ is a function with domain F and codomain F, and since $\circ$ is associative in $M(F)$, we have that $\circ$ is associative for elements of $\text{Aff}(F)$. (Recall that $(M(F), \circ)$ is a monoid.)

If $[b,m]$, $[b',m'] \in \text{Aff}(F)$, then

$$\begin{aligned}[b,m] \circ [b',m'](x) &= [b,m](b' + m'x) = b + m(b' + m'x) \\ &= (b + b'm) + (mm')x = [b + b'm,\ mm'](x).\end{aligned}$$

So, we get the rule

$$[b,m] \circ [b',m'] = [b + mb',\ mm'] \tag{4:26}$$

for arbitrary $[b,m]$, $[b',m'] \in \text{Aff}(F)$, and so $\text{Aff}(F)$ is closed with respect to $\circ$.

Note that $[0,1](x) = 0 + 1x = x$ for each $x \in F$. Hence $[0,1] = 1_F$, the identity of $M(F)$, and so $\text{Aff}(F)$ has the identity $[0,1]$ with respect to $\circ$.

There remains to show that $[b,m]^{-1} \in \text{Aff}(F)$ for any $[b,m] \in \text{Aff}(F)$. To do this, look at the equation

$$[b,m] \circ [b',m'] = [b + mb', mm'] = [0,1].$$

We see that we need $m' = m^{-1}$ and $b' = -m^{-1}b$. Let's try it!

$$[b,m] \circ [-m^{-1}b, m^{-1}] = [b + m(-m^{-1}b). mm^{-1}] = [b - b,\ 1] = [0,1].$$

Also

$$[-m^{-1}b,\ m^{-1}] \circ [b,m] = [-m^{-1}b + m^{-1}b,\ m^{-1}m] = [0,1].$$

Hence, $[b,m]^{-1} = [-m^{-1}b, m^{-1}] \in \mathrm{Aff}(F)$. (Note how we have needed that m^{-1} exists. That is why we had $m \in F^*$ in the definition of $\mathrm{Aff}(F)$.) So $(\mathrm{Aff}(F), \circ)$ is a group.

But is it nonabelian? We already have $[b,m] \circ [b',m'] = [b+mb', mm']$. So $[b',m'] \circ [b,m] = [b'+m'b, m'm]$. We certainly have no problem with $mm' = m'm$ for all $m, m' \in F^*$, but is

$$(*) \qquad b + mb' = b' + m'b$$

for all $m, m' \in F^*$ and for all $b, b' \in F$? If we take $b \neq 0$, $b' = 0$, and $m' \neq 1$, then $(*)$ becomes

$$b = m'b,$$

which is not true if $|F| > 2$. We have now finished the proof of the theorem. □

We will use the complex number field $(\mathbb{C}, +, \cdot)$ frequently as a model. Perhaps we should review a few facts. For i being a solution to the equation $x^2 = -1$, we have $u \in \mathbb{C}$ if and only if $u = a + ib$ for some $a, b \in \mathbb{R}$. The *absolute value* of u is $|u| = \sqrt{a^2 + b^2}$. The *real part* of u is a, and the *imaginary part* of u is ib. We identify the complex number $a + ib$ with the point $(a,b) \in \mathbb{R} \times \mathbb{R}$. If $u = a + bi \neq 0$, then the ray from $(0,0)$ through (a,b) makes an angle θ with the ray from $(0,0)$ through $(1,0)$. It is elementary to see that $(a,b) = (|u|\cos\theta, |u|\sin\theta)$, so $a + bi = |u|(\cos\theta + i\sin\theta)$ is a point on the circle or radius $r = |u|$ centered at the origin $(0,0)$.

If $v = c + id = |v|(\cos\phi + i\sin\phi)$, then using elementary trigonometry, one gets that $uv = |u||v|(\cos(\theta+\phi) + i\sin(\theta+\phi))$. It is customary to use the notation

$$e^{i\theta} = \cos\theta + i\sin\theta.$$

So we have $u = |u|e^{i\theta}$, $v = |v|e^{i\phi}$, and so $uv = |u||v|e^{i(\theta+\phi)}$.

Take any point u on the circle of radius $r = 1$ and center $(0,0)$. Then $|u| = 1$, so $u = e^{i\theta}$ for some θ. Certainly, if $\phi \in \mathbb{R}$, then $e^{i\phi}$ is a point on the unit circle, that is, the circle of radius $r = 1$ centered at the origin $(0,0)$. Let

$$T = \{e^{i\phi} \mid \phi \in \mathbb{R}\}.$$

Then it is easy to see that $(T, \cdot)$ is an abelian group.

For any positive integer n, consider the complex number

$$\gamma = e^{\frac{2\pi i}{n}}.$$

Then

$$\gamma^k = e^{\frac{2\pi k i}{n}},$$

and so

$$C_n = \{\gamma, \gamma^2, \ldots, \gamma^{n-1}, \gamma^n = 1\}$$

is a set of n distinct values in T. In fact C_n is a subgroup of T. The group C_n is closely related to $(\mathbb{Z}_n, +)$, the additive group of the ring $(\mathbb{Z}_n, +, \cdot)$. In fact, algebraists often think of them as the same group. The reason for this is that they are isomorphic as groups.

Theorem A4.7. *The map $\phi : \mathbb{Z}_n \to C_n$ defined by $\phi(k) = \gamma^k$ is a group isomorphism.*

Proof. Certainly, ϕ is a bijection. Also, $\phi(s+t) = \gamma^{s+t} = \gamma^s \gamma^t = \phi(s)\phi(t)$. This says that ϕ is a group homomorphism. Putting everything together allows us to conclude that ϕ is a group isomorphism. □

Definitions A4.8. Any group isomorphic to $(\mathbb{Z}_n, +)$ is a *finite cyclic group of order n*. Any group isomorphic to $(\mathbb{Z}, +)$ is an *infinite cyclic group*. A group $(G, \cdot)$ is *finite* if $|G| < \infty$, and we say that the *order of G is finite*. Otherwise, $|G| = \infty$, and we say that G is an *infinite* group, or that G has *infinite order*.

Let

$$\tau = e^{i\sqrt{2}\pi},$$

and $V = \{\tau^n | n \in \mathbb{Z}\}$. (Actually, we could use any number $\lambda \in \mathbb{R} \setminus \mathbb{Q}$ for $\sqrt{2}$, as the reader should check as we go along.) We want to see that V is a subgroup of T isomorphic to $(\mathbb{Z}, +)$. That is, we want to prove that V is an infinite cyclic group. Certainly $\tau^m\tau^n = \tau^{m+n} \in V$, and $\tau^{-n}\tau^n = \tau^n\tau^{-n} = 1$. Now $(\tau^n)^{-1} = \tau^{-n} \in V$, and so V is a subgroup of T. The mapping $F : \mathbb{Z} \to V$ defined by $F(m) = \tau^m$ has the property that $F(m+n) = \tau^{m+n} = \tau^m\tau^n = F(m)F(n)$, and so F is a homomorphism. Easily, F is surjective, and so F is an epimorphism. But is F injective? If $F(m) = F(n)$, then $\tau^m = \tau^n$, and so $\tau^{m-n} = 1$, which is to say that

$$e^{i\sqrt{2}(m-n)\pi} = 1.$$

This means, of course, that $\sqrt{2}(m-n)\pi = 2k\pi$ for some integer k. Solving for k, we get $k = \sqrt{2}(m-n)/2$, which implies that $\sqrt{2} = (m-n)/k$, if $k \neq 0$. But if $k = 0$, then $m = n$. Otherwise, we have that $\sqrt{2} \in \mathbb{Q}$, which is false. In any event, we must have $m = n$, which means that F is injective, and this assures us that F is an isomorphism. Summarizing, we have

Theorem A4.9. *The group T has the subgroup*

$$V = \{\tau^k \mid k \in \mathbb{Z}\},$$

where

$$\tau = e^{i\sqrt{2}\pi},$$

and V is an infinite cyclic group. (Actually, any irrational number λ could be used in place of $\sqrt{2}$ to get the same result.)

With Theorem A4.7 and Theorem A4.9, we have seen that the group T has an isomorphic copy of all the finite cyclic groups $\mathbb{Z}_n$ as well as a subgroup isomorphic to the infinite cyclic group $(\mathbb{Z}, +)$. But every group $(G, \cdot)$ has isomorphic copies of some of the cyclic groups. Take any $g \in G$. If $g, g^2, g^3, \ldots, g^n, \ldots$ are all distinct, then $1, g^{-1}, g^{-2}, \ldots, g^{-n}, \ldots$ are also all distinct, and so $\langle g \rangle = \{g^k \mid k \in \mathbb{Z}\}$ is an infinite cyclic subgroup of G which is therefore isomorphic to $(\mathbb{Z},+)$ using the map $F:\mathbb{Z}\to\langle g\rangle$ defined by $F(k)=g^k$. On the other hand, there may be distinct positive integers m and n such that $g^m=g^n$. Suppose k is the smallest positive integer for which there is an integer $m>k$ such that $g^m=g^k$. Then $g^{m-k}=e$. With $m-k>0$, the set $\{n \mid g^n=e\}\neq\varnothing$, and as such, it has a smallest member s. That is, $g^s=e$, and if $t>0$ and $g^t=0$, then $t\geq s$. For such a t, divide t by s and get $t=sq+r$ for some r, $0\leq r<s$. Then $e=g^t=g^{sq+r}=g^{sq}g^r=(g^s)^q g^r=e^q g^r=eg^r=g^r$. Since s is the smallest positive integer x so that $g^x=e$, and since $0\leq r<s$, we must conclude that $r=0$. Hence, $t=sq$, and so s divides t, and we sometimes write $s|t$. This reasoning can be repeated to show that $g^t=g^r$ where $t=sq+r$ and $0\leq r<s$. This means that $\{g,g^2,\ldots,g^n,\ldots\}=\{g,g^2,\ldots,g^{s-1},g^s=e\}$, a finite set, and so $\langle g\rangle=\{g,g^2,\ldots,g^{s-1},g^s=e\}$ is a subgroup of G of order s. We also say that the *order of* g is s, and write $|g|=s$. The map $\phi:\mathbb{Z}_s\to\langle g\rangle$ defined by $\phi(k)=g^k$ is an isomorphism. We say that $g\in G$ *generates the cyclic group* $\langle g\rangle$ and that g is a *generator* of $\langle g\rangle$.

We summarize some of our points in

Theorem A4.10. *Let $(G,\cdot)$ be a group. Take any $g\in G$. If $\langle g\rangle$ is infinite, then $\langle g\rangle$ is an infinite cyclic subgroup of G generated by g, and g has infinite order. If $\langle g\rangle$ is finite, then there is a smallest positive integer k for which $|\langle g\rangle|=k$. Also, $\langle g\rangle$ is a finite cyclic subgroup of G isomorphic to $(\mathbb{Z}_k,+)$. Hence, $\langle g\rangle$ and g both have order k.*

We want to show that if G is a finite group and H is a subgroup, then $|H|$ divides $|G|$. So, in particular, $|g|$ divides $|G|$ for every $g\in G$, if G is finite. But to show this property of any subgroup H, we need a very important concept.

Definition A4.11. Let H be a subgroup of a group G. For any $g\in G$, define

$$gH=\{g\}H=\{gh \mid h\in H\},$$

and call gH the *left coset of H in G with representative g*.

A subgroup H of a group G defines an equivalence relation $\equiv_H$ on G. For $x, y \in G$, we write $x \equiv_H y$ if and only if $x^{-1}y \in H$.

Theorem A4.12. *For any subgroup H of a group G, $\equiv_H$ is an equivalence relation.*

Proof. Take any $x \in G$. Then $x^{-1}x = e \in H$, and so $x \equiv_H x$. Hence, $\equiv_H$ is reflexive. If $x \equiv_H y$, then $x^{-1}y \in H$, and so $(x^{-1}y)^{-1} = y^{-1}(x^{-1})^{-1} = y^{-1}x \in H$, which means that $\equiv_H$ is symmetric. Finally, suppose $x \equiv_H y$ and $y \equiv_H z$. Then $x^{-1}y$, $y^{-1}z \in H$, and so $(x^{-1}y)(y^{-1}z) = x^{-1}z \in H$. Thus, $x \equiv_H z$, and so $\equiv_H$ is transitive. We have now what we need to conclude that $\equiv_H$ is an equivalence relation. □

Theorem A4.13. *For a subgroup H of a group G, we have $xH = yH$ if and only if $x \equiv_H y$.*

Proof. If $xH = yH$, then $ye = xh$ for some $h \in H$, that is, $y = xh$ for some $h \in H$. So $x^{-1}y = h \in H$, and then $x \equiv_H y$. Conversely, if $x \equiv_H y$, then $x^{-1}y = h \in H$ for some h. This means that $y = ye = xh$, and so $y \in xH$. There remains to prove that $xH = yH$ if $y \in xH$.

Suppose $y \in xH$. Then $y = xh$ for some $h \in H$. Take any $yh' \in yH$. Then $yh' = xhh' = x(hh')$ and $hh' \in H$. Hence $yH \subseteq xH$. With $y \in xH$, and $y = xh$ with $h \in H$, we have $x = yh^{-1}$ and $x \in yH$. By what we have just seen, we conclude that $xH \subseteq yH$. This all comes from $y \in xH$. So, $y \in xH$ implies $yH \subseteq xH$ and $xH \subseteq yH$. Now we may conclude that $xH = yH$.

So, $x \equiv_H y$ implies that $y \in xH$, and $y \in xH$ implies $xH = yH$. Hence, $x \equiv_H y$ implies that $xH = yH$. This completes the proof of the converse, and also this proof of the theorem. □

Corollary A4.14. *For a subgroup H of a group G, we have $xH = [x]_{\equiv_H}$. That is, each left coset xH is the equivalence class $[x]_{\equiv_H}$ of the equivalence relation $\equiv_H$.*

Proof. If $x \equiv_H y$, then $xH = yH$, and so $y = ye = xh$ for some $h \in H$. Hence, $y \in xH$. Now we have $[x]_{\equiv_H} \subseteq xH$. Conversely, if $t = xh' \in xH$ for some $h' \in H$, then $x^{-1}t = h' \in H$, and so $x \equiv_H t$. This means that $t \in xH$ implies $t \in [x]_{\equiv_H}$, and we have $xH \subseteq [x]_{\equiv_H}$. What can we do but conclude that $[x]_{\equiv_H} = xH$? □

Lemma A4.15. *For a subgroup H of a group G, we have for any two left cosets xH and yH of H in G, a map* $\phi : xH \to yH$ *defined by* $\phi(xh) = yh$. *This mapping* ϕ *is a bijection.*

Proof. We will first show that $\mu_x : H \to xH$ defined by $\mu_x(h) = xh$ is a bijection. Easily, μ_x is surjective. If $\mu_x(h) = \mu_x(h')$, then $xh = xh'$, and so $h = h'$. This means that μ_x is injective. Now we know that μ_x is a bijection.

Since $\mu_x : H \to xH$ and $\mu_y : H \to yH$ are bijections, $\mu_y \circ \mu_x^{-1} : xH \to yH$ is also a bijection. Now $\mu_y \circ \mu_x^{-1}(xh) = \mu_y(h) = yh$. So, define $\phi = \mu_y \circ \mu_x^{-1}$. □

This lemma assures us that any two cosets of the subgroup H in G have the same cardinality. From Corollary A4.14, we have that the left cosets of H in G are the equivalence classes of $\equiv_H$. Let $[G : H]$ denote the number of left cosets of H in G, and call it the *index of H in G*. We have that $|G| = [G : H]|H|$, which completes the proof of

Theorem A4.16. *Let H be a subgroup of a finite group G. Then the order of H and the index of H in G both divide the order of G. In fact,*

$$|G| = [G : H]|H|. \tag{4:27}$$

Corollary A4.17. *If g is any element of a finite group G, then* $|g|$ *divides* $|G|$.

Proof. Certainly, $|g| = |\langle g \rangle|$, the order of the cyclic subgroup generated by g. By Theorem A4.16, $|\langle g \rangle|$ divides $|G|$, so we are permitted to conclude that $|g|$ divides $|G|$. □

Many years ago, you learned about the *greatest common divisor* and the *least common multiple* of two positive integers m and n. Every positive integer has a unique factorization into primes, so

$$m = p_1^{e_1} \cdots p_s^{e_s},$$

where each p_i is a prime, and each e_i is a positive integer. If

$$n = q_1^{f_1} \cdots q_t^{f_t}$$

is the factorization of n into primes q_i with each f_i a positive integer, and if

$$\{p_1, \ldots, p_s\} \cup \{q_1, \ldots, q_t\} = \{r_1, \ldots, r_u\},$$

then $\{r_1, \ldots, r_u\}$ is the set of primes which divide m or n. So we can write

$$m = r_1^{a_1} \cdots r_u^{a_u}$$

and

$$n = r_1^{b_1} \cdots r_u^{b_u}$$

where each a_i and each b_i is a nonnegative integer. (So $a_i = 0$ if and only if $r_i \notin \{p_1, \ldots, p_s\}$, and $b_i = 0$ if and only if $r_i \notin \{q_1, \ldots, q_t\}$.) For each integer i, $1 \le i \le u$, let $c_i = \min\{a_i, b_i\}$, that is, c_i is the smaller of a_i and b_i. Similarly, let $d_i = \max\{a_i, b_i\}$, so d_i is the larger of a_i and b_i. Then the *greatest common divisor of m and n* is

$$\gcd\{m, n\} = r_1^{c_1} \cdots r_u^{c_u},$$

and the *least common multiple of m and n* is

$$\operatorname{lcm}\{m, n\} = r_1^{d_1} \cdots r_u^{d_u}.$$

For example, if $m = 2^3 \cdot 3^5$ and $n = 3^5 \cdot 7^2$, then $(p_1, p_2) = (2, 3)$, $(q_1, q_2) = (3, 7)$, $(e_1, e_2) = (3, 5)$, and $(f_1, f_2) = (5, 2)$. Hence,

$$\{p_1, p_2\} \cup \{q_1, q_2\} = \{r_1, r_2, r_3\} = \{2, 3, 7\}.$$

So we have $m = 2^3 \cdot 3^5 \cdot 7^0$ and $n = 2^0 \cdot 3^5 \cdot 7^2$. With $(c_1, c_2, c_3) = (0, 5, 0)$ and $(d_1, d_2, d_3) = (3, 5, 2)$, we have $\gcd\{m, n\} = 2^0 \cdot 3^5 \cdot 7^0$ and $\operatorname{lcm}\{m, n\} = 2^3 \cdot 3^5 \cdot 7^2$.

These next three lemmas will play an important role for two of the main results we need for our development.

Lemma A4.18. *Let $a, b \in G$, where $(G, \cdot)$ is an abelian group. Suppose $|a| = m$, $|b| = n$, and $\gcd\{m,n\} = 1$. Then $|ab| = |a||b|$.*

Proof. With $m = |a| = p_1^{e_1} \cdots p_s^{e_s}$ and $n = |b| = q_1^{f_1} \cdots q_t^{f_t}$, the factorization of m and n into primes p_i and q_j, and with each $e_i > 0$ and each $f_j > 0$, then $\{p_1, \ldots, p_s\}$ and $\{q_1, \ldots, q_t\}$ are disjoint as $\gcd\{m,n\} = 1$. Hence,

$$\operatorname{lcm}\{m,n\} = p_1^{e_1} \cdots p_s^{e_s} q_1^{f_1} \cdots q_t^{f_t} = mn.$$

With $\langle a \rangle = \{a, a^2, \ldots, a^{m-1}, a^m = e\}$ and $\langle b \rangle = \{b, b^2, \ldots, b^{n-1}, b_n = e\}$, we have $\langle a \rangle \cap \langle b \rangle = \{e\}$. For if $a^i = b^j \in \langle a \rangle \cap \langle b \rangle$, then $|a^i|$ divides m and $|b^j|$ divides n. But the $\gcd\{m,n\} = 1$, and so $|a^i| = |b^j| = 1$. Hence, $a^i = b^j = e$, the only element of order 1.

Let $u = |ab|$. Then $1 = (ab)^u = a^u b^u$. Equivalently, $a^u = b^{-u} \in \langle a \rangle \cap \langle b \rangle$, and so $a^u = b^{-u} = e$. Hence, $m|u$ and $n|u$ by Corollary A4.17. Hence, $mn|u$ since $m = p_1^{e_1} \cdots p_s^{e_s}$, $n = q_1^{f_1} \cdots q_t^{f_t}$, and $\{p_1, \ldots, p_s\} \cap \{q_1, \ldots, q_t\} = \varnothing$. Since $mn|u$, we have $mn \leq u = |ab|$. But $(ab)^{mn} = a^{mn} b^{mn} = (a^m)^n (b^n)^m = e^n e^m = e$. Again by Corollary A4.17, we have $u|mn$, and so $u \leq mn$. With $mn \leq u \leq mn$, we must conclude that $u = mn$. That is, $|ab| = |a||b|$, and this completes our proof. □

An immediate application of Lemma A4.18 is in the next lemma.

Lemma A4.19. *For an abelian group G, suppose $a, b \in G$, $|a| = m$, and $|b| = n$. Then there is a $c \in G$ such that $|c| = \operatorname{lcm}\{m,n\} = \operatorname{lcm}\{|a|, |b|\}$.*

Proof. Factor $m = p_1^{e_1} \cdots p_s^{e_s}$ and $n = p_1^{f_1} \cdots p_s^{f_s}$ where each p_i is a prime, each $e_i \geq 0$, and each $f_i \geq 0$. (Here, $\{p_1, \ldots, p_s\}$ is the $\{r_1, \ldots, r_u\}$ discussed earlier.) We order the primes p_i in such a way that for $1 \leq k \leq t$, we have $p_k^{e_k} < p_k^{f_k}$, and for $t+1 \leq j \leq s$, we have $p_j^{f_j} \leq p_j^{e_j}$. (For each i, either $p_i^{e_i} < p_i^{f_i}$ or $p_i^{f_i} \leq p_i^{e_i}$.) So,

$$\operatorname{lcm}\{m,n\} = p_1^{f_1} \cdots p_t^{f_t} p_{t+1}^{e_{t+1}} \cdots p_s^{e_s}$$

and

$$\gcd\{p_{t+1}^{e_{t+1}} \cdots p_s^{e_s},\, p_1^{f_1} \cdots p_t^{f_t}\} = 1.$$

Let $m_1 = p_1^{e_1} \cdots p_t^{e_t}$ and $n_1 = p_{t+1}^{f_{t+1}} \cdots p_s^{f_s}$. Then $\gcd\{m_1, n_1\} = 1$, $|a^{m_1}| = p_{t+1}^{e_{t+1}} \cdots p_s^{e_s}$, and $|b^{n_1}| = p_1^{f_1} \cdots p_t^{f_t}$. We apply Lemma A4.18 to a^{m_1} and b^{n_1} to get

$$|a^{m_1}b^{n_1}| = |a^{m_1}||b^{n_1}| = p_1^{f_1} \cdots p_t^{f_t} p_{t+1}^{e_{t+1}} \cdots p_s^{e_s} = \operatorname{lcm}\{m, n\}.$$

So, if we let $c = a^{m_1}b^{n_1}$, then we may conclude that $|c| = \operatorname{lcm}\{|a|, |b|\}$. □

Lemma A4.20. *If G is a finite group with $|G| = n$, and if $g \in G$, then $g^n = e$.*

Proof. Suppose $|g| = k$. Then $|g| = |\langle g \rangle|$, and $|\langle g \rangle|$ divides n, so $n = km$ for some integer m. Hence, $g^n = g^{km} = (g^k)^m = e^m = e$. □

We are ready now for two of our more important results needed for the development of the ideas we present within these notes.

Theorem A4.21. *Let C_n be a cyclic group of order n. Let k be a divisor of n. Then C_n has a cyclic subgroup C_k of order k.*

Proof. Suppose $C_n = \{g, g^2, \ldots, g^{n-1}, g^n = e\}$. Also suppose $n = km$. Then

$$\begin{aligned}\langle g^m \rangle &= \{g^m, (g^m)^2, \ldots, (g^m)^{k-1}, (g^m)^k\} \\ &= \{g^m, g^{2m}, \ldots, g^{mk-m}, g^{mk} = g^n = e\}\end{aligned}$$

is a cyclic subgroup of order k, and so we let $C_k = \langle g^m \rangle$. □

Theorem A4.22. *Let $(F, +, \cdot)$ be a finite field. Then the multiplicative group $(F^*, \cdot)$ is cyclic.*

Proof. Let $|F| = q$, and so $|F^*| = q - 1$. Take $a \in F^*$. Then $|a|$ divides $q - 1$ by Corollary A4.17, and so $1 \leq |a| \leq q - 1$. Let

$$M = \{|b| \mid b \in F^*\},$$

a subset of $\{1, 2, \ldots, q-1\}$, so M has a largest element s and there is a $u \in F^*$ for which $|u| = s$.

For $a \in F^*$, and for u as just described, we have, by Lemma A4.19, an element $c \in F^*$ so that $|c| = \text{lcm}\{|a|, |u|\} = \text{lcm}\{|a|, s\}$. Hence, $s \leq |c|$ and $|c| \in M$, so $|c| \leq s$. Thus $|c| = |u| = s$, and s is a multiple of $|a|$ for any $a \in F^*$. If $s = k|a|$, then $a^s = (a^{|a|})^k = 1^k = 1$. We have just seen that any of the $q-1$ elements $a \in F^*$ satisfies the equation $x^s - 1 = 0$. From Theorem A2.36, we know that $q - 1 \leq s$. But, by Lemma A4.20, we also have that $|u| = s$ is a divisor of $q-1$, and so $s \leq q-1$. What are we to do but to conclude that $s = q-1$?

With $u \in F^*$ such that $|u| = s$, and with $s = q-1$, we have $|u| = q-1$, and so $\langle u \rangle = F^*$. Now we have finished a proof of this very important theorem. □

Corollary A4.23. *If F is as finite field with $|F| = q$, and if k divides $q-1$, then F^* has a cyclic subgroup of order k.*

Proof. From Theorem A4.22, we have $F^* = \langle u \rangle$ is cyclic of order $q-1$. From Theorem A4.21, we know that $\langle u \rangle$ has a subgroup C_k of order k. □

Exercise A4.24. (You may want to keep your results of these exercises for future reference.)

(1) For the finite field $(\mathbb{Z}_{13}, +, \cdot)$, $(\mathbb{Z}_{13}^*, \cdot)$ has cyclic subgroups of orders 1, 2, 3, 4, 6, and 12. Find these subgroups.

(2) For the finite field $(F, +, \cdot) = (\mathbb{Z}_3[x]/(1+x^2), +, \cdot)$ of 9 elements, F^* has subgroups of orders 1, 2, 4, and 8. Find these subgroups.

(3) For the finite field $(F, +, \cdot) = (\mathbb{Z}_5[x]/(2+x^2), +, \cdot)$ of 25 elements, F^* has subgroups of orders 1, 2, 4, 6, 8, 12, and 24. Find these subgroups.

A4.1 Morphisms of groups

We have discussed morphisms of semigroups in Definitions A3.9, and our morphisms of groups are really the same thing, but we will formalize these ideas now.

Definitions A4.25. Let $(G,\cdot)$ and $(H,*)$ be groups. A function $f: G \to H$ is a *group homomorphism* if $f(x\cdot y) = f(x) * f(y)$ for every $x, y \in G$. If $\text{Im} f = H$, then f is as *group epimorphism*. If f is injective, then f is a *group monomorphism*. If f is both surjective and injective, then f is a *group isomorphism*, and we say that G and H are *isomorphic*. (In this case, $f^{-1}: H \to G$ is also a group isomorphism.) If $(G,\cdot) = (H,*)$, then f is a *group endomorphism*. If f is a group endomorphism and is also a group isomorphism, then f is a *group automorphism*.

Definition A4.26. Let $f: G \to H$ be a group homomorphism. The *kernel of* f, denoted by $\ker f$, is defined by $\ker f = \{g \in G \mid f(g) = e\}$. (Compare this with Definition A2.22.)

Theorem A4.27. *Let $f: G \to H$ be a group homomorphism. If e_G and e_H are the identities of G and H, respectively, then $f(e_G) = e_H$. If $f(x) \in \text{Im} f$, then $f(x)^{-1} \in \text{Im} f$ and $f(x)^{-1} = f(x^{-1})$. In addition, $\text{Im} f$ is a subgroup of H and $\ker f$ is a subgroup of G.*

Proof. First of all, in a group $(T,\circ)$, there is exactly one element $u \in T$ for which $u \circ u = u$ is true. For $u^{-1} \in T$ exists, and from $u \circ u = u$, we get

$$e = u^{-1} \circ u = u^{-1} \circ (u \circ u) = (u^{-1} \circ u) \circ u = e \circ u = u.$$

But $f(e_G) = f(e_G \cdot e_G) = f(e_G) * f(e_G)$, and so $f(e_G) = e_H \in \text{Im} f = f(G)$. So $f(G) \neq \varnothing$. Also,

$$e_H = f(e_G) = f(x \cdot x^{-1}) = f(x) * f(x^{-1}),$$

and

$$e_H = f(e_G) = f(x^{-1} \cdot x) = f(x^{-1}) * f(x).$$

Hence, $f(x^{-1}) = f(x)^{-1}$.

Now take $f(x), f(y) \in f(G)$. Then $f(x) * f(y) = f(x \cdot y) \in f(G)$, and so $f(G)$ is closed with respect to $*$. We've already seen that $f(x)^{-1} = f(x^{-1}) \in f(G)$ for any $f(x) \in f(G)$. Hence, $\text{Im}f = f(G)$ is a subgroup of H.

Take $x, y \in \ker f$. Then $f(x \cdot y) = f(x) * f(y) = e_H * e_H = e_H$, and so $x \cdot y \in \ker f$, and we have that $\ker f$ is closed with respect to multiplication $\cdot$. Also, we have that $f(e_G) = e_H$, and so $e_G \in \ker f$, and we know that $\ker f \neq \varnothing$. If $x \in \ker f$, then $f(x^{-1}) = f(x)^{-1} = e_H^{-1} = e_H$, and so $x^{-1} \in \ker f$. This makes $\ker f$ a subgroup of G, and finishes our proof of the theorem. □

Examples A4.28. (1) Define $f : \mathbb{R} \to GL(2, \mathbb{R})$ by $f(a) = \begin{pmatrix} 1 & a \\ 0 & 1 \end{pmatrix}$. Then f is a monomorphism from $(\mathbb{R}, +)$ into $(GL(2, \mathbb{R}), \cdot)$.

(2) For a field $(F, +, \cdot)$, define $g : F \to \text{Aff}(F)$ by $g(b) = [b, 1]$. Then g is a monomorphism from $(F, +)$ into $(\text{Aff}(F), \circ)$.

(3) Define $\phi : \mathbb{R} \to \mathbb{R}_+$ by $\phi(x) = e^x$. Then ϕ is an isomorphism from $(\mathbb{R}, +)$ onto $(\mathbb{R}_+, \cdot)$, and $\phi^{-1}(x) = \ln x$ is an isomorphism from $(\mathbb{R}_+, \cdot)$ onto $(\mathbb{R}, +)$.

(4) For any group $(G, \cdot)$, let $\text{Aut}G = \{\phi : G \to G \mid \phi \text{ is an automorphism}\}$. Then $(\text{Aut}G, \circ)$ is a group. Define $T : G \to \text{Aut}G$ by $T(g) = \phi_g$, the inner automorphism of G induced by g. Then T is a group homomorphism. (See Theorem A3.28.)

(5) For any field $(F, +, \cdot)$, define $f : F^* \to \text{Aff}(F)$ by $f(m) = [0, m]$. Then f is a group monomorphism.

(6) Define $f : \mathbb{R}^* \to GL(2, \mathbb{R})$ by $f(a) = \begin{pmatrix} a & 0 \\ 0 & a \end{pmatrix}$. Then f is a group monomorphism from $(\mathbb{R}^*, \cdot)$ into $GL(2, \mathbb{R})$.

(7) In (6) of Examples A4.2, we defined a subgroup T of $GL(2, \mathbb{R})$. Now define $f : \mathbb{R} \to T$ by $f(\theta) = \begin{pmatrix} \cos\theta & \sin\theta \\ -\sin\theta & \cos\theta \end{pmatrix}$. Then f is an epimorphism form $(\mathbb{R}, +)$ onto $(T, \cdot)$.

(8) Just before Theorem A4.7, we defined $T = \{e^{i\theta} \mid \theta \in \mathbb{R}\}$. Define $g : \mathbb{R} \to T$ by $g(\theta) = e^{i\theta}$. Then g is an epimorphism from $(\mathbb{R}, +)$ onto T. (How are the T's of (7) and (8) related?)

There are two main things we want to do with group morphisms. One is to select certain subgroups of Aut$G$, and the other is to select certain subgroups of $(F^*, \cdot)$, where $(F, +, \cdot)$ is a field. In (4) of Examples A4.28, we had the group homomorphism $T : G \to \text{Aut}G$, and from Definitions A4.25, we know that $\text{Im}T = \text{Inn}G < \text{Aut}G$. As long as $(G, \cdot)$ is nonabelian, we get $\text{Inn}G \neq \{1_G\}$. As long as $(G, \cdot)$ is nonabelian, we get $Z(G) \neq G$. But, the subgroups of $(\text{Aut}G, \circ)$ of interest to us are not as easily obtained, but certain ones are even easier.

Definitions A4.29. For a group $(G, \cdot)$, an automorphism $\phi \in \text{Aut}(G, \cdot)$ is *fixed point free* if the only $x \in G$ for which $\phi(x) = x$ is $x = e$. That is, $\phi(x) = x$ implies $x = e$. A subgroup $\Phi < \text{Aut}(G, \cdot)$ is *fixed point free* or *regular* if $\Phi \neq \{1_G\}$, and every $\phi \in \Phi$, $\phi \neq 1_G$ is fixed point free.

Examples A4.30. (1) Let $(A, +)$ be an abelian group without any element of order 2. So, if $a \in A$, and $a \neq 0$, then $a + a \neq 0$. Define $m : A \to A$ by $m(a) = -a$. Then $m(a+b) = -(a+b) = (-b) + (-a) = (-a) + (-b) = m(a) + m(b)$. This is exactly what is needed to conclude that m is an endomorphism. If $m(a) = m(b)$, then $-a = -b$, and so $a = b$. Now we know that m is a monomorphism. If $b \in A$, then $-b \in A$ and $m(-b) = -(-b) = b$. This makes m an epimorphism. Altogether, m is an automorphism. Finally, if $m(a) = a$, then from $m(a) = -a$, we get $-a = a$, and so $0 = a + a$. This cannot happen in $(A, +)$ unless $a = 0$. So m is fixed point free. We also have that $\Phi = \{1_A, m\}$ is a regular subgroup of $\text{Aut}(A, +)$.

(2) For any field $(F, +, \cdot)$, take any subgroup $\Phi < F^*$, $\Phi \neq \{1\}$. For each $a \in \Phi$, define $a^\bullet : F \to F$ by $a^\bullet(x) = ax$. Then

$$a^\bullet(x+y) = a(x+y) = (ax) + (ay) = a^\bullet(x) + a^\bullet(y).$$

So each $a^\bullet$ is an endomorphism of $(F, +)$. If $a^\bullet(x) = a^\bullet(y)$, then $ax = ay$, and so $a^{-1}(ax) = a^{-1}(ay)$, which means that $x = y$. Hence, $a^\bullet$ is a monomorphism. From $a^\bullet(a^{-1}x) = a(a^{-1}x) = x$, we learn that $a^\bullet$ is an epimorphism. So, each

$a^\bullet \in \mathrm{Aut}(F,+)$. If $a^\bullet(x) = x$, then $ax = x$. If $x \neq 0$, then $(ax)x^{-1} = xx^{-1}$, and so $a = 1$. Hence, $a^\bullet = 1^\bullet = 1_F$. So $\{a^\bullet \mid a \in \Phi,\ a \neq 1\}$ is a set of fixed point free automorphisms of $(F,+)$. Since

$$(ab)^\bullet(x) = (ab)x = a(bx) = a^\bullet(b^\bullet x) = (a^\bullet \circ b^\bullet)(x),$$

we have that the map $\lambda : \Phi \to \mathrm{Aut}(F,+)$ defined by $\lambda(a) = a^\bullet$ is a homomorphism, and $\lambda(\Phi) = \{a^\bullet \mid a \in \Phi\}$. Hence,

$$\Phi^\bullet = \{a^\bullet \mid a \in \Phi\} < \mathrm{Aut}(F,+).$$

Now we have that $\Phi^\bullet$ is a regular group of automorphisms. The reader may want to prove that λ is an isomorphism. For this reason, we identify Φ with $\Phi^\bullet$.

(3) Let $(R,+,\cdot,1)$ be a ring with identity 1. So $(R,\cdot,1)$ is a monoid and $\mathscr{U}(R)$ is its group of units. For each $u \in \mathscr{U}(R)$, we define $u^\bullet : R \to R$ by $u^\bullet(x) = ux$. As in (2) above, each $u^\bullet \in \mathrm{Aut}(R,+)$. But what does it take for $u^\bullet$ to be fixed point free? If $x \neq 0$ and $u^\bullet(x) = x$, then $ux = x$, and so $-x + ux = (-1+u)x = 0$. This means that $-1+u \notin \mathscr{U}(R)$. Conversely, if $-1+u \in \mathscr{U}(R)$, then $(-1+u)x \neq 0$ if $x \neq 0$. Hence, $ux \neq x$, and so $u^\bullet$ is fixed point free. So, if we select $\Phi < \mathscr{U}(R)$ so that $-1+u \in \mathscr{U}(R)$ for each and every $u \in \Phi \setminus \{1\}$, then Φ will yield $\Phi^\bullet = \{u^\bullet \mid u \in \Phi\}$, a regular group of automorphism of $(R,+)$. For example, in $(\mathbb{Z}_{25},+,\cdot,1)$, $\mathscr{U}(\mathbb{Z}_{25}) = \mathbb{Z}_{25} \setminus \{0,5,10,15,20\}$, and if $\Phi = \langle 7 \rangle = \{7,24,18,1\}$, then $-1+u \in \mathscr{U}(\mathbb{Z}_{25})$ for every $u \in \Phi \setminus \{1\}$. Hence, $\Phi = \{7^\bullet, 24^\bullet, 18^\bullet, 1^\bullet\}$ is a regular group of automorphisms. We also have here that Φ and $\Phi^\bullet$ are isomorphic, and so we tend to identify $u \in \Phi$ with $u^\bullet \in \Phi^\bullet$.

Getting regular groups of automorphisms is rather easily done, but how to find all regular groups of automorphisms is not so easy. Nevertheless, regular groups of automorphisms play a central role in our development.

Our next point is really very easy, but it also is very important.

Theorem A4.31. *Let $(F, +, \cdot)$ be a field, and let n be any positive integer. Define $\phi_n : F^* \to F^*$ by $\phi_n(x) = x^n$. Then ϕ_n is an endomorphism of $(F^*, \cdot)$.*

Proof. Take $x, y \in F^*$. Then $\phi_n(xy) = (xy)^n = x^n y^n = \phi_n(x)\phi_n(y)$. That's all there is too it. □

As an easy application of Theorem A4.31, consider a finite field $(F, +, \cdot)$ where $|F| = q$ is an odd number. Then $|F^*| = q - 1$ is even, and so $\phi_2 : F^* \to F^*$ has $\ker \phi_2 = \{\pm 1\}$. This means that $\phi_2(F^*)$ contains half of the elements of F^*, and this in turn means that exactly half of the equations $x^2 - a = 0$, $a \in F^*$ have solutions, and the other half do not. So, there are usually numerous irreducible polynomials $b + x^2 \in F[x]$. The fields $(F[x]/(b + x^2), +, \cdot)$, where $b + x^2 \in F[x]$ is irreducible, each has q^2 elements. So, if there is a field F with q elements, q odd, then there is a field $(F[x]/(b + x^2), +, \cdot)$ with q^2 elements. Hence

Theorem A4.32. *If $(F, +, \cdot)$ is a finite field with q elements, and if q is odd, then exactly half of the elements $b \in F^*$ make $b + x^2$ irreducible, and each of these yield a field $(F[x]/(b + x^2), +, \cdot)$ with exactly q^2 elements.*

A4.2 Direct products

We remarked earlier how we identify $a + ib \in \mathbb{C}$ with the ordered pair (a, b) in $\mathbb{R} \times \mathbb{R}$. From $(a + ib) + (c + id) = (a + c) + i(b + d)$, we would then add $(a, b) + (c, d) = (a + c, b + d)$. We could do the same in $(\mathbb{R}[x]/(1 + x + x^2), +, \cdot)$. See Theorem A2.32. If we identify $a + bx + (1 + x + x^2)$ with (a, b), we have

$$[a + bx + (1 + x + x^2)] + [c + dx + (1 + x + x^2)]$$
$$= (a + c) + (b + d)x + (1 + x + x^2).$$

This would again tell us to add $(a, b) + (c, d) = (a + c, b + d)$. So, if we define $+$ on $\mathbb{R} \times \mathbb{R}$ by $(a, b) + (c, d) = (a + c, b + d)$, we get a group $(\mathbb{R} \times \mathbb{R}, +)$, which is isomorphic to $(\mathbb{C},+)$ as well as to $(\mathbb{R}[x]/(1+x+x^2),+)$. From this we make two observations. The first is that $(\mathbb{R}\times\mathbb{R},+)$ is in many ways simpler than either $(\mathbb{C},+)$ or $(\mathbb{R}[x]/(1+x+x^2),+)$, and stands as a rather simple representative of the three isomorphic groups. The second is that if we have a group $(A,+)$, then we should be able to make another group $(A\times A,+)$, where $+$ is defined on $A\times A$ by $(a,b)+(c,d)=(a+c,b+d)$. In fact, we can do that with two different groups.

Theorem A4.33. *Let $(A,+)$ and $(B,*)$ be two groups. On $A\times B$ define $\cdot$ by $(a,x)\cdot(b,y)=(a+b,x*y)$. Then $(A\times B,\cdot)$ is a group.*

Proof. For (a,x), (b,y), $(c,z)\in A\times B$, we have $(a,x)\cdot(b,y)=(a+b,x*y)\in A\times B$, so $\cdot$ is a binary operation. Easily,

$$\begin{aligned}(a,x)\cdot[(b,y)\cdot(c,z)] &= (a,x)\cdot(b+c,y*z)=(a+(b+c),x*(y*z))\\ &= ((a+b)+c,(x*y)*z)=(a+b,x*y)\cdot(c,z)\\ &= [(a,x)\cdot(b,y)]\cdot(c,z).\end{aligned}$$

So $\cdot$ is associative. If $0\in A$ and $1\in B$ are the identities, then $(0,1)\cdot(a,x)=(0+a,1*x)=(a,x)=(a+0,x*1)=(a,x)\cdot(0,1)$. Hence $(0,1)\in A\times B$ is the identity for $\cdot$. For $(a,x)\in A\times B$, we have $(-a,x^{-1})\in A\times B$, and so

$$\begin{aligned}(-a,x^{-1})\cdot(a,x) &= (-a+a,x^{-1}*x)=(0,1)\\ &= (a+(-a),x*x^{-1})=(a,x)\cdot(-a,x^{-1}).\end{aligned}$$

This makes $(a,x)^{-1}=(-a,x^{-1})$, and completes our proof. □

Definitions A4.34. Let $(A,+)$ and $(B,*)$ be groups. Then the group $(A\times B,\cdot)$ of Theorem A4.33 is the *(external) direct product* of A and B. In general, if $(A_1,*_1)$, $(A_2,*_2)$, …, $(A_n,*_n)$ are groups, then $(A_1\times A_2\times\cdots\times A_n,*)$ is a group and is the *(external) direct product* of $A_1,A_2,\cdots,A_n$, where

$$(a_1,\ldots,a_n)*(b_1,\ldots,b_n)=(a_1*_1b_1,\ldots,a_n*_nb_n).$$

Direct products contain isomorphic copies of their factors.

Theorem A4.35. *Let* $(A \times B, \cdot)$ *be the direct product of* A *and* B. *Let*

$$A' = \{(a,e) \mid a \in A\} \text{ and } B' = \{(1,b) \mid b \in B\}.$$

Define $e_A : A \to A \times B$ *and* $e_B : B \to A \times B$ *by* $e_A(a) = (a,e)$ *and* $e_B(b) = (1,b)$. *Then* e_A *and* e_B *are monomorphisms with* $e_A(A) = A'$ *and* $e_B(B) = B'$. *Hence,* A' *and* B' *are isomorphic copies of* A *and* B, *respectively.*

Proof. The proof is easy, and we encourage the reader to provide it. □

In Theorem A4.33, the subgroups A' and B' of $A \times B$ have the following properties. First, notice that $A' \cap B' = \{(1,e)\}$. Also, if $(a,x) \in A \times B$, then $(a,x)(b,e)(a,x)^{-1} = (aba^{-1},e) \in A'$ and $(a,x)(1,y)(a,x)^{-1} = (1,xyx^{-1}) \in B'$. In addition, $(a,x) = (a,e)(1,x)$, so $A'B' = A \times B$. From these, we isolate two concepts.

Definition A4.36. A subgroup H of a group G is a *normal* or *invariant* subgroup of G if $ghg^{-1} \in H$ for every $g \in G$ and for every $h \in H$.

So A' and B' are normal subgroups of $(A \times B, \cdot)$.

Definition A4.37. Let A and B be normal subgroups of a group G with the properties that $A \cap B = \{e\}$ and $AB = G$. Then G is the *(internal) direct product* of A and B.

Examples A4.38. (1) For the ring $(\mathbb{Z}_8, +, \cdot)$, the monoid $(\mathbb{Z}_8, \cdot, 1)$ has its group of units $\mathscr{U}(\mathbb{Z}_8) = \{1,3,5,7\}$. Then $A = \{1,3\}$ and $B = \{1,5\}$ are normal subgroups of $\mathscr{U}(\mathbb{Z}_8)$, $A \cap B = \{1\}$ and $AB = \mathscr{U}(\mathbb{Z}_8)$. Hence, $\mathscr{U}(\mathbb{Z}_8)$ is the (internal) direct product of A and B. Note also that $\mathscr{U}(\mathbb{Z}_8)$ is isomorphic to $(\mathbb{Z}_2 \times \mathbb{Z}_2, +)$.

(2) The field $(\mathbb{C}, +, \cdot)$ has multiplicative group $\mathbb{C}^*$. If $z \in \mathbb{C}^*$, then $z = |z|e^{i\theta}$ for some $e^{i\theta} \in T$. Note that $\mathbb{R}_+$ and T are normal subgroups of $\mathbb{C}^*$, that $\mathbb{R}_+ \cap T = \{1\}$, and that $\mathbb{C}^* = \mathbb{R}_+ T$. Hence, $\mathbb{C}^*$ is the (internal) direct product

of $\mathbb{R}_+$ and T. This example will be used to motivate some very interesting observations.

(3) Consider the abelian group $(\mathbb{Z}_6, +)$. We note that $\langle 2 \rangle = \{0, 2, 4\}$ and $\langle 3 \rangle = \{0, 3\}$ are normal subgroups of $\mathbb{Z}_6$. Also, $\langle 2 \rangle \cap \langle 3 \rangle = \{0\}$ and $\langle 2 \rangle + \langle 3 \rangle = \mathbb{Z}_6$. Hence, $\mathbb{Z}_6$ is the internal direct product (sum) of $\langle 2 \rangle$ and $\langle 3 \rangle$. Since $\langle 2 \rangle$ is isomorphic to $(\mathbb{Z}_3, +)$ and $\langle 3 \rangle$ is isomorphic to $(\mathbb{Z}_2, +)$, we have that $\mathbb{Z}_6$ is isomorphic to $\mathbb{Z}_2 \times \mathbb{Z}_3$.

(4) For the real numbers $\mathbb{R}$, note that $\mathbb{R}_+$ and $\{\pm 1\}$ are normal subgroups of $(\mathbb{R}^*, \cdot)$, and that $\mathbb{R}_+ \cap \{\pm 1\} = \{1\}$. Certainly, $\mathbb{R}^* = \{\pm 1\}\mathbb{R}_+$, so $\mathbb{R}^*$ is the internal direct product of $\{\pm 1\}$ and $\mathbb{R}_+$. Now $\{\pm 1\}$ is isomorphic to $\mathbb{Z}_2$ and $\mathbb{R}_+$ is isomorphic to $(\mathbb{R}, +)$, so $\mathbb{R}^*$ is isomorphic to $\mathbb{Z}_2 \times \mathbb{R}$.

A4.3 Normal Subgroups

Normal subgroups play a very important role in group theory. We have just focused on normal subgroups in describing direct products, and direct products are used to describe more complicated groups in terms of more familiar groups. Another role of normal subgroups is in constructing quotient groups. The role of an ideal in ring theory is somewhat parallel to the role of a normal subgroup in group theory. Just as every ideal is the kernel of a ring homomorphism, so is every normal subgroup the kernel of a group homomorphism.

Theorem A4.39. *Let $f : G \to H$ be a homomorphism from the group G to the group H. Then* $\ker f$ *is a normal subgroup of G.*

Proof. From Theorem A4.27, we know that $\ker f$ is a subgroup of G. Take $h \in \ker f$ and $g \in G$. Then $f(ghg^{-1}) = f(g)e_H f(g^{-1}) = f(g)f(g)^{-1} = e_H$. Hence, $ghg^{-1} \in \ker f$, and so $\ker f$ is a normal subgroup. □

With an ideal I of a ring R, we can make $(R/I, +)$ an abelian group with the addition $(x+I)+(y+I)=(x+y)+I$ for cosets $x+I, y+I \in R/I$. Let K be a normal subgroup of a group $(G,\cdot)$. Let G/K be all the left cosets xK as x varies through G. For $xK, yK \in G/K$, define

$$(xK)\cdot(yK)=xyK.$$

This will make $(G/K,\cdot)$ a group with identity $K=eK$, where e is the identity of G.

Theorem A4.40. *Let K be a normal subgroup of a group $(G,\cdot)$ with identity e. Then $(xK)\cdot(yK)=xyK$ is a binary operation on G/K, and $(G/K,\cdot)$ is a group with identity $eK=K$. For $gK \in G/K$, we have $(gK)^{-1}=g^{-1}K$.*

Proof. First, we must be assured that multiplication $\cdot$ is well defined. Assume that $xK=x'K$ and $yK=y'K$. We must show that $xyK=x'y'K$. There are $h_x, h_y \in K$ such that $x'=xh_x$ and $y'=yh_y$, so $x'y'=xh_xyh_y=xyy^{-1}h_xyh_y$. Since $y^{-1}h_xy \in K$, let $h'=y^{-1}h_xy$, and then $x'y'=xyh'h_y$. With $h'h_y \in K$, we now have that $xyK=x'y'K$, and so $\cdot$ is a well defined binary operation on G/K.

It is easy to see that $\cdot$ is associative on G/K, that $eK=K\in G/K$ is the identity for $\cdot$, and that $(gK)^{-1}=g^{-1}K$ for any $gK \in G/K$. Hence, $(G/K,\cdot)$ is a group. □

Theorem A4.41. *Let K be a normal subgroup of a group $(G,\cdot)$. Define the function $p_K : G \to G/K$ by $p_K(g)=gK$. Then p_K is an epimorphism, and p_K is called the natural map. The kernel of p_K is $\ker p_K = K$.*

Proof. Easily, $p_K(xy)=xyK=(xK)(yK)=p_K(x)p_K(y)$. If $xK \in G/K$, then $x\in G$ and $p_K(x)=xK$. Hence, p_K is an epimorphism.

With $K=eK$ being the identity of G/K, and with $xK=eK=K$ if and only if $x \in K$, we easily have $x \in \ker p_K$ if and only if $x\in K$. Hence, $\ker p_K = K$. □

With Theorem A4.39 and Theorem A4.41, we have that normal subgroups are kernels of homomorphisms, and conversely. With these, and the next theorem, the reader is encouraged to re-read Theorem A2.23

Theorem A4.42. *Let $f : G \to H$ be a group epimorphism. Let $K = \ker f$. Then there is a unique isomorphism $F : G/K \to H$ such that $F \circ p_K = f$.*

Proof. Define $F : G/K \to H$ by $F(xK) = f(x)$. If $xK = x'K$, then $x^{-1}x' \in K$, and so $e_H = f(x^{-1}x') = f(x^{-1})f(x') = f(x)^{-1}f(x')$. Consequently, $f(x) = f(x')$. This makes F well defined.

For $x \in G$, $(F \circ p_K)(x) = F(xK) = f(x)$, and so $F \circ p_K = f$. On the other hand, $F((xK)(yK)) = F(xyK) = f(xy) = f(x)f(y) = F(xK)F(yK)$, so F is a homomorphism. Since f is an epimorphism, each $h \in H$ has an $x \in G$ such that $f(x) = h$. Hence, $F(xK) = f(x) = h$, and so F is also an epimorphism. If $F(xK) = F(yK)$, then $f(x) = f(y)$, and so $e_H = f(y)^{-1}f(x) = f(y^{-1})f(x) = f(y^{-1}x)$. This assures us that $y^{-1}x \in \ker f = K$, and so $xK = yK$. Hence, F is an isomorphism.

Suppose $F' \circ p_k = f$. Then for any $x \in G$, we have

$$F(xK) = f(x) = F'(p_K(x)) = F'(xK).$$

This makes $F = F'$, and gives us uniqueness. □

As an application of Theorem A4.42, recall that any group $(G, \cdot)$ is a monoid with the property that $\mathscr{U}(G) = G$. Apply Theorem A3.29 to get the $T : G \to \mathrm{Inn}(G)$ defined by $T(g) = \phi_g$, where $\phi_g(x) = gxg^{-1}$. Then T is a group epimorphism. What is the kernel of T, $\ker T$? Now $g \in \ker T$ if and only if $T(g) = \phi_g = 1_G$. So, for $x \in G$, $x = 1_G(x) = \phi_g(x) = gxg^{-1}$. Hence, $xg = gx$ for every $x \in G$. This means that $g \in Z(G)$, the center of G. So, $\ker T \subseteq Z(G)$. Conversely, if $g \in Z(G)$, then $T(g) = \phi_g$ where $\phi_g(x) = gxg^{-1} = xgg^{-1} = x$ for every $x \in G$. Hence, $\phi_g = 1_G$, and we have $Z(G) \subseteq \ker T$. This makes $Z(G) = \ker T$, and so $G/Z(G)$ is isomorphic to $\mathrm{Inn}(G)$ by Theorem A4.42. Summarizing, we have

Theorem A4.43. *For a group $(G, \cdot)$, consider the mapping $T : G \to \mathrm{Inn}(G)$ defined by $T(g) = \phi_g$, where $\phi_g(x) = gxg^{-1}$ for every $x \in G$. Then T is a group epimorphism*

with kernel $\ker T = Z(G)$, *the center of* G. *Consequently, the quotient group* $G/Z(G)$ *is isomorphic to* $\mathrm{Inn}(G)$, *the group of inner automorphisms of* G.

Exercise A4.44. Refer to Exercise A3.21, Exercise A3.22, and Exercise A3.31. If $|S| \in \{3,4,5\}$, show that the center of $\mathscr{U}(M(S))$ consists of the identity 1_S only. Hence, $\mathscr{U}(M(S))$ is isomorphic to $\mathrm{Inn}(\mathscr{U}(M(S)))$. Can you generalize your results for $|S| \geq 6$?

Theorem A4.45. *Suppose a group* $(G,\cdot)$ *is a direct product of two of its subgroups* H *and* K. *Then:*

(a) *each* $g \in G$ *has a unique factorization*

$$g = g_H g_K$$

with $g_H \in H$ *and* $g_K \in K$;

(b) *each* $h \in H$ *commutes with each* $k \in K$.

Proof. Suppose $g = g_H g_k = x_H x_K$ with $g_H, x_h H \in H$ and $g_K, x_K \in K$. Then we have $g_H g_K = x_H x_K$ and so $g_K x_K^{-1} = g_H^{-1} x_K \in H \cap K = \{1\}$. From this we can conclude that $g_H = x_H$ and $g_K = x_K$. Hence, (a).

By saying that each $h \in H$ commutes with each $k \in K$, we mean that $hk = kh$ for arbitrary $h \in H$ and arbitrary $k \in K$. From $hk = kk^{-1}hk$ and $h_1 = k^{-1}hk \in H$, we see that $hk = kh_1$ $hkh^{-1} = kh_1h^{-1} \in K$. Also, from $k \in K$, we have $k^{-1}kh_1h^{-1} = h_1h^{-1} \in H \cap K = \{1\}$. Therefore, $h_1 = h$, or $k^{-1}hk = h$, which gives us the desired result that $hk = kh$. □

Bibliography

[B & F] Berman, G., and Fryer, K. D.
(1972) *Introduction to Combinatorics*, Academic Press, New York-London.

[C] Clay, J. R.
(1992) *Nearrings: Geneses and Applications*, Oxford University Press, Oxford.

[Di] Dickson, L. E.
(1905) Definitions of a group and a field by independent postulates, *Trans. Amer. Math. Soc.*, **6**, 198–204.

[E] Eggetsburger, Roland A.
(1992) *Codes from Field Generated Finite Planar Nearrings*, Diplomarbeit, Johannes Kepler Universität Linz, Linz.

[FHP] Fuchs, P., Hofer, G. and Pilz, G.
(1990) Codes from planar near-rings, *IEEE Trans. Inform. Theory*, **36**, 647–651.

[H] Hungerford, T. W.
(1980) *Algebra*, Graduate Texts in Mathematics, 73. Springer-Verlag, New York-Berlin.

[G] Gallian, J. A.
(1990) *Contemporary Abstract Algebra*, D. C. Heath, Lexington.

[Gr] Gorenstein, D.
(1980) *Finite groups*. Second edition. Chelsea Publishing Co., New York.

[I-H] Chen, I-Hsing
(1991) *Some Combinatorial Structures Arising from Finite Planar Near-rings*, Thesis, National Chiao Tung University, Hsin-chu.

[K] Kline, M.
(1967) *Mathematics for Liberal Arts*, Addison–Wesley, Reading.

[Ke] Ke, Wen-Fong
(1992) *Structures of Circular Planar Nearrings*, Ph.D. Dissertation, University of Arizona, Tucson.

[L] Lüneburg, H.
(1965) Über projektive Ebene, in denen jede Fahne von einer nicht-trivialen Elation invariant gelassen wird, *Abh. Math. Sem. Uni. Hamburg*, **29**, 37–76.

[M & S] McWilliams, F. J. and Sloane, N. J. A.
(1977) *The Theory of Error-correcting Codes*, North-Holland, Amsterdam,

[R] Ross, S.
(1988) *A First Course in Probability, Third Edition*, Macmillan, New York

[S] Stevenson, F.
(1972) *Projective Planes*, Freeman, San Francisco.

[VW] Veblen, O. and MacLagan-Wedderburn, J. H.
(1907) Non-desarguesian and non-pascalian geometries, *Trans. Amer. Math. Soc.*, **8**, 379–388.

[W] Wagner, Gerhard
(1992) *On Constucting BIB-Designs and Constant Weight Codes from Nearfield-Generated Planar Nearrings*, Diplomarbeit, Johannes Kepler Universität Linz, Linz.

[Wä] Wähling, H.
(1987) *Theorie der Fastkörper*, Thales Verlag, Essen.

[Ze] Zemmer, J. L.
(1964) Near-fields, planar and non-planar, *Math. Student*, **32**, 145–150.

[Z] Zassenhaus, H.
(1936) Über endliche Fastkörper, *Abh. Math. Sem. Hamburg*, **11**, 187–220.

Index

k-transitive, 49
    sharply, 49, 102

absolute value
    of complex number, 50
addition, 310
adjacent, 164
affine plane, 6, 33, 192
    order, 38
association scheme, 151
automorphism
    fixed point free, 95
    inner, 307

balanced incomplete block design, 2
    circular, 4
    efficiency, 128
BIBD, 2
binary operation, 253
    associative, 254
    commutative, 254
    identity, 254
    inverse, 255
    left distributive, 255
    right distributive, 255
binomial coefficient, 19
block, 3

cardinality, 252
cartesian product, 244
center, 306, 313
characteristic, 85
circle, 4, 8, 51, 56
closed, 260
code, 171
    column, 172
    constant weight, 172
    FHP-procedure, 189
    row, 172
codeword, 172
    weight, 172
coefficient, 273
complex number
    absolute value, 315

conjugate, 49
imaginary part, 315
real part, 315
coset, 260, 318
representative, 260
cross ratio, 48

desarguesian, 104
determinant, 82
direct product
external, 330
internal, 331
disjoint union, 244
Division Algorithm
for $\mathbb{Z}$, 279
of $F[x]$, 280
division ring, 77
divisor, 294
double nearring, 215
double planar nearring, 114, 215
absolute value, 221
alternate measurements of a triangle, 226
angle, 221
angular measurement, 226
circle, 220
congruent triangles, 227
diameter of a circle, 229
dilatation, 223
equiangular triangle, 226
equilateral triangle, 226
has rays, 220
isosceles triangle, 226
lateral measurement, 226
length, 221
measurements of a triangle, 225
orientation of a triangle, 225
right angle, 226
rotation, 223
translation, 223
triangle, 225

edge, 164
empty set, 243
equivalence class, 245

Ferrero pair, 115
field, 256
characteristic, 86
extension, 285
rational function, 98
fixed point free automorphism, 327
fixed point free group of automorphisms, 327
formal power series, 273
function, 248
bijective, 249
codomain, 248
composition, 249
domain, 248

identity, 249
image, 250
inclusion, 250
injective, 249
insertion, 250
inverse, 249
inverse image, 250
Kronecker delta, 275
pre-image, 250
restriction, 250
surjective, 249

graph, 164
cartesian product, 166
complete, 164
null, 164
regular, 164
strongly regular, 164
greatest common divisor, 294, 320
computing, 294
group, 304, 310
abelian, 313
automorphism, 325
commutative, 313
cyclic, 316
epimorphism, 325
finite, 316
fixed point free, 95
generator, 318
homomorphism, 325
infinite, 316
isomorphism, 325
monomorphism, 325
nonabelian, 313
noncommutative, 313
of automorphisms, 233
of linear fractional transformations, 46
of units, 304
order, 316

ideal, 260
maximal, 266
minimal, 266
principal, 266
proper, 266
trivial, 266
identity, 256, 299
incidence matrix, 172
incidence relation, 137
incidence structure, 105, 137
dual, 137
indeterminate, 273
index, 320
index set, 246
intersection, 243, 246
inversive plane, *see* Möbius plane
IP-nearring, 109
Ferrero Planar Nearring Factory, 116

zero-symmetric, 110
isomorphic, 269, 325

Johnson's bound, 179

kernel, 270, 325

Latin square, 10
mutually orthogonal, 11, 135
orthogonal, 11
least common multiple, 320
left identity, 85
length
of complex number, 50
line, 2, 33, 39
parallel, 33
linear fractional transformations, 46

Möbius plane, 8, 51
monoid, 299
multinomial coefficient, 21
multiplication, 253, 310

natural map, 252
nearfield, 84
planar, 97
nearring, 108
abelian, 108
distributive, 108
integral planar, 109
left, 108
right, 108
net, 129, 130, 150
null set, 243

orbit, 95, 115
trivial, 95, 115
order
of a group element, 318
ordered pair, 244

pairwise disjoint, 246
partially balanced incomplete block design, 158
partition, 245
PBIBD, 158
permutation, 17
point, 2
polynomial, 273
constant, 282
degree, 277
irreducible, 282
leading coefficient, 277
reducible, 282
power set, 244
principle of mathematical induction for integers, 278
product, 253
project plane, 192
projective plane, 7, 39, 138

quadratic extension, 290
quaternion

absolute value, 81
conjugate, 80
length, 81
norm, 81
quaternions, 79
quotient, 281

regular group of automorphisms, 327
relation, 245
domain, 245
equivalence, 245
range, 245
reflexive, 245
symmetric, 245
transitive, 245
relatively prime, 296
remainder, 281
right identity, 111
right inverse, 112
ring, 256
commutative, 256
endomorphism, 269
epimorphism, 269
homomorphism, 269
isomorphism, 269
monomorphism, 269
of integers modulo *n*., 259
quotient, 262
ring automorphism, 269
root, 284
semigroup, 297, 299
automorphism, 302
endomorphism, 302
epimorphism, 302
homomorphism, 302
monomorphism, 302
set difference, 243
skew field, *see* division ring
subgroup, 312
invariant, 331
normal, 331
proper, 313
subgroups
trivial, 313
subring, 260
subset, 243
proper, 243

tactical configuration, 151
transversal design, 149
plots, 149
size, 149

union, 246
unit, 304

valency, 164
vertex, 164

well-ordering principle, 278

zero, 300

A mathematician, like a painter or a poet, is a maker of patterns. If his patterns are more permanent than theirs, it is because they are made with ideas. The mathematician's patterns, like the painter's or the poet's, must be beautiful; the ideas, like the colours or the words, must fit together in a harmonious way. Beauty is the first test: there is no permanent place in the world for ugly mathematics.

—G. H. Hardy (A Mathematician's Apology)

FSC
www.fsc.org
MIX
Papier | Fördert
gute Waldnutzung
FSC® C083411

Kolibri 360 GmbH
Ferdinand-Jühlke-Straße 7
99095 Erfurt, Deutschland
produktsicherheit@kolibri360.de

Druck:
CPI Druckdienstleistungen GmbH
im Auftrag der
Kolibri 360 GmbH
Ferdinand-Jühlke-Str. 7
99095 Erfurt